American Institute of Chemical Engineers

Topical Conference on Refinery Processing 2005

Held at the 2005 AIChE Spring National Meeting

April 10-14, 2005
Atlanta, Georgia, USA

Printed from e-media with permission by:

Curran Associates, Inc.
57 Morehouse Lane
Red Hook, NY 12571
www.proceedings.com

ISBN: 978-1-60423-473-2

ISBN: 978-1-60423-473-2

American Institute of Chemical Engineers
Proceedings
Three Park Avenue
New York, NY 10016-5991
Phone: 212-591-8100

www.aiche.org

TC - International Conference on Refinery Processing

Session 60 - Hydrotreating Catalyst Tutorial
Chair: Stuart Shih
Vice Chair: Peter Kokayef

Session 61 - Advances in Hydrotreating and Ultra-Clean Fuel Production
Chair: Jinwen Chen
Vice Chair: Mike Schultz

Session 62 - Refinery Reliability, Safety and Protection
Chair: Tim Olsen
Vice Chair: James W. Harris

Hydrotreating Catalyst Loading, Activation, & Conditioning

L.S. Kraus
Albemarle Catalysts Company LP
Houston, Texas

Abstract

To achieve maximum hydrotreating catalyst performance, proper loading, activation, and start-of-run conditioning procedures must be employed. The catalyst loading method (sock or dense) employed has a significant impact on start-of-run pressure drop, loaded catalyst activity, and catalyst/feed contacting. Catalyst loading quality (uniform bed density, proper particle orientation) critically impacts feed distribution and catalyst performance. The impact of catalyst loading on hydrotreater performance is discussed. Proper catalyst activation procedures must be employed to achieve maximum catalyst activity and to avoid catalyst damage during activation. The major catalyst activation steps are: drying, oil wetting, low temperature sulfiding, and high temperature sulfiding. The important aspects of each step are reviewed. The relative advantages and disadvantages of typical sulfiding methods (sulfur-spiked oil feed, oil feed only, sour gas, and ex-situ sulfiding) are discussed. A detailed example of sulfur-spiked feed sulfiding is presented to illustrate key concepts in catalyst activation. Catalyst conditioning after activation influences catalyst start-of-run performance and cycle length. The impact of premature introduction of cracked feedstocks following catalyst activation and best practices for catalyst conditioning are discussed.

60b Hydrotreating Catalysts for Ultra Low-Sulfur Naphtha and Diesel Production

Henrik W Rasmussen

The tutorial paper convers catalysts for production of ultra low-sulfur naphtha and diesel fuels. For the ultra low-sulfur diesel, the current and future regulations for ULSD around the world will be briefly described. The different reaction mechanisms and process parameters and their impact on ULSD production will be discussed and highlighted with commercial examples.

For the ultra low-sulfur naphatha, the hydroprocessing principles and objectives for SR naphtha will be discussed. The main focus will be hydroprocessing of coker naphtha feedstocks containing silicon from the delayed coking operation. The mechanism for silicon deposition on the catalyst, the impact on catalyst activity, and the optimum catalyst selection will be presented.

The Principles and Practices of FCC Feed Pretreating

FCC Feed Pretreat Units were originally constructed and operated for compliance with environmental regulations primarily related to FCC regenerator sulfur emissions. Fortunately, sulfur removal is not the only benefit of FCC feed pretreatment. Reductions in the FCC feed nitrogen, metals, and carbon residue contents and saturation of aromatic species result in improved FCC yields and product properties.

The tutorial session on The Principles and Practices of FCC Feed Pretreating will cover the basics of FCC Feed Pretreating which include the chemistry, kinetic responses, and various modes of operation. The session will explore the benefits of feed pretreatment on FCC yields and product properties with an emphasis on FCC Pretreater catalyst types and selection. Finally, the future role of the FCC Feed Pretreater in a Clean Fuels Refinery will be defined with comparisons to other options.

60d Residue Hydroprocessing Catalysts – Function and Application

Woodrow K Shiflett

The requirements for various catalysts in the hydrotreating of petroleum residues are set by the nature of the process involved. This tutorial will describe how catalysts used in ebullated-bed processes must be multifunctional. In single-stage designs and many double-stage designs, the catalysts employed must simultaneously facilitate metals removal (HDM), hydrodesulfurization (HDS), carbon residue conversion (HDCR) and vacuum bottoms hydrocracking. Examples will be giving showing how this sets rigorous requirements for chemical and physical properties employed. Some descriptions will be given showing how these catalytic functions enable the production of clean petroleum fuels.

The tutorial will also describe the systemic approach used in specifying catalysts for fixed-bed residue processes and how this provides greater flexibility in catalyst design as well as in the variety of catalysts that can be used. The chemical and physical catalyst needs for HDM, HDS and HDCR in this service will be described.

DEVELOPMENT OF NEW CATALYST SYSTEMS AND INNOVATIVE SOLUTIONS FOR MEETING NEW DIESEL SPECIFICATIONS

Madhusudan Sau. Sau, Ujjwal Manna, Ganesh V. Butley, Hillol Biswas, Manoranjan Santra and Ram Prakash Verma [+]
Indian Oil Corporation Limited
R&D Centre, Sector-13, Faridabad 121 007, India

ABSTRACT

There has been a tremendous upheaval in refining industry worldwide to reduce the sulfur (S) level in diesel for abatement of vehicular pollution. The quality of the crude oil is deteriorating day by day and the targets for 'diesel sulfur' have been receding continuously and have come down to the level of 10-50 ppm in many parts of the world. In near future, most of the countries would follow the same. Refiners are in the process of making considerable investment and would require more for meeting the timetable laid by various Governments in their respective countries. As the quanta of investments are huge and they have considerable impact on refinery profitability, refiners are under tremendous pressure to utilize existing resources and to minimize additional investment for meeting these targets. In countries like India, where 'S' in diesel is currently at the level of 350/500 ppm, bringing down the sulfur level to the required range of 10-50 ppmw levels and conforming to other key specifications (e.g. 95% boiling point, poly aromatic hydrocarbon etc.) is a formidable challenge - especially using the existing low to medium hydrogen pressure units. However, more fundamental understanding of catalytic functions, reaction mechanisms and effect of other micro-constituents have led to various solutions for achieving these targets even at relatively lower pressures.

Improvement of today's technology for achieving ultra low sulfur target in diesel fuel requires overcoming the difficulty of desulfurizing 4-substituted and 4,6-disubstituted dibenzothiophenes (β-DBTs). The presence of nitrogen species (e.g., substituted carbazoles) also plays a crucial role in the ultra-low sulfur regime.

This paper discusses various research studies conducted at our Research centre and, based upon these studies, proposes novel catalytic solutions to achieve these goals in the existing low/medium pressure units, particularly, with respect to meeting diesel S target of 50 ppm and reduction of 95% boiling point as per Euro-IV specification. In many parts of the world, the target of 50 ppm 'S' is achieved with relative ease by hydrotreating lighter feed meeting the reduced ASTM 95% boiling point specification. However, in countries like India, disposal of heavy diesel would cause problem as most of the middle distillate is used for automotive applications. In this context, a new hybrid catalyst system has been developed with high desulfurisation function and selective cracking activity. A stacked catalyst bed of conventional diesel hydrodesulphurization (DHDS) catalyst and this hybrid catalyst can effectively convert the heavy end of diesel back to diesel in the same DHDS reactor. It has been observed that by controlling the bed temperature of catalyst having cracking activity, substantial boiling point shift can be achieved with generation of minimum amount of wild naphtha. In this paper, experimental results are presented for producing ultra low sulfur diesel (ULSD) based on the improved catalyst formulations and use of hybrid catalyst system. Additionally, experimental results are also discussed in the paper on the inhibition effect of nitrogen compounds on DHDS reactions and their removal from diesel stream prior to desulphurization.

On the basis of the pilot studies conducted, it is seen that it would be possible to produce 50 ppm 'S' diesel with proper selection of feed, catalysts and operating conditions. The use of hybrid catalyst system can potentially reduce the 95% boiling point by about 15°C without producing much of lighter product (wild naphtha yield increase is about +2wt%). Experimental results also show that the stacked bed of two catalyst systems can produce better desulfurisation activity than individual catalyst systems. The adsorption of nitrogen compounds by adsorbents prior to desulfurisation can achieve 50% more desulfurisation compared to that with untreated actual refinery stream.

KEYWORDS: DHDS, hydrotreating, catalytic process, stacked bed, nitrogen inhibition.

+ Author to whom the correspondence to be made (vermarp@iocrd.co.in)

1. INTRODUCTION

There has been a tremendous upheaval in refining industry worldwide to reduce the sulfur (S) level in diesel for abatement of vehicular pollution. The quality of the crude oil is deteriorating day by day and the targets for 'diesel S' have been receding continuously and have come down to the level of 10-50 ppm in many parts of the world. In near future, most of the countries would follow the same. Refiners are in the process of making considerable investment and would require more for meeting the timetable laid by various Governments in their respective countries. As the quanta of investments are huge and they have considerable impact on refinery profitability, refiners are under tremendous pressure to utilize existing resources and to minimize additional investment for meeting these targets. In Indian scenario, the specification of future diesel fuel is shown in **Table 1**.

Table 1. The specification of diesel fuel in Indian Scenario.

Parameter		BIS-2000	Bharat-II	Bharat-III (EURO-III Equivalent)	Bharat-IV (EURO-IV Equivalent)
Sulfur, ppm		2500	500	350	50
Cetane Number		48	48	51	51
Density, Kg/m^3		820-860	820-860	820-845	820-845
95% Recovery, ^{0}C		370	370	360	360
PAH*, wt%				11	11
Year of Imple-mentation	**Metros**		2000-2001	April 1, 2005	April 1, 2010
	Mega Cities		April 1, 2003	April 1, 2005	April 1, 2010
	Rest of India	April 1, 2000	April 1, 2005	April 1, 2010	

*PAH – Poly Aromatic Hydrocarbon

The diesel hydrodesulphurization (DHDS) units in Indian refineries were originally built for producing 2500 ppm sulfur diesel using one reactors and subsequently another reactor was used for reducing the product sulfur level to 500 ppm. The question before the refining industry is whether these existing units can produce diesel of Euro III/IV quality or there is a need for new investment?

Bringing down the current sulfur content of diesel (500 ppmw) to the required range of 10-50 ppmw levels is a formidable challenge - especially at low to medium hydrogen pressures. However, more fundamental understanding of catalytic functions, reaction mechanisms and effect of other compounds have led to various options for achieving these targets even at relatively lower pressures.

The sulfur compounds present in the diesel fuel are thiophene, thiophene derivatives, benzothiophenes, benzothiophene derivatives etc. Aliphatic sulfur compounds i.e. thiophene and thiophene derivatives are relatively easy to desulfurize than benzothiophenes, dibenzothiophenes and alkylated benzothiophenes. Improvement of today's technology for achieving ultra low sulfur target in diesel fuel requires overcoming the difficulty of desulfurizing 4-substituted and 4,6-disubstituted dibenzothiophenes (β-DBTs). These species are the most refractory sulfur heterocyclics. The various refractive sulphur compounds with TBP cut ranges present in the diesel fuel [2] is shown in **Figure 1**.

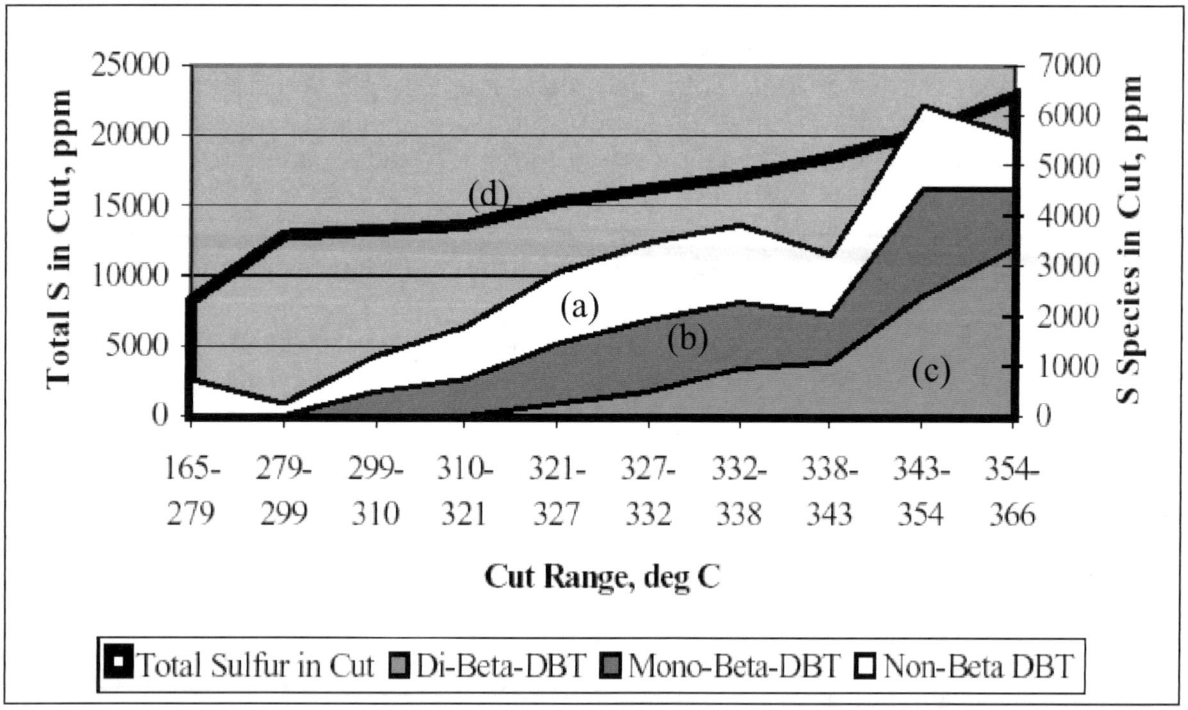

Figure 1. Sulfur compound class and total sulfur in various cut is shown [2]. (a): Non beta dibenzothiophene (DBT), (b): Mono-beta Dibenzothiophene (DBT), (c): Di- beta dibenzothiophene (DBT), (d) : Total sulfur in Cut.

The presence of polycondensed aromatics and nitrogen species in the diesel fractions adversely affects the desulphurization reactions. It has also been found that nitrogen species (e.g., substituted carbazoles) play a crucial role in the ultra-low sulfur regime [3] as can be seen in **Figure 2**. It may be observed that the presence of basic and nonbasic nitrogen compounds affect the desulphurization differently. The catalysts of high hydrogenation functionality which are good for removal of refractory sulfur and nitrogen compounds are not ideally suited for the HDS of non-β-DBTs (viz. thiophenes, benzothiophenes etc.), which can be more efficiently desulfurized over catalysts with strong hydrogenolysis functionality. Therefore, an obvious question is whether these two catalyst types can be used together in a synergistic manner. Hence, there is a need to develop catalytic strategies for producing Ultra Low Sulphur Diesel (ULSD).

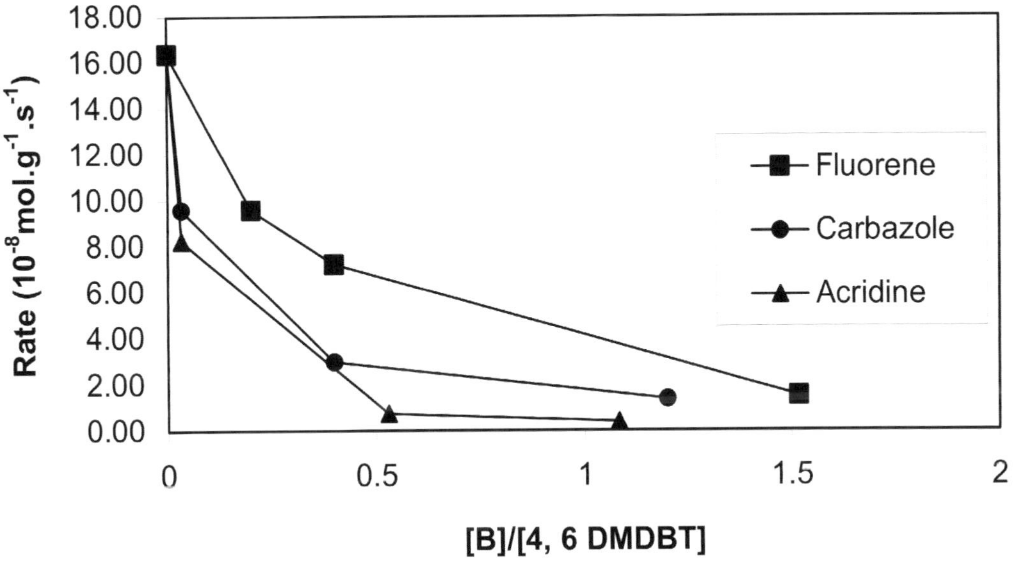

Figure 2. Effect of Inhibition with nitrogen compounds on rate of transformation of 4, 6 DMDBT [3].

This paper deals with following catalytic solutions explored at our Research Centre to achieve these goals in the existing units, particularly, with respect to meeting Diesel 'S' target of 50 ppm and 95% boiling point of 360°C as required in Euro-IV specifications. The options discussed in the paper for ULSD production based on pilot plant results are:

i. Reduction of ASTM 95% boiling point
ii. Use of high activity catalyst systems and use of stacked bed.
iii. Effects of removal of nitrogen compound from diesel stream prior to desulphurization

2. EXPERIMENTS

Experiments were conducted for hydrotreating and hydrocracking reactions in hydroprocessing pilot plant reactors shown in **Figure 3**. This pilot plant consists of two reactors in series and both are in downflow mode configuration. These reactors can be operated up to pressure of 180 kg/cm^2 and temperature of 450°C. One can see from **Figure 3** that the feed is pumped to the reactor1 (R1) by a piston type of pump (P) from feed tank (T1). The partially converted feed from Reactor 1 (R1-ex) goes to reactor2 (R2). The total products and unconverted feed is then routed to a high pressure separator (HPS). The gas from high pressure separator meet other gas (to be described later) to a gas line. The liquid from HPS goes to low pressure separator (LPS). The gas from LPS joins with the gas from HPS and the combined gas is passed through a caustic scrubber and vented through metering by a weight gas meter (WGM). The liquid from LPS to collected into product tank (T2).

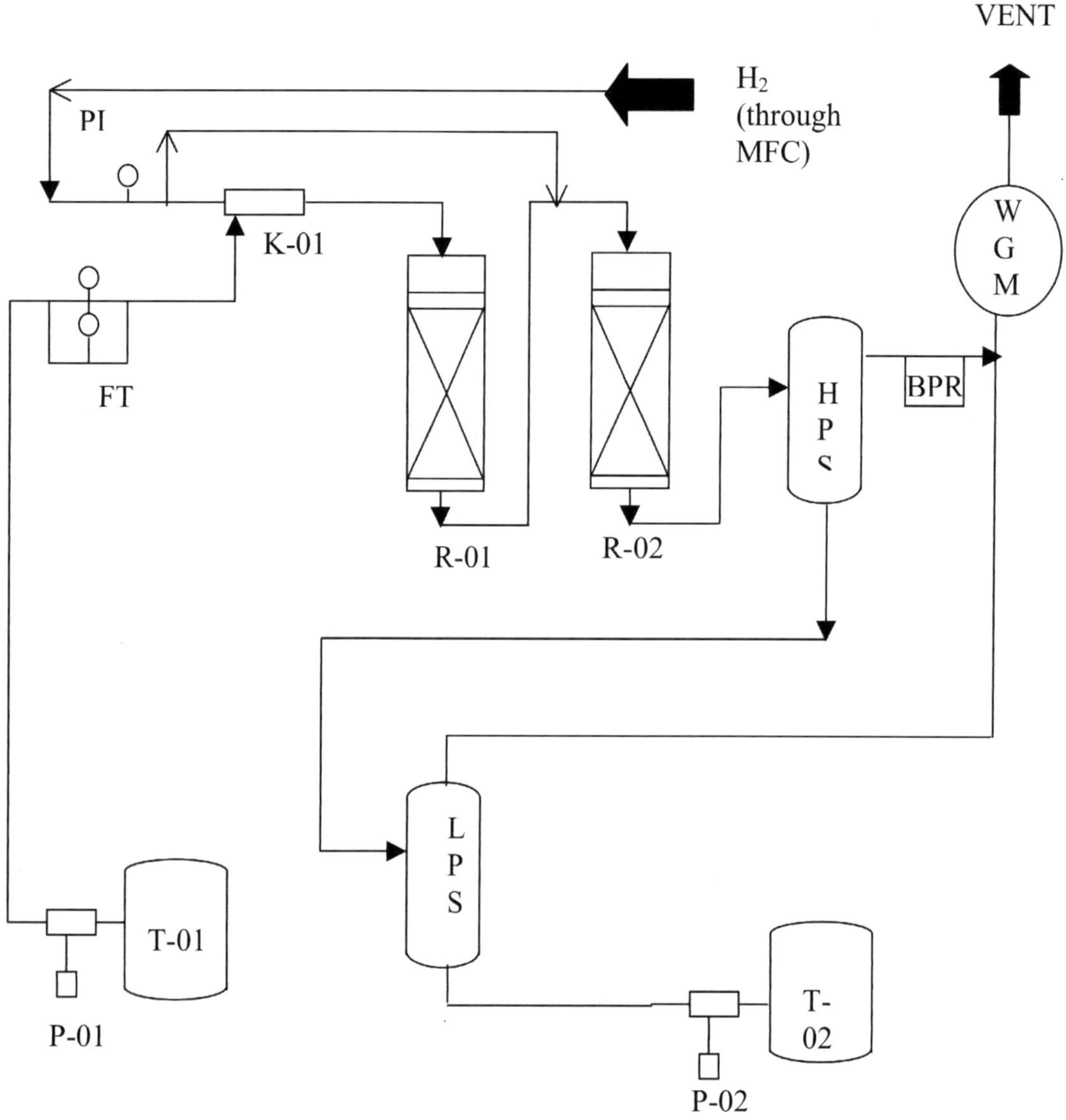

MFC: Mass flow controller, PI: Pressure indicator, T-01 and T-02: Feed and product tank, P-01 and P-02: Feed and product pump, FT: Flow transmitter, K-01: Knok out drum, RB-01: Reactor 1, RB-02: Reactor 2, HPS: High pressure separator, LPS: Low pressure separator, BRP: Back pressure regulator, WGM: Wet gas meter

Figure 3. A schematic of the hydroprocessing pilot plant reactors used for experimentations.

Experiments were conducted in pilot plant reactors using various types of diesel stream and various operating conditions. The diesel feeds used in this study are referred as "Light feed" and "Full range diesel feed". The properties of both feeds are given in **Table 2a**. The catalysts used in this study are eleven in numbers and these are hydrotreating catalyst and hybrid catalyst systems. The description of each catalyst is given in **Table 2b**. The operating conditions mostly used in the experiments are: Pressure = 49 bar, Temperature =338- 350 °C, LHSV (liquid hourly space velocity)= 1.0-1.5 h^{-1}, H2/HC (hydrogen/diesel feed) ratio = 267-400

(Nm^3/m^3). Diluents such as silicon carbide were used to have comparable kinetic data with commercial plant.

Table 2a. Properties of diesel feeds used in the study.

Description	Sulphur (ppmw)	Nitrogen (ppmw)	SIMTBP (D2887) IBP* (^0C)	SIMTBP (D2887) FBP* (^0C)	Density (Kg/m^3)	Saturates (wt%)	Aromatics (wt%)
Light feed	9228	64	125	370	0.8351	72.7	27.3
Full range feed	11800	174	136	447	0.8478	71.0	29.0

*IBP: Initial boiling point and FBP: Final boiling point

Table 2b. Description of catalysts used in the study.

Catalyst name	**Catalyst description**
CAT-A, CAT-B	CoMo based hydrotreating catalyst
CAT-C, CAT-D, CAT-E, CAT-F	Isomerization catalyst
CAT-GH	Hybrid catalyst system
CAT-I	NiMo based hydrotreating catalyst
CAT-J	CoMo based hydrotreating catalyst
CAT-K	CoMo based hydrotreating catalyst
CAT-L	NiMo based hydrotreating catalyst

3. RESULTS AND DISCUSSION

3.1 Reduction of 95% ASTM boiling point

ASTM 95% boiling points is to be reduced from 370^{0}C to 360^{0}C in Euro-III and Euro-IV diesel specification. This can be achieved by adjusting the cut points in the distillation columns. This also has a beneficial impact on the desulphurization, as most of the refractory material present in the heavy end of the feed would be out of diesel feed. In many parts of the world, the target of 50 ppm 'S' is achieved with relative ease by hydrotreating lighter feed meeting the reduced ASTM 95% boiling point specification. However, in countries like India, disposal of heavy diesel would cause problem as most of the middle distillate is used for automotive applications. So, the ideal solution in this case would be employment of some special catalysts in the existing DHDS reactors, which can selectively convert this heavy end back to diesel fraction. There are two options of doing this, which have been explored in our study.

The first option is the saturation of aromatic compounds and isomerization of n-paraffinic compounds, which can shift the boiling point towards lower side as these compounds have lower boiling point. Depending upon the level of aromatic saturation taking place in the DHDS reactor, boiling range is shifted to a certain degree. Isomerisation of n-paraffins can cause this shift further away from heavy end. In this pilot plant study, four isomerization catalysts have been used along with hydrodesulphurization catalyst for estimating this shift. The results are shown in **Table 3a**. It can be seen that in case of some catalysts, a fair amount of lighter material in naphtha range (called the wild naphtha, boiling point 127°C-) has also been generated. This indicates that the success of this scheme would depend on proper calibration of catalyst functions.

Table 3a. Pilot plant results of different Isomerization catalysts for reduction of ASTM 95% boiling point.

Catalyst	Feed ASTM 95% boiling point (D2887) (°C)	Operating conditions				Decrease in ASTM 95 % (D2887) boiling point (°C)	Increase in quantity of 127 °C- hydro-carbon (wt%)
		Temp (°C)	Press (bar)	LHSV (h^{-1})	H2/HC ratio (Nm^3/m^3)		
CAT-C	337	350	49	1.0	400	4	2
CAT-D	337	350	49	1.0	400	22	17
	349	350	49	1.0	400	9	7
CAT-E	349	350	49	1.0	400	43	7
CAT-F	349	350	49	1.0	400	8	3

The other option is to use hybrid catalyst with enhanced desulfurization function and selective cracking function for selective and controlled cracking of heavy hydrocarbon present in diesel. In the current study, hybrid catalysts have been used under same operating conditions of DHDS units. The results are shown in **Table 3b**. It is very interesting to observe that by controlling the hybrid catalyst bed temperatures, substantial boiling point shift can be achieved with generation of minimum amount of naphtha. We can expect the diesel quality also to be better.

3.2 Use of high activity catalyst systems for ULSD production in pilot plant reactors

A larger number of catalyst systems both commercial and laboratory scale have been evaluated in our pilot plant using different feedstocks and results of achieving ultra low 'S' target are discussed in details. The results of the commercial catalysts are shown in **Table 4** and the results of laboratory scale catalyst are shown in **Table 6**. It can be observed that with a proper choice of catalyst and operating conditions, it is possible to produce less than 50 ppm

'S' diesel in the units originally designed for production of 500 ppm 'S'. It can be interesting to note that Ni-Mo catalyst show better results in this regime with full range of feedstock. In case the feed is cut to lighter fraction, Co-Mo catalyst can be equally effective.

Table 3b. Pilot plant results of combine catalyst system (Hydrotreating and hybrid catalyst) for reduction of diesel ASTM 95% boiling point.

Catalyst	Feed ASTM 95% boiling point (D2887) (°C)	Operating conditions				Decrease in ASTM 95 % (D2887) boiling point (°C)	Increase in quantity of 127 °C- hydro-carbon (wt%)
		Temp. Bed I/ Bed II * (°C)	Press (bar)	LHSV (h^{-1})	H2/HC ratio (Nm^3/m^3)		
CAT-G&H	390	345/360	49	1.5	566	2	2
	390	345/375	49	1.5	566	10	2
	390	345/380	49	1.5	566	10	2
	390	345/390	49	1.5	566	15	2

* Bed I is hydrotreating catalyst & Bed II is hybrid catalyst

Table 4. Pilot plant results of high activity commercial catalysts for production of Ultra Low Sulphur Diesel (ULSD).

Catalyst	Feed description (Defined in Table 2a)	Feed Sulphur (ppmw)	Operating conditions				Product Sulphur (ppmw)
			Temp (°C)	Press (bar)	LHSV (h^{-1})	H_2/HC ratio (Nm^3/m^3)	
CAT-I	Light feed	9228	345	49	1.5	250	10
	Full range feed	11800	345	49	1.5	250	41
CAT-J	Light feed	9228	345	49	1.5	250	11
	Full range feed	11800	345	49	1.5	250	131

The results of the stacked beds are shown in **Table 5**. One can see in **Table 5** that the Catalyst-L and Catalyst-K combination can perform far better than Catalyst K as well as the catalyst combination of Catalyst -K and Catalyst-L. As described in **Table 2b**, the Catalyst-K is CoMo based and Catalyst-L is NiMo based hydrotreating catalyst.

Pilot plant results show that the stacked bed of two catalyst systems can produce better desulphurization activity than individual catalyst systems. Since most of the Indian refineries have hydrodesulphurization units with two reactors, reactors can be loaded with combination of catalyst systems in stacked manner to achieve the desired target of ultra low sulfur diesel (ULSD).

Table 5. Pilot plant results of stacked bed of commercial catalysts for production of Ultra Low Sulphur Diesel.

Catalyst	Feed description (Defined in Table 2a)	Feed Sulphur (ppmw)	Operating conditions				Product Sulphur (ppmw)
			Temp (^{0}C)	Press (bar)	LHSV (h^{-1})	H2/HC Ratio (Nm^3/m^3)	
CAT-K	Light feed	9228	345	49	1.5	250	75
	Full range feed	11800	345	49	1.5	250	413
CAT-K + CAT-L	Light feed	9228	345	49	1.5	250	22
	Full range feed	11800	345	49	1.5	250	203
CAT-L + CAT-K	Light feed	9228	345	49	1.5	250	20
	Full range feed	11800	345	49	1.5	250	131

Table 6. Pilot plant results of laboratory scale catalysts for production of Ultra Low Sulphur Diesel (ULSD). Effects of nitrogen removal prior to desulphurization are also shown.

Catalyst	Feed description (Defined in Table 2a)	Feed Nitrogen (ppmw)	Feed Sulphur (ppmw)	Operating conditions				Product Sulphur (ppmw)
				Temp (°C)	Press (bar)	LHSV (h^{-1})	H2/HC ratio (Nm^3/m^3)	
CAT-A	Light feed	64	9228	338	49	1.5	267	97
	Light feed (after nitrogen removal)	15	9228	338	49	1.5	267	54
CAT-B	Light feed	64	9228	345	49	1.5	267	60
	Light feed (after nitrogen removal)	15	9228	345	49	1.5	267	25
	Full range feed	174	11800	345	49	1.5	267	128
	Full range feed (after nitrogen removal)	50	11800	345	49	1.5	267	78
	Full range feed (after nitrogen removal)	15	11800	345	49	1.5	267	68

3.3 Effects of removal of nitrogen from diesel stream prior to desulphurization

It has been reported [4] that presence of nitrogen acts as hindrance to removal of sulfur compounds, especially in the low ppm range, due to competitive adsorption. Therefore, if nitrogen compounds are selectively removed by means of adsorption, extraction, etc., prior to hydrodesulphurization, performance of DHDS catalysts can be enhanced.

In this study, nitrogen compounds were selectively removed from partially hydrotreated diesel feed by adsorption using silica and alumina in varying ratios. Partially hydrotreated diesel feedstocks having different nitrogen content were subjected to pilot plant testing using DHDS catalysts and the results are shown in **Table 6**. These results show that removal of nitrogen compounds prior to desulphurization can achieve 40-60% or even more desulphurization compared to that with untreated feed. This additional step of nitrogen removal can be installed in an existing DHDS reactor to achieve more desulphurization or can be designed with the grass roots units. This, in addition, reduces the consumption of precious hydrogen.

4. CONCLUSIONS

On the basis of the pilot studies conducted, it is seen that it would be possible to produce new diesel specifications with the new catalyst system and innovative solutions described in the paper. It would be possible to achieve the target of 50 ppm 'S' very easily with lighter feed meeting the ASTM 95% boiling point specification of Euro-III and IV. However, this would require separate disposal routes for heavy diesel. It may be possible to effectively convert this heavy end back to diesel in the same DHDS reactor with employment of hybrid catalyst or isomerization catalysts. The use of hybrid catalyst system can potentially reduce the 95% boiling point by about 15°C without producing much of lighter product (wild naphtha yield increase is about +2wt%). Stacked bed of Ni-Mo and Co-Mo catalyst has been shown to give better performance than the individual catalyst. The units having problem to meet the 50 ppm 'S' specification target may remove nitrogen selectively prior to desulphurization. The adsorption of nitrogen compounds by adsorbents prior to desulfurisation can achieve 50% more desulfurisation compared to that with untreated actual refinery stream. Moreover, this also would reduce the hydrogen consumption.

5. REFERENCES

1. Report of the Expert Committee on *Auto-fuel Policy* (2002).
2. Bjorklund, B. L., Heckel, T. L., Howard, N. D., Lindsay, D. A., and Piasecki, D. J., UOP LLC, Des Plaines, Illinois, *"The Lower It Goes, The Tougher It Gets (The Practical Implications of Producing Ultra-Low Sulfur Diesel)"*, AM-00-16, NPRA Annual Meeting.
3. Breysse, M., Djega-Mariadassou, G., Pessayre, S., Geantet, C., Vrinat, M., Perot, G., Lemaire, M., *"Deep Desulphurization: Reactions, Catalysts and Technological Challenges", Catalysis Today*, Vol. 84, 129-138, 2003.
4. Min, W. S., Choi, K. I., Khang, S. Y., Min, D.S., Ryu, J. W., Yoo, K. S., Kim, J. H., *"Method of manufacturing cleaner diesel", US Patent No. 6,248,230.*

A short Bio-data of the Authors

Madhusudan Sau, PhD., UA, OHIO (USA) is working as *Deputy Manager (Research)* in IndianOil Corporation Research and Development Centre.

Ujjwal Manna, PhD., UMIST, (UK) is working as *Senior Research Manager* in IndianOil Corporation Research and Development Centre.

Ganesh V. Butley, M. Tech, LIT Nagpur is working as *Research Officer (Chemical Engineering)* in IndianOil Corporation Research and Development Centre.

Hillol Biswas, M. Tech, I.I.T. Kharagpur is working as *Research Officer (Chemical Engineering)* in IndianOil Corporation Research and Development Centre.

Manoranjan Santra, PhD, I.I.T. Kharagpur is working as *Chief Research Manager* in IndianOil Corporation Research and Development Centre.

Ram Prakash Verma, PhD., UDCT Bombay, is working as *Executive Director in* IndianOil Corporation Research and Development Centre.

Experimental Study on Vapour-Liquid Phase Equilibrium of LCO Hydrotreating Systems at High Temperatures and Pressures

Jinwen Chen, Hong Yang and Zbigniew Ring
National Centre for Upgrading Technology (NCUT), One Oil Patch Drive
Devon, Alberta, Canada, T9G 1A8

Prepared for Presentation at the 2005 AIChE National Spring Meeting, Session 61: Advances in Hydrotreating and Ultra-Clean Fuel Production, April 10-14, 2005, Atlanta, GA, USA

APRIL 2005

Abstract

In this study, a continuous flow unit was established to conduct vapour-liquid equilibrium (VLE) measurement with systems containing hydrogen and petroleum fractions. VLE data were measured with two systems containing hydrogen and light cycle oil. The experiments were conducted at high temperatures (250-400°C) and pressures (50-100 atm.) that are typically used in commercial petroleum hydrotreating reactors. It was found that the relative amount of LCO, the total sulphur and individual sulphur compounds in the vapour phase increased significantly with the increase in temperature and decrease in pressure. The VLE ratios, the ratios of vapour to liquid mole fraction at equilibrium, for individual sulphur compounds increased with the increase in temperature. Under the same temperature, sulphur compounds with low molecular weight/size had higher equilibrium ratios than those with higher molecular weight/size.

1. INTRODUCTION

Hydrotreating is a widely used petroleum refining process that reduces sulphur, nitrogen, aromatics, and heavy metal contents in intermediate streams and final products. In a hydrotreating process, petroleum fractions and hydrogen react on catalyst surface at relatively high temperatures and pressures (300-400°C and 30-100 atm.)(Froment et al., 1994; Mayers, 1996; Chen et al., 2003). Under such conditions, the reaction system (hydrogen and hydrocarbons) is at vapour-liquid equilibrium (VLE) with both the vapour and liquid phases containing hydrogen and hydrocarbons (including sulphur and nitrogen compounds). In a trickle-bed hydrotreating reactor, under the conditions of complete catalyst wetting, only the components present in the liquid phase can come in contact with the catalyst surface and undergo hydrotreating reactions. The flow rate of the liquid phase and the concentrations of various reacting species in the liquid phase may be significantly different from those used by assuming no liquid vaporization in the reactor, which ultimately affects kinetics analysis and the predictive ability of the resulting kinetics models. The understanding of hydrotreating reactions and, consequently, the predictive ability of process models can be dramatically improved by including VLE calculation in the kinetics analysis and reactor modeling to determine the composition reacting components in the vapour and liquid phases.

Although a limited number of papers have been published on the effect of VLE on gas-liquid-solid three phase reactors (Akgerman et al., 1985; Smith and Satterfield, 1986; LaVopa and Satterfield, 1988; Jong, 1994; Tesser et al., 2003), very few studies have been published to account for VLE effect during hydrotreating of petroleum fractions (Kocis and Ho, 1986; Bellos and Papayannakos, 2003). VLE is often ignored in practice, and HDS kinetics modeling is usually based on the assumption that all components in the feedstock are present in the liquid phase regardless of the operating conditions. Akgerman et al. (1985) discussed in general the VLE effect on reaction conversion in trickle-bed reactors and found that the difference in conversion could be up to 38% with and without VLE consideration when the same kinetics model was used. Smith and Satterfield (1986) studied the VLE effect on the performance of a trickle-bed reactor used to conduct catalytic hydrodenitrogenation (HDN) of quinoline. They observed that the increase in gas-to-oil ratio and, therefore, the increase in liquid vaporization, caused the decrease in quinoline conversion and this effect became more pronounced with the increase in reaction order. LaVopa and Satterfield (1988) investigated the

VLE effect on hydrodeoxygenation of dibenzofuran and hydrogenation of n-butylbenzene. They concluded that the greatest VLE effect on conversion occurred when liquid volatility was low and when the absolute difference in volatilities between reactant and un-reacting liquid was large. Kocis and Ho (1986) observed higher conversion of dibenzothiophene (DBT) with $C_{16}H_{34}$ as solvent than with $C_{24}H_{50}$ as solvent in a trickle-bed reactor, which was attributed to the stronger partial evaporation of $C_{16}H_{34}$. Bellos and Papayannakos (2003) studied hydrodesulphurization (HDS) kinetics of a straight run heavy gas oil in a microreactor by taking VLE effect into account. In their study, the gas oil compounds as well as the sulphur compounds were represented by five pseudo-components. Only the reactants present in the liquid phase were considered to participate in catalytic reactions. Since moderate temperature and pressure (320-350°C and 54 atm., respectively) were used in the experiments and since heavy gas oil was used as the feed, the VLE effect on HDS kinetics was not expected to play a significant role in their study.

The virtual lack of research on VLE effect on HDS is probably due to the significant difficulty of experimental measurements (or simulation) plus the extra complexity in kinetics analysis that accounts for VLE. Experimental data on VLE of hydrogen-petroleum fraction systems obtained under commercial hydrotreating conditions are rarely published in the open literature because of the poor definition of the problem, stemming from the complex composition of those systems. In addition, high temperature and pressure, and sample characterization and analysis, add more difficulties to the experiments. Instead, systems consisting of hydrogen and a single hydrocarbon or a simulated hydrocarbon mixture are often studied. VLE data for hydrogen-toluene system were reported by Simnick et al. (1978) and Laugier et al. (1983). Ramanujam et al. (1985) reported VLE data for a system consisting of hydrogen and simulated coal-derived liquid in the temperature range of 427.8 ~ 541.1 K and pressure range of 23.3 ~ 100 atm. The simulated aromatic oil was composed of benzene (54.8%), toluene (18.6%), and 14 other aromatic compounds. Connolly and Kandalic (1986) measured VLE data with the system of hydrogen in 2,3-dimethylbutane, cyclohexane, m-xylene and 1,4-diethylbenzene in the temperature range of 308 ~ 593 K and pressure range of 13.6 ~ 163 atm. Wiegand et al. (1989) measured VLE data for a hydrogen-coal liquid system in the temperature range of 452 ~ 612K and the pressure range of 50 ~ 302 atm. The coal liquid contained 87.2% carbon, 11.9% hydrogen and 0.34% nitrogen, oxygen and sulphur.

This paper reports our VLE experimental data obtained with systems containing hydrogen and light cycle oil (LCO), a fraction of the liquid product from the fluid catalytic cracking (FCC) process. The operating conditions of interest were those typically used during hydrotreating of LCO to remove sulphur and nitrogen compounds, and to saturate aromatic compounds. Unlike the other published studies, our experiments uses a real system of significant importance and looks in detail at the distributions of various sulphur compounds between the vapour and liquid phases in a bid to determine the intrinsic kinetics of hydrotreating reactions free of the effect of phase equilibrium. In addition, our experimental conditions cover the range specific for the production of ultra low sulphur diesel.

2. EXPERIMENTAL

The experimental setup used in our VLE experiments was based on a bench-scale continuous hydrotreating unit. A continuous-flow VLE cell and a microreactor are both integral parts of the unit so it can be used for either VLE or hydrotreating studies, or both depending on he need. The diagram is shown in Figure 1 where the reactor section that was not used in this study is shown in dashed lines.

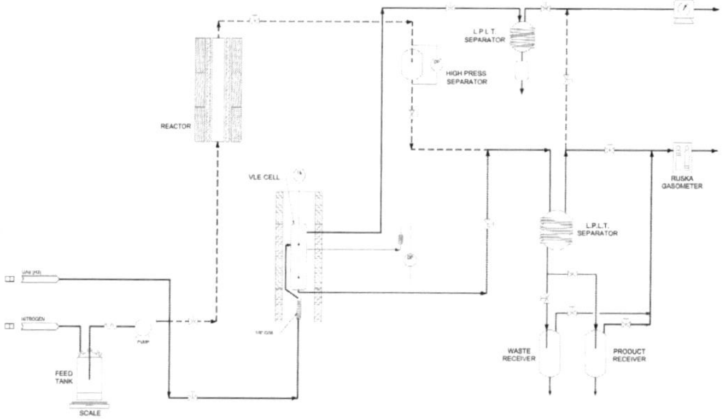

Figure 1. Diagram of the VLE measurement unit

The feedstocks, the original or hydrotreated LCO, were pumped from the feed tank and then mixed with hydrogen (99.99% purity supplied by PRAXAIR Inc.). The resulting mixture, under the desired pressure, went through a stainless steel coil pre-heater (12-m long stainless steel tube of 3-mm I.D.). The coil and the VLE cell were both situated inside a 4-zone blanket furnace to ensure that the mixture reached the desired temperature before reaching the VLE cell and maintained at that temperature in the cell. The preheated mixture entered the VLE cell at a middle location to undergo flash operation. The cell, made of stainless steel, had an I.D. of 2.54 cm and a height of 30 cm, was arranged as a high-pressure high-temperature phase separator. A differential pressure (DP) cell (DP2000, FoxBoRo) was used to control the liquid level in the VLE cell. A two-point thermocouple was positioned at the center of the cell to measure the temperatures of the liquid and vapour phase. At the top of the cell, stainless steel wool was packed to capture any liquid droplets carried by the vapour phase. The vapour phase, leaving through the top of the cell, went into a low-temperature low-pressure phase separator where the hydrocarbons in the vapour were condensed into liquid while the uncondensed hydrogen went to a dry test meter (DTM) for metering. The liquid phase, leaving through the bottom of the cell, entered another low-temperature low-pressure phase separator where the hydrocarbons were condensed and the dissolved hydrogen was released and introduced to a high-precision gasmeter (Ruska 2331-801, Technel Engineering Inc.) for accurate volume measurement at closely monitored temperature and atmospheric pressure (a relatively small amount of hydrogen was dissolved in the liquid at the operating conditions). The condensed liquid samples from both vapour and liquid phase separators were collected, weighed and analyzed separately. A detailed overall mass balance calculation of the phase separation process was performed to ensure that the experiments were properly conducted. The two liquid samples were analysed for simulated distillation (SimDis), total sulphur content, and for the content of a number of individual sulphur species of vital interest in hydrotreating.

In middle distillate hydroprocessing, dibenzothiophenic compounds are normally considered as refractory sulphur species to which more attention has to be paid in order to produce ultra low sulphur diesel fuels. In this study, 27 dibenzothiophenic compounds were positively identified and quantified in both the hydrotreated and the original LCO. The speciation of these sulphur compounds was achieved with an HP6890-series gas chromatograph, equipped with an atomic emission detector (AED, sulfur channel 181 nm) and a 60 m PIONA silica capillary column. Programmed temperature control was employed during the GC analysis with initial temperature of 40°C and ramping rate of 2°C/min to 285°C, at which the temperature was held for 20 min. Sulphur compound identification was achieved by either matching the retention times of the GC peaks with those of standard compounds, or by using GC-MS spectra (HP5890-series II with 5972-series MS detector) obtained separately, or by using literature data. External-standard method was employed to quantify sulfur content in the samples. The standard solutions were either a NIST diesel or a standard solution of benzothiophene (BT), depending on the need for total sulphur or for peak-by-peak analysis. More details on sulphur speciation and analysis can be found elsewhere (Chen et al., 2003; Chen and Ring, 2004).

The key physical properties and chemical compositions of the two feedstocks used in our study are listed in Table 1. The hydrotreated LCO had lower density, and sulphur and nitrogen content than the original LCO, consistent with partial hydrogenation of aromatics and removal of sulphur and nitrogen during hydrotreating.

Table 1. Physical properties and chemical compositions of the LCO samples

	Original LCO	Hydrotreated LCO
Density (15.6°C), g/mL (ASTM D4052)	0.9380	0.9201
Carbon, wt% (ASTM D5291)	88.25	88.34
Hydrogen, wt% (ASTM D5291)	10.54	11.07
Total sulphur, wppm (GC-AED)	11704	7200
Total nitrogen, wppm (ASTM D4629)	665.8	87.97
SimDis, °C (ASTM D2887)		
IBP	128.4	115.5
10wt%	225.0	220.6
30wt%	255.3	252.5
50wt%	288.5	280.0
70wt%	326.4	314.8
90wt%	373.9	362.8
FBP	430.8	425.8

In all experimental runs conducted in this study, the liquid feed and gas rate were fixed at 10 g/h and 10 NL/h, respectively. The resulting gas-to-oil ratio of 1000 NL/kg of feed has been typically used in commercial hydrotreating and our previous LCO hydrotreating studies (Chen et al., 2003; Chen and Ring, 2004)

3. RESULTS AND DISCUSSION

3.1. Verification of the Experimental Setup and Procedures

So far, no experimental data have been reported with systems consisting of hydrogen and petroleum fractions by using a continuous flow unit like the one used in this study. However, a similar unit was used by Simnick et al. (1978) to measure the VLE data of different systems consisting of pure hydrocarbons and hydrogen. To verify the reliability of the data obtained with our VLE cell and to establish an efficient experimental procedure, VLE data were measured for a hydrogen-toluene system at the temperature of 188.7°C and pressures ranging from 50 to 100 atm. Our data closely follow the data measured by Simnick et al (1978) as shown in Figure 2, implying that the VLE measurement unit used in this study is reliable and suitable for high-temperature high-pressure VLE measurements.

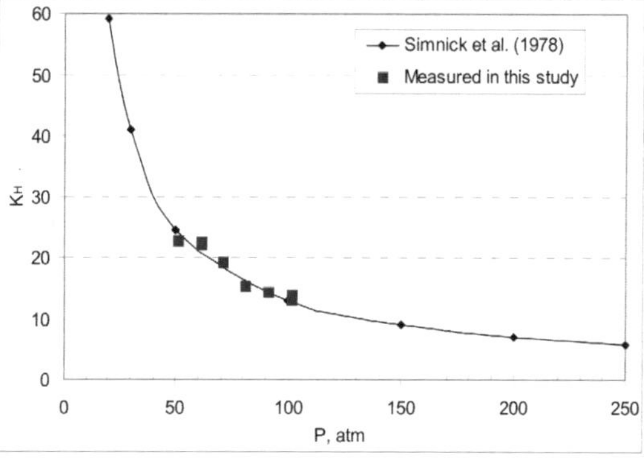

Figure 2. Comparison of VLE data for toluene-hydrogen system at 188.7°C

3.2. VLE measurement with hydrogen-LCO system

VLE experiments were then carried out with the two LCO feedstocks and hydrogen under different temperatures and pressures of practical significance in petroleum hydrotreating processes. Figure 3 shows the change of LCO amount (wt% of feed) in the -vapour phase in the temperature range of 250-400°C under constant pressure of 70 atm. The amount of LCO in the vapour phase increased dramatically with temperature. At 350°C, the amount of LCO in the vapour phase was over 40wt% for the original LCO and over 50wt% for the hydrotreated LCO (indicating greater volatility of the hydrotreated material). At 400°C, it was over 80wt% for the original LCO and over 90wt% for the hydrotreated LCO. In this case the commercial hydrotreater might have been operated in partial wetting mode. It was also noted that generally the amount of LCO in the vapour phase for the original LCO was about 10wt% less than that for the hydrotreated LCO. In another view of our data, Figure 4 shows decreasing amount of LCO in the vapour phase with increased pressure. Again, the hydrotreated LCO is more volatile than the original LCO and the difference at each pressure is 10wt% on average.

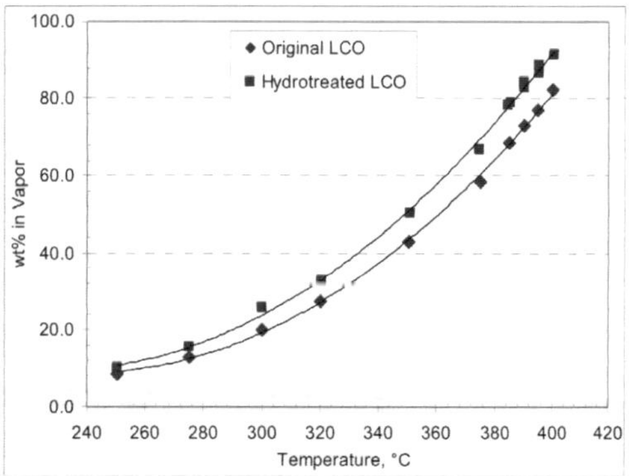

Figure 3. Wt% of LCO in the vapour phase at different temperatures

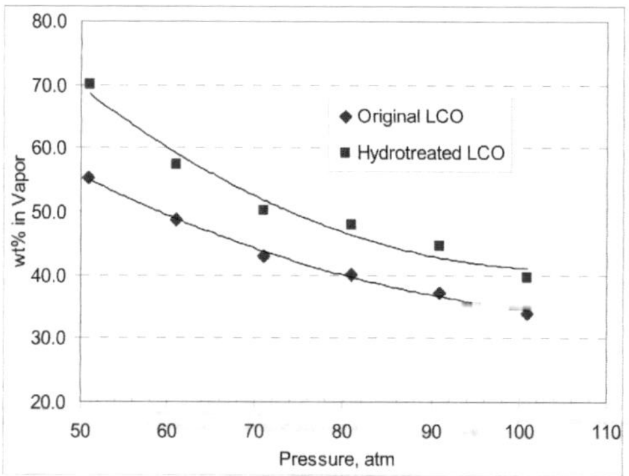

Figure 4. Wt% of LCO in the vapour phase at different pressures

One of the purposes of LCO hydrotreating is to reduce the sulphur content to a desired level so it could be used as one of the diesel-blending components. Therefore the variation of total sulphur concentration in the vapour and liquid phase with temperature and pressure, as well as the concentrations of each individual sulphur compound in the two phases, is an important information in precise process design and reactor simulation. Figures 5 and 6 show the total sulphur concentrations in the vapour and liquid phase at different temperatures and pressures, respectively, for the original LCO (the hydrotreated LCO showed similar trends). It should be noted that here the concentrations of total sulphur (and the individual sulphur compounds as will be discussed later) in the vapour and liquid phase do not represent those in the VLE cell (or hydrotreating reactor) where hydrogen is present in both phases. In stead, these concentrations represent those measured after the removal of hydrogen from the liquid samples of the two phases. As seen in Figures 5 and 6, the total sulphur concentration in the liquid phase decreases with the increase in temperature and the decrease in pressure, while the total sulphur concentration in the vapour phase increases with the increase in temperature and the decrease in pressure. The reason for such a trend is that most of the sulphur species in the LCO have relatively low boiling points. With increased temperature and decreased pressure, they are easily vaporized into the vapour phase together with other hydrocarbons in the LCO. At high temperatures and low pressures, the sulphur compounds left in the liquid phase are mostly those with higher boiling points, such as alkyl substituted dibenzothiophenes and naphthothiophenes.

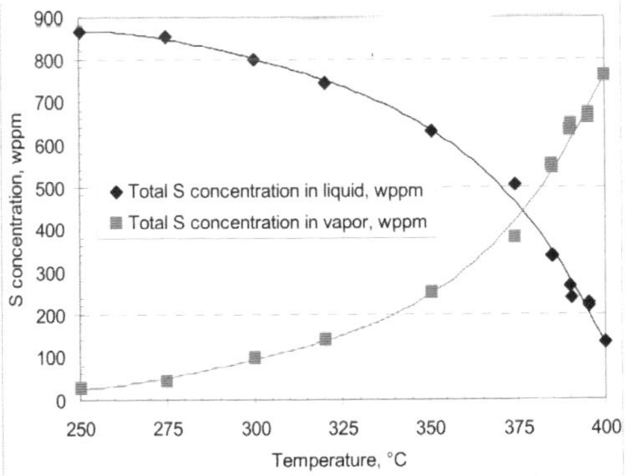

Figure 5. Total S concentration in the vapour and liquid phase at different temperatures

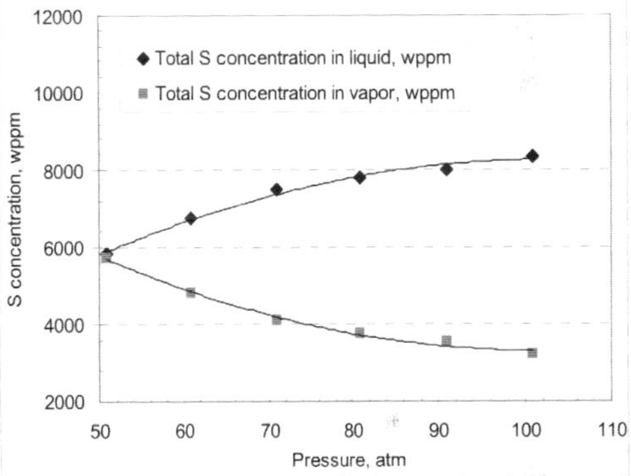

Figure 6. Total S concentration in the vapour and liquid phase at different pressures

Figures 7 and 8 show the 4,6-dimethyldibenzothiophene (4,6-DMDBT) concentrations in the vapour and liquid phase at different temperatures and pressures, respectively, obtained with the original LCO. Unlike the total sulphur concentrations, the 4,6-DMDBT concentrations in the vapour and liquid phase both increase with the increase in temperature and the decrease in pressure. All other dibenzothiophenic compounds investigated in this study in the hydrotreated LCO and original LCO showed similar trends. It is also noted in Figure 7 that the increase in 4,6-DMDBT's concentration in the vapour phase with temperature is more dramatic than that in the liquid phase. Although the 4,6-DMDBT's concentration in liquid phase slightly increases with the increase in temperature and decrease in pressure, the amount of 4,6-DMDBT in liquid phase decreases with the increase in temperature and decrease in pressure, as shown in Figures 9 and 10, since the total amount of LCO in liquid phase decreases dramatically (see Figures 3 and 4) with the increase in temperature and decrease in pressure. This finding might partially explain the observed reduced temperature sensitivity of HDS reactivity at high temperatures (Chen et al., 2003). According to the present observation, dibenzothiophenic compounds, such as 4,6-DMDBT, would be mostly present in the vapour phase at high temperatures. If the catalyst bed in the hydrotreating reactor were still in complete catalyst wetting, these sulphur compound molecules in the vapour phase would not have a chance to directly contact with catalyst surface for hydrotreating reactions. This is equivalent to partial catalyst by-passing, resulting in a reduced HDS conversion. Such a detrimental effect of phase equilibrium on HDS becomes more significant at higher temperatures and lower pressures. It will also depend on the boiling point distribution of the hydrocarbon matrix of the feed being processed.

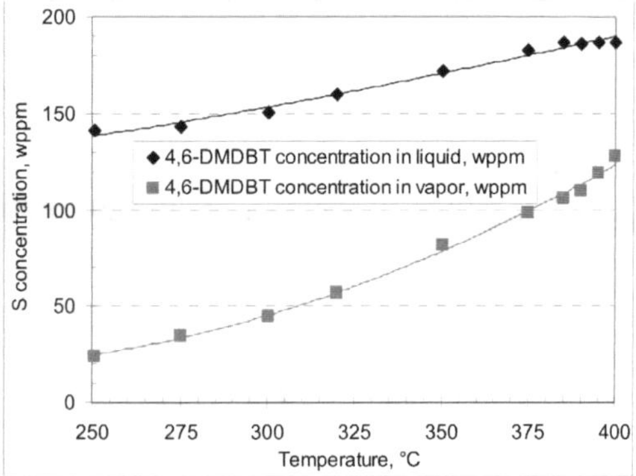

Figure 7. 4,6-DMDBT concentrations in vapour and liquid phase at different temperatures

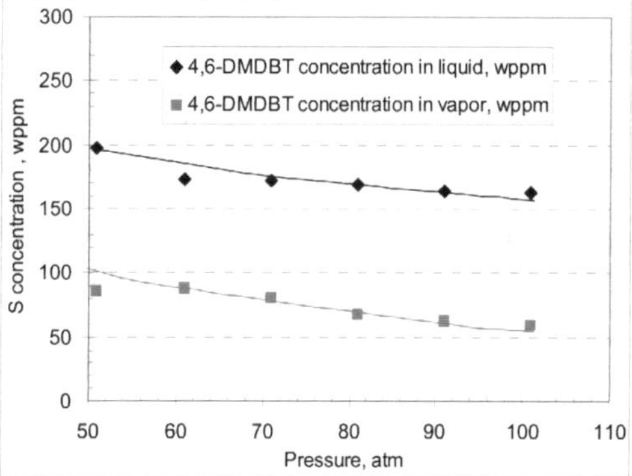

Figure 8. 4,6-DMDBT concentrations in vapour and liquid phase at different pressures

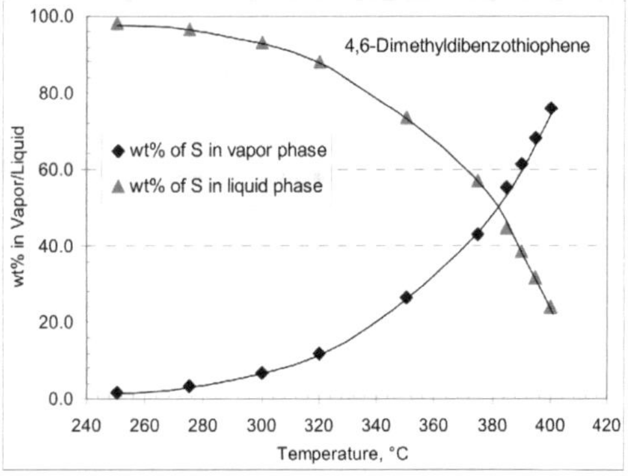

Figure 9. Wt% of total 4,6-DMDBT in vapour and liquid phase at different temperatures

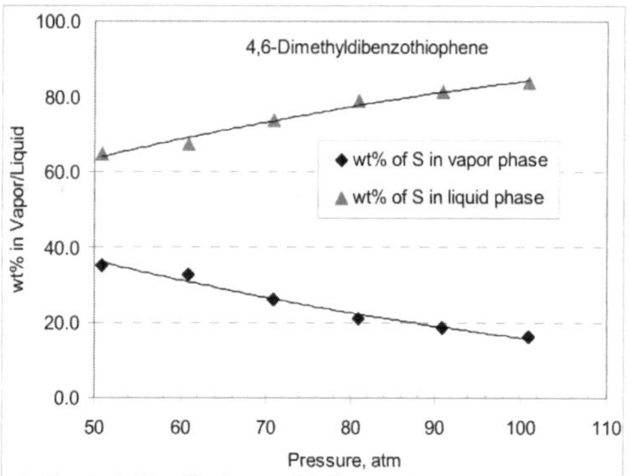

Figure 9. Wt% of total 4,6-DMDBT in vapour and liquid phase at different pressures

In vapour-liquid phase equilibrium, the equilibrium ratio or the K factor represents the volatility of a component in a mixture. The equilibrium ratio of a component is defined as the ratio of the mole faction in the vapour phase to that in the liquid phase. The differences in equilibrium ratios of the sulphur compounds investigated in this study represent their relative volatilities. Table 2 and Table 3 list the equilibrium ratio values for the 27 dibenzothiophenic compounds at different temperatures (at the pressure of 70 atm.) and pressures (at the temperature of 350°C), respectively. These sulphur compounds represent dibenzothiophene and its methyl and/or ethyl substitutes at different positions. Since the equilibrium ratios of the same component in the hydrotreated LCO and original LCO are quite close to each other, the averaged values between these two are listed in Tables 2 and 3. As seen in the tables, under the same pressure (Table 2) and temperature (Table 3) the equilibrium ratios increased dramatically with the increase in temperature (Table 2) and decrease in pressure (Table 3).

Also it is observed that smaller molecules (with no or fewer attaching methyl/ethyl groups) have higher K values than larger molecules (with more attaching methyl/ethyl groups).

The data acquired in this study are being used to estimate the interaction coefficients between hydrocarbons (including organic sulphur and nitrogen compounds) and hydrogen. These interaction coefficients are required parameters during flash calculation using equation of state. These results will be published in the future.

4. CONCLUSIONS

1) The continuous flow VLE cell used in this study provided reliable VLE experimental data for systems containing hydrogen and petroleum fractions. In this study VLE data were accurately measured with systems consisting of hydrogen and petroleum light cycle oil at different temperatures and pressures.
2) The relative amount of LCO, total sulphur, and individual sulphur compounds in the vapour phase significantly increased with temperature and decreased with pressure. However, the rates of increase differed which has implications for the kinetics analysis of HDS data.
3) The total sulphur concentration in the liquid phase decreased with the increase in temperature and decrease in pressure while that in the vapour phase increased with the increase in temperature and decrease in pressure. However, the concentrations of dibenzothiophenic compounds, such as 4,6-DMDBT, in both the liquid and the vapour phase increased with the increase in temperature and decrease in pressure.
4) Generally smaller dibenzothiophenic compounds have higher volatility than larger dibenzothiophenic compounds as represented by the K values.

5. ACKNOWLEDGEMENT

The authors are grateful to the following individuals for their help and support: Dennis Carson for conducting the pilot plant experiments, Yevgenia Briker for sulphur speciation analysis, and Norm Sacuta for editing this paper. Partial funding for NCUT has been provided by the Canadian Program for Energy Research and Development (PERD), the Alberta Research Council (ARC) and the Alberta Energy Research Institute (AERI).

Table 2. Equilibrium ratios for individual sulphur compounds at different temperatures

Temperature, °C	250.45	300.04	350.51	400.30
Dibenzothiophene	0.0045	0.0129	0.0354	0.0738
4-Methyldibenzothiophene	0.0029	0.0106	0.0287	0.0644
2-Methyldibenzothiophene	0.0032	0.0092	0.0285	0.0609
3-Methyldibenzothiophene	0.0028	0.0093	0.0281	0.0702
4-Mthyldibenzothiophene	0.0024	0.0085	0.0249	0.0567
4,6-Dimethyldibenzothiophene	0.0021	0.0081	0.0244	0.0594
2,4-Dimethyldibenzothiophene	0.0019	0.0077	0.0222	0.0573
2,6-Dimethyldibenzothiophene	0.0019	0.0079	0.0231	0.0572
3,6-Dimethyldibenzothiophene	0.0021	0.0075	0.0220	0.0587
2-Ethyldibenzothiophene	0.0018	0.0071	0.0216	0.0707
2,8-Dimethyldibenzothiophene	0.0019	0.0088	0.0243	0.0552
2,7-Dimethyldibenzothiophene	0.0022	0.0078	0.0259	0.0727
3,7-Dimethyldibenzothiophene	0.0026	0.0095	0.0321	0.0828
1,8-Dimethyldibenzothiophene	0.0023	0.0161	0.0416	0.0682
1,4-Dimethyldibenzothiophene	0.0020	0.0076	0.0207	0.0594
1,6-Dimethyldibenzothiophene	0.0020	0.0071	0.0209	0.0430
3,4-Dimethyldibenzothiophene	0.0020	0.0071	0.0205	0.0530
4-Ethyl-6-mehyldibenzothiophene	0.0007	0.0066	0.0204	0.0486
1,7-Dimethyldibenzothiophene	0.0012	0.0072	0.0243	0.0701
2,3-Dimethyldibenzothiophene	0.0010	0.0072	0.0353	0.0729
2,4,6-Trimethyldibenzothiophene	0.0015	0.0062	0.0190	0.0497
2,4,8-Trimethyldibenzothiophene	0.0019	0.0061	0.0189	0.0464
2,4,7-Trimethyldibenzothiophene	0.0018	0.0057	0.0179	0.0487
1,4,6-Trimethyldibenzothiophene	0.0016	0.0055	0.0173	0.0482
1,4,7-Trimethyldibenzothiophene	0.0022	0.0067	0.0176	0.0484
3,4,7-Trimethyldibenzothiophene	0.0014	0.0060	0.0162	0.0409
1,3,7-Trimethyldibenzothiophene	0.0014	0.0071	0.0212	0.0465

Table 3. Equilibrium ratios for individual sulphur compounds at different pressures

Temperature, °C	50.92	70.93	90.93	100.92
Dibenzothiophene	0.0423	0.0354	0.0255	0.0258
4-Methyldibenzothiophene	0.0336	0.0287	0.0237	0.0222
2-Methyldibenzothiophene	0.0355	0.0285	0.0210	0.0190
3-Methyldibenzothiophene	0.0310	0.0281	0.0229	0.0192
4-Mthyldibenzothiophene	0.0263	0.0249	0.0184	0.0174
4,6-Dimethyldibenzothiophene	0.0273	0.0244	0.0182	0.0169
2,4-Dimethyldibenzothiophene	0.0257	0.0222	0.0170	0.0156
2,6-Dimethyldibenzothiophene	0.0251	0.0231	0.0177	0.0157
3,6-Dimethyldibenzothiophene	0.0242	0.0220	0.0165	0.0156
2-Ethyldibenzothiophene	0.0265	0.0216	0.0167	0.0139
2,8-Dimethyldibenzothiophene	0.0243	0.0243	0.0164	0.0133
2,7-Dimethyldibenzothiophene	0.0231	0.0259	0.0186	0.0153
3,7-Dimethyldibenzothiophene	0.0273	0.0321	0.0178	0.0147
1,8-Dimethyldibenzothiophene	0.0554	0.0416	0.0217	0.0218
1,4-Dimethyldibenzothiophene	0.0244	0.0207	0.0156	0.0156
1,6-Dimethyldibenzothiophene	0.0224	0.0209	0.0161	0.0144
3,4-Dimethyldibenzothiophene	0.0225	0.0205	0.0154	0.0148
4-Ethyl-6-methyldibenzothiophene	0.0254	0.0204	0.0154	0.0141
1,7-Dimethyldibenzothiophene	0.0293	0.0243	0.0154	0.0136
2,3-Dimethyldibenzothiophene	0.0236	0.0353	0.0190	0.0117
2,4,6-Trimethyldibenzothiophene	0.0237	0.0190	0.0155	0.0137
2,4,8-Trimethyldibenzothiophene	0.0204	0.0189	0.0144	0.0128
2,4,7-Trimethyldibenzothiophene	0.0210	0.0179	0.0138	0.0122
1,4,6-Trimethyldibenzothiophene	0.0208	0.0173	0.0134	0.0113
1,4,7-Trimethyldibenzothiophene	0.0212	0.0176	0.0160	0.0129
3,4,7-Trimethyldibenzothiophene	0.0192	0.0162	0.0122	0.0120
1,3,7-Trimethyldibenzothiophene	0.0207	0.0212	0.0156	0.0122

6. REFERENCES

Akgerman, A., Collins, G. M., Hook, B. D., "Effect of Feed Volatility on Conversion in Trickle-bed reactors", Ind. Engng. Chem. Fundam., 24, 398-401, 1985.

Bellos, G. D., and Papayannakos, N. G., "The Use of a Three Phase Microreactor to Investigate HDS Kinetics", Catal. Today, 79&80, 349-355, 2003.

Chen, J., Te, M., Yang, H., and Ring, Z., "Hydrodesulfurization of Dibenzothiophenic Compounds in a Light Cycle Oil", Petrol. Sci. and Technol., 21 (5&6), 911-935, 2003.

Chen, J. and Ring, Z., "HDS Reactivities of Dibenzothiophenic Compounds in a LC-Finer LGO and H_2S/NH_3 Inhibition Effect", Fuel, 83,305-313, 2004.

Connolly, J. and Kandalic, G., "Gas Solubilities, Vapour-Liquid Equilibria, and Partial Molar Volumes in Some Hydrogen-Hydrocarbon Systems", J. Chem. Eng. Data, 31, 396-406, 1986.

Froment, G. F., Depauw, G. A., and Vanrysselberghe, V., "Kinetic Modeling and Reactor Simulation in Hydrodesulphurization of Oil Fractions", Ind. Engng. Chem. Res., 33, 2975-2988, 1994.

de Jong, K. P., "Effect of Vapour-Liquid Equilibrium on Coke Deposition in Trickle-Bed Reactors during Heavy Oil Processing. 2. Modeling", Ind. Engng. Chem. Res. 33, 3141-3145, 1994.

Kocis, G. R. and Ho, T. C., "Effects of Liquid Evaporation on the Performance of Trickle-Bed Reactors", Chem. Engng. Res. Des., 64, 288-291, 1986.

Laugier, S., Richon, D., and Renon, H., "Vapour-Liquid Equilibria of Hydrogen-Hydrocarbon Mixtures Up to 40 MPa, 400°C: Measurements and Representation", Chemical Engineering Thermodynamics, paper presented at the Thermodynamics Sessions of the 2nd World Congress on Chemical Engineering, 1983, Montreal, Que., Canada.

LaVopa, V., and Satterfield, C. N., "Some Effects of Vapour-Liquid Equilibria on Performance of a Trickle-Bed Reactor", Chem. Engng. Sci., 43, 2175-2180, 1988.

Meyers, R. A., Handbook of Petroleum Refining Processes, 2nd ed., New York, McGraw-Hill, 1996.

Ramanujam, S., Leipziger, S., and Well, S., "Vapour-Liquid Equilibrium for a Hydrogen/ Simulated Coal-Derived Liquid System", Ind. Engng. Chem. Proc. Des. Dev., 24, 364-368, 1985.

Simnick, J., Sebastian, H., Li, H-M, and Chao, K-C, "Solubility of Hydrogen in Toluene at Elevated Temperatures and Pressures", J. Chem. Eng. Data, 23, 339-340, 1978.

Smith, C. M. and Satterfield, C. N., "Some Effects of Vapour-Liquid Flow Ratio on Performance of a Trickle-bed reactor", Chem. Engng. Sci. 41, 839-843 1986.

Tesser, R., Di Serio, M., and Santacesaria, E., "Influence of the Vapour-Liquid Equilibrium (VLE) on the Kinetics in Gas-Liquid and Gas-Liquid-Solid Systems", Catal. Today, 79&80, 323-331, 2003.

Wiegand, K., Strobel, B. and Hofmann, H., "Vapour-Liquid Equilibrium of Hydrogen/Coal Liquid and Methane/Coal Liquid Systems at Elevated Temperatures and Pressures", Chem. Engng. Technol., 12, 280-288, 1989.

Yang, H., Chen, J., Fairbridge, C., Yevgenia, B., Zhu. Y., and Ring, Z.," Inhibition of Nitrogen Compounds on the Hydrodesulfurization of Substituted Dibenzothiophenes in Light Cycle Oil", Fuel Proc. Technol., 85, 1415-1429, 2004.

Catalytic Hydrodesulphurization of DBTs in the Presence of Aromatics

Jinwen Chen[a], Nobumasa Nakajima[b], Hong Yang[a], Zbigniew Ring[a]
Tao Song[c], Jason Zhang[d]

[a] *National Center for Upgrading Technology,*
One Oil Patch Drive, Devon, AB, T9G 1A8, Canada
[b] *Cosmo Oil Co. Ltd. 1134-2, Gongendo, Satte-Shi, Saitama-Ken 340-0193, Japan*
[c] *Currently in Nova Chemicals, 2928 16 Ave., NE., Calgary, AB, T2E 7K7, Canada*
[d] *Department of Chemical Engineering, University of Ottawa, 161 Louis Pasteur St., Ottawa, Canada K1N 6N5*

Prepared for Presentation at the 2005 AIChE National Spring Meeting, Session 61: Advances in Hydrotreating and Ultra-Clean Fuel Production, April 10-14, 2005, Atlanta, GA, USA

April 2005

Abstract

This study quantifies the effect of aromatics on the HDS reaction mechanism through experiments in a bench-scale trickle-bed reactor over a commercial hydrotreating $NiMo/Al_2O_3$ catalyst. A special feed was designed and prepared with model sulphur and aromatics compounds, and solvent to simulate the chemical composition of light cycle oil. The HDS kinetics of dibenzothiophene (DBT) and 4,6-dimethyldibenzothiophene (4,6-DMDBT) were assumed to follow Langmuir-Hinshelwood (L-H) type kinetics, with the aromatic compounds being the competitive adsorption species. To simplify the problem, the hydrogenation kinetics of the aromatic compounds were lumped into 5 different groups according to the number of aromatic rings in the compounds, namely naphthenes, mono-aromatics, di-aromatics, tri-aromatics and tetra-aromatics. It was further assumed that within each aromatic group, the compounds had the same reactivity and adsorption equilibrium constant. The hydrogenation kinetics of these 5 aromatic groups was assumed to be of power law kinetics, and the kinetics parameters were estimated prior to the evaluation of the HDS kinetics of DBT and 4,6-DMDBT. The L-H type of HDS kinetics for DBT and 4,6-DMDBT were then evaluated.

1. INTRODUCTION

In the last several years, deep hydrodesulphurization (HDS) has received more attention due to the increasingly stringent environmental regulations on sulphur content in transportation fuels (Chen et al., 2003, 2004; Song, 2003). Since hundreds of sulphur compounds are present in middle distillates (components for diesel blending) and since these HDS reactions are further affected by hydrodenitrogenation (HDN) and hydrodearomatization (HDA) reactions, the mechanism of HDS is still not fully understood, although it has been a research subject for many years. For example, the quantitative effect of aromatics content on the HDS reaction mechanism is rarely studied and reported in the open literature. Understanding this effect is especially important when middle distillates rich in aromatics, such as light cycle oil (LCO) from the fluid catalytic cracking (FCC) process, are hydrotreated. Lo (1981) studied the inhibition effect of naphthalene on HDS of DBT over $NiMo/Ai_2O_3$ and concluded that naphthalene inhibited both the hydrogenation and hydrogenolysis reactions of the DBT HDS reaction network mildly. Nagai and Kabe (1983) observed that phenanthrene affected HDS of DBT over Mo/Al_2O_3 only through the hydrogenation reaction. However, this effect was much stronger than that by naphthalene. Isoda et al. (1996) conducted a 4,6-DMDBT HDS study in the presence of naphthalene and found that a mixture of $NiMo/Al_2O_3$ and Ru/Al_2O_3 catalyst could minimize the inhibition effect of naphthalene on the HDS of 4,6-DMDBT. More recently, Farag et al. (1999) studied the inhibition effect of naphthalene on the HDS of 4,6-DMDBT over CoMo/C catalyst. They observed that increasing naphthalene concentration in the feed significantly decreased the HDS conversion, through both hydrogenation and hydrogenolysis reactions with slightly less effect on hydrogenolysis. Koltai et al. (2002) investigated the comparative inhibiting effect of anthracene, phenantrene and fluorene on the HDS of 4,6-DMDBT over $NiMo/Al_2O_3$ catalyst. Among these 3 aromatic compounds, fluorene was the strongest HDS inhibitor while anthracene was the weakest one. A Langmuir-Hinshelwood (L-H) type of kinetics model was used to qualitatively describe the inhibition effect. Ho (2003) conducted dynamic experiments to investigate the naphthalene inhibition effect on the HDS of 4,6-diethyldibenzothiophene (4,6-DEDBT) over $CoMo/Al_2O_3$ - SiO_2 catalyst. He observed a significant decrease in HDS conversion when feed containing 10wt% of naphthalene was used. However, this inhibition effect could be completely reversed after striping the catalyst with non-naphthalene-containing feed for 200 hours. Egorova and Prins (2004) also used naphthalene to study the inhibition effect of aromatics on the HDS of DBT and 4,6-DMDBT over NiMo/Al_2O_3. They concluded that the HDS of both DBT and 4,6-DMDBT were inhibited by naphthalene with a stronger inhibition effect measured for 4,6-DMDBT. The authors also observed that naphthalene inhibited both hydrogenation and hydrogenolysis pathways to a similar extent.

As discussed above, the aromatic inhibition studies reported so far were all limited to feeds with a single sulphur and a single aromatic compound at relatively low concentrations of aromatics in the feed (less than 10wt%). Middle distillates used as diesel blending components normally contain much more aromatics (>30w%) with a different number of aromatic rings, from one to four. It is therefore important and informative to study the aromatics inhibition effect on HDS of dibenzothiophenic compounds by using feeds with higher aromatics contents and multiple aromatic compounds of similar compositions to those in middle distillates. This is the objective of the present study.

2. EXPERIMENTAL

Details of the experimental setup, procedures, conditions and related analysis have been reported elsewhere (Song et al., 2003). Only a brief description is given below.

The experimental setup used was a micro fixed-bed reactor (0.64 mm in diameter and 30 cm in length) packed with 6 ml (6.2 g) of crushed commercial NiMo/Al_2O_3 hydrotreating catalyst (35-60 mesh). A movable thermocouple, housed in a centrally located thermowell, was used to measure the reactor bed temperature in the axial direction. The reactor tube was heated by a computer-controlled three-zone soft furnace to ensure isothermal operation both axially and radially. After heated to the desired temperature in a pre-heater, the mixture of feed and treat gas (hydrogen) was introduced into the reactor from the bottom in up-flow mode. The reactor effluent went through a high-pressure (HP) separator where hydrogen and liquid product were separated. The liquid product was further introduced into a low-pressure (LP) separator to remove any dissolved hydrogen and then collected in a product receiver. The two gas streams from the HP and LP separator were mixed and sent to vent after metering.

To study the aromatics inhibition effect, special feeds were designed and prepared. DBT and 4,6-DMDBT were used as sulphur model compounds to represent relatively easy-to-remove and most difficult-to-remove dibenzothiophenic compounds in middle distillates. N-hexadecane was used as solvent. 6 compounds were selected as aromatic model compounds, which included two mono-aromatics, two di-aromatics, one tri-aromatic and one tetra-aromatic: 1,3,5-thimethylbenzene, tetralin, 2-methylnaphthalene, fluorene, phenanthrene and pyrene, respectively. The relative amounts of these four types of aromatic compounds in the designed feeds were set to match those in the similar boiling range of a real LCO sample determined through GC-MS by-boiling-point-distribution analysis. The total content of aromatics in the designed feeds varied between 0 and 40wt% (5 different levels). Other experimental conditions, such as temperature, pressure, gas-to-oil ratio and liquid hourly space velocity (LHSV), also covered wide ranges to obtain meaningful kinetics data. The identification of each individual aromatic compound and its corresponding hydrogenated products were achieved with GC-MS, while the quantification of each species present in the feed and hydrotreated products was achieved with GC-AED.

It should be noted that, throughout the experimental program, catalyst deactivation was monitored by periodically conducting check-back experiments with the same feed and under the same conditions. The data reported in this paper were already corrected according to the catalyst deactivation behavior.

3. EXPERIMENTAL RESULTS

3.1. Catalyst Deactivation

As mentioned above, the catalyst deactivation was monitored during the experimental program. The feed used for the check-back experiments contained only DBT and 4,6-DMDBT with no aromatics. The check-back conditions were: temperature of 200°C, pressure of 70 atm., LHSV of 1.58 h^{-1}, and gas/oil of 1000 NL/kg feed. The catalyst deactivations for DBT and 4,6-DMDBT, in terms of pseudo-first order rate constants, are shown in Figure 1. The catalyst activity dropped significantly during the first 1250 h of catalyst age. After that, it tended to stabilize until the end of the program (2600 h of catalyst age). Overall, the catalyst activity decreased by about 40% for DBT and 60% for 4,6-DMDBT. Another immediate observation from Figure 1 is that 4,6-DMDBT is much less reactive than DBT, which is usually attributed to the steric hindrance effect of the two methyl groups in 4,6-DMDBT (Lecrenay et al., 1997; Shin et al. 2001; Chen and Ring 2003; Chen et al., 2003, 2004).

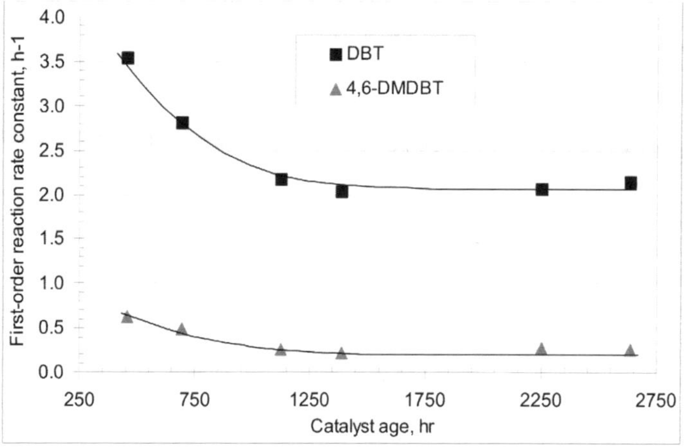

Figure 1. Catalyst deactivation over time

3.2. Aromatics Inhibition Effect on HDS

In this work, the total aromatics contents in the feed was varied at 5 different levels: 0w%, 5w%, 10w%, 20w%, and 40w%. Figures 2(a) and 2(b) show the conversions of DBT and 4,6-DMDBT, respectively, along with the variation of aromatics content in the feed. As seen in these figures, the HDS of both DBT and 4,6-DMDBT was significantly inhibited by aromatics in the feed. With the increase in aromatics content from 0 to 40wt% in the feed, DBT conversion decreased from 90% to 15% and 4,6-DMDBT conversion decreased from 35% to 3% at the reaction temperature of 473 K. At higher temperatures of 513K and 553K, similar observations can be made, although DBT was completely converted (no DBT was detected in the effluent) at a temperature of 553K. This observation is consistent with those reported in the literature (Farag et al., 1999; Koltai et al., 2002). The two figures note that as the aromatics content increases, the increase in inhibition effect become weaker, which follows the Langmuir-Hinshelwood type of equation as discussed below.

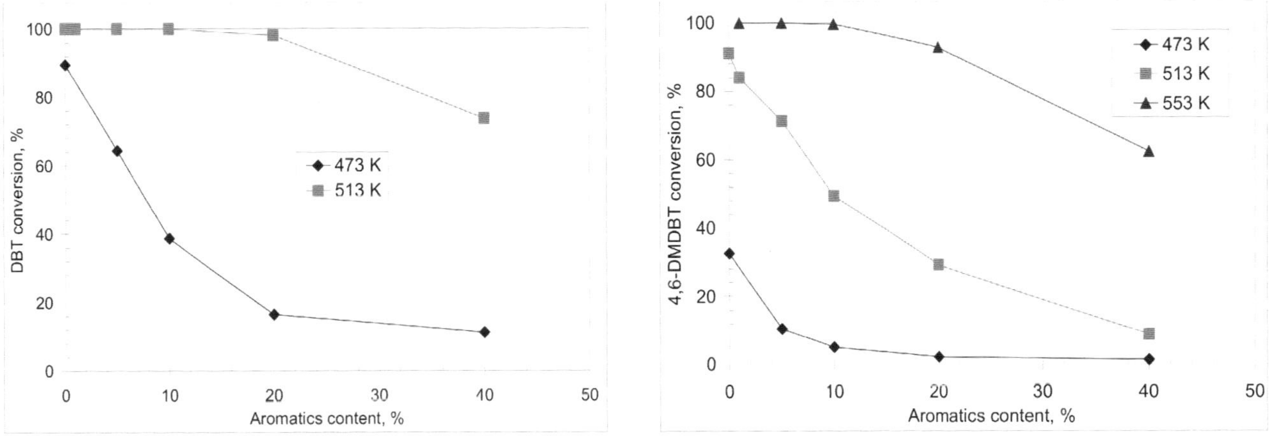

(a) DBT (b) 4,6-DMDBT

Figure 2. DBT and 4,6-DMDBT conversions at different aromatics content in the feed

As already reported in an earlier publication (Song et al., 2003), at certain aromatics contents in the feed, the HDS of both DBT and 4,6-DMDBT could be approximated by pseudo-first order kinetics. The pseudo-first order reaction rate constants for DBT and 4,6-DMDBT obtained under different temperatures and with different feeds are shown in Figures 3(a) and 3(b), respectively. The straight lines of Ln(k) vs. 1/T indicate the pseudo-first order assumption is acceptable in this work.

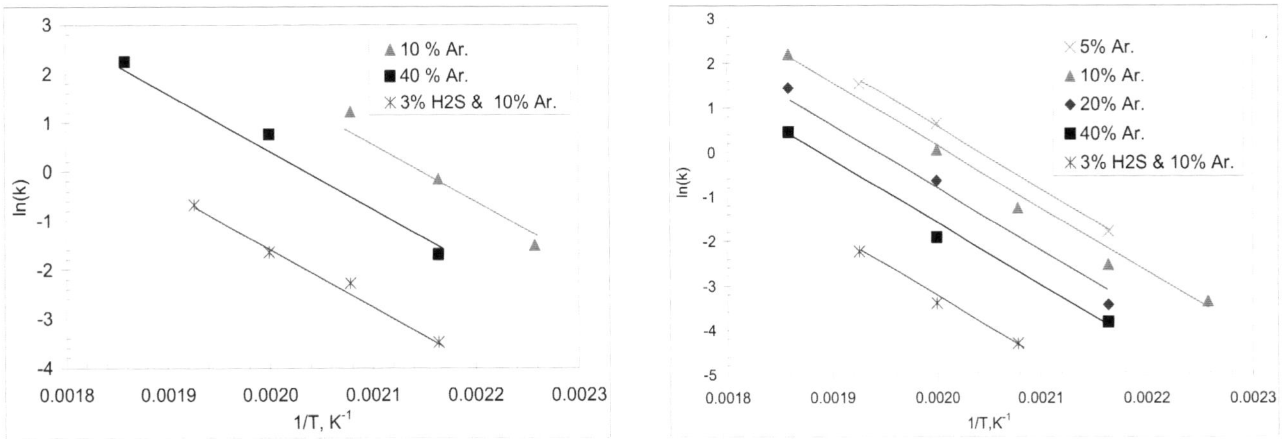

(a) DBT (b) 4,6-DMDBT

Figure 3. Pseudo-first order reaction rate constants at different temperatures

4. KINETICS MODELING

To properly model the HDS kinetics of DBT and 4,6-DMDBT with the significant inhibition effect of aromatics, a Langmuir-Hinshelwood type of kinetic model was proposed. Since there are too many individual aromatic compounds in the system (those present in the feed and their hydrogenated intermediate products) and since a limited number of experimental data points are available, it is impractical to account for all the aromatic compounds individually in the kinetics model. Instead, a lumping method was used in this study to lump these aromatics compounds into 5 different groups according to the number of aromatic rings. These lumps are named as naphthene, mono-, di-, tri-, and tetra- aromatics (Naph, Mono, Di, Tri, and Tetra, for short). Each lump was represented by a single concentration, reaction kinetics and adsorption equilibrium constant. It was further assumed that the inhibition effect of H_2S (generated during HDS) and of the DBT and 4,6-DMDBT themselves are negligible due to their much lower concentrations in the liquid phase, compared to that of aromatics. Therefore, the HDS reaction rates for DBT and 4,6-DMDBT are written as follows:

$$r_{DBT} = \frac{k_{DBT}C_{DBT}}{1 + K_{Mono}C_{Mono} + K_{Di}C_{Di} + K_{Tri}C_{Tri} + K_{Tetra}C_{Tetra}} \quad (1)$$

$$r_{4,6-DMDBT} = \frac{k_{4,6-DMDBT}C_{4,6-DMDBT}}{1 + K_{Mono}C_{Mono} + K_{Di}C_{Di} + K_{Tri}C_{Tri} + K_{Tetra}C_{Tetra}} \quad (2)$$

To further simplify the modeling work, a power-low type of kinetics was applied to the hydrogenation of each aromatic lump. In addition, it was assumed that DBT and 4,6-DMDBT concentrations did not affect these hydrogenation reactions. The hydrogenation of the 4 aromatic groups can be represented by the following set of irreversible reactions in series:

$$A4 \xrightarrow{k_4} A3 \xrightarrow{k_3} A2 \xrightarrow{k_2} A1 \xrightarrow{k_1} A0 \quad (3)$$

The reaction rates for the 4 aromatic groups are written as:

$$r_{Mono} = k_1 C_{Mono}^{n_1} - k_2 C_{Di}^{n_2} \quad (4)$$

$$r_{Di} = k_2 C_{Di}^{n_2} - k_3 C_{Tri}^{n_3} \quad (5)$$

$$r_{Tri} = k_3 C_{3Tri}^{n_3} - k_4 C_{Tetra}^{n_4} \quad (6)$$

$$r_{Tetra} = k_4 C_{Tetra}^{n_4} \quad (7)$$

The reactor used in this study can be reasonably described by a plug-flow model. For DBT and 4,6-DMDBT,

$$\frac{dC_{DBT}}{dx} = -\frac{1}{Sv}\frac{k_{DBT}C_{DBT}}{1 + K_{Mono}C_{Mono} + K_{Di}C_{Di} + K_{Tri}C_{Tri} + K_{Tetra}C_{Tetra}} \quad (8)$$

$$\frac{dC_{4,6-DMDBT}}{dx} = -\frac{1}{Sv}\frac{k_{4,6-DMDBT}C_{4,6-DMDBT}}{1 + K_{Mono}C_{Mono} + K_{Di}C_{Di} + K_{Tri}C_{Tri} + K_{Tetra}C_{Tetra}} \quad (9)$$

For the 4 aromatics groups,

$$\frac{d\vec{C}}{dx} = M\vec{C}_p \quad (10)$$

where

$$\vec{C} = \begin{bmatrix} C_1 \\ C_2 \\ C_3 \\ C_4 \end{bmatrix} \quad \vec{C}_p = \begin{bmatrix} C_1^{\ n_1} \\ C_2^{\ n_2} \\ C_3^{\ n_3} \\ C_4^{\ n_4} \end{bmatrix} \quad M = \begin{bmatrix} -\frac{k_1}{Sv} & \frac{k_2}{Sv} & 0 & 0 \\ 0 & -\frac{k_2}{Sv} & \frac{k_3}{Sv} & 0 \\ 0 & 0 & -\frac{k_3}{Sv} & \frac{k_4}{Sv} \\ 0 & 0 & 0 & -\frac{k_4}{Sv} \end{bmatrix} \tag{11}$$

In Equations (1) to (11):

C_{DBT}, $C_{4,6\text{-}DMDBT}$: Concentration of DBT and 4,6-DMDBT.

C_{Mono}, C_{Di}, C_{Tri}, C_{Tetra}: Concentration of Mono, Di, Tri and Tetra.

k_{DBT}, $k_{4,6\text{-}DMDBT}$: Reaction rate constants for DBT and 4,6-DMDBT.

K_{Mono}, K_{Di}, K_{Tri}, K_{Tetra}: Adsorption equilibrium constants for Mono, Di, Tri and Tetra.

k_1, k_2, k_3, k_4: Reaction rate constants for Mono, Di, Tri and Tetra.

n_1, n_1, n_1, n_1: Reaction order for Di, Tri and Tetra.

Sv : Liquid hourly space velocity

x : Dimensionless axial position

Since Equations (4) to (7) are independent from Equations (1) and (2), the kinetics parameters (k_1 to k_4 and n_1, to n_4) of the hydrogenation reactions can be evaluated prior to the evaluation of the kinetics parameters for the HDS reactions. In this case, the aromatics concentrations in Equations (1) and (2) can be calculated independently during the estimation of the HDS parameters (k_{DBT}, $k_{4,6\text{-}DMDBT}$, K_{Mono}, K_{Di}, K_{Tri}, K_{Tetra}), which makes the parameter estimation much easier. At this time, the kinetics models described above do not account for the temperature effect since the experimental data used for the parameter estimation were obtained under the same temperature.

The parameter estimation of the aromatics hydrogenation kinetics model was performed by fitting the experimentally measured concentrations in the reactor effluent with those calculated from the reactor model [Equations (10) and (11)]. Figures 4(a) and 4(b) shows the fitting results. Clearly, the fit is reasonably good, indicating that this simple model can be used to calculate the aromatics concentration profiles along the reactor (the sole purpose of this exercise). The model parameters are:

$k_1 = 1.35\ (m^3/mol)^{-0.3}\ h^{-1}$, $k_2 = 0.027\ (m^3/mol)^{1.1} h^{-1}$, $k_3 = 4.10\ h^{-1}$, $k_4 = 5.47\ (m^3/mol)^{0.3}\ h^{-1}$

$n_1 = 0.7$, $n_2 = 2.1$, $n_3 = 1.0$, $n_4 = 1.3$

As seen above, the hydrogenation reaction orders are in the range from 0.7 to 2.1 for the 4 different aromatics groups. One should not assume that this model fully represents the real hydrogenation mechanism. Again, the purpose of this model is to independently calculate the axial concentration profiles of the 4 aromatics groups that are required during the HDS kinetics model parameter estimation.

With the aromatics concentration profiles modeled, the kinetics parameters in Equations (1) and (2), k_{DBT}, $k_{4,6\text{-}DMDBT}$, K_{Mono}, K_{Di}, K_{Tri}, K_{Tetra}, were estimated by fitting the experimentally measured concentrations of DBT and 4,6-DMDBT in the reactor effluent with those calculated from the reactor model (Equations 8 and 9). The fitting results are shown in Figure 5. Again, in the figure the fitting is very good. The kinetics parameters are:

$k_{DBT} = 4024.3\ h^{-1}$, $k_{4,6\text{-}DMDBT} = 222.4\ h^{-1}$
$K_{Mono} = 1.27\ m^3/mol$, $K_{Di} = 0.75\ m^3/mol$, $K_{Tri} = 0.26\ m^3/mol$, $K_{Tetra} = 0.1323\ m^3/mol$

It is clear from these numbers that the intrinsic reactivity of DBT is much higher than that of 4,6-DMDBT. As discussed by many researchers (Lecrenay et al., 1997; Shin et al., 2001; Chen et al., 2004), the HDS of DBT proceeds mainly through the hydrogenolysis route, while that of 4,6-DMDBT proceeds mainly through the hydrogenation route due to the steric

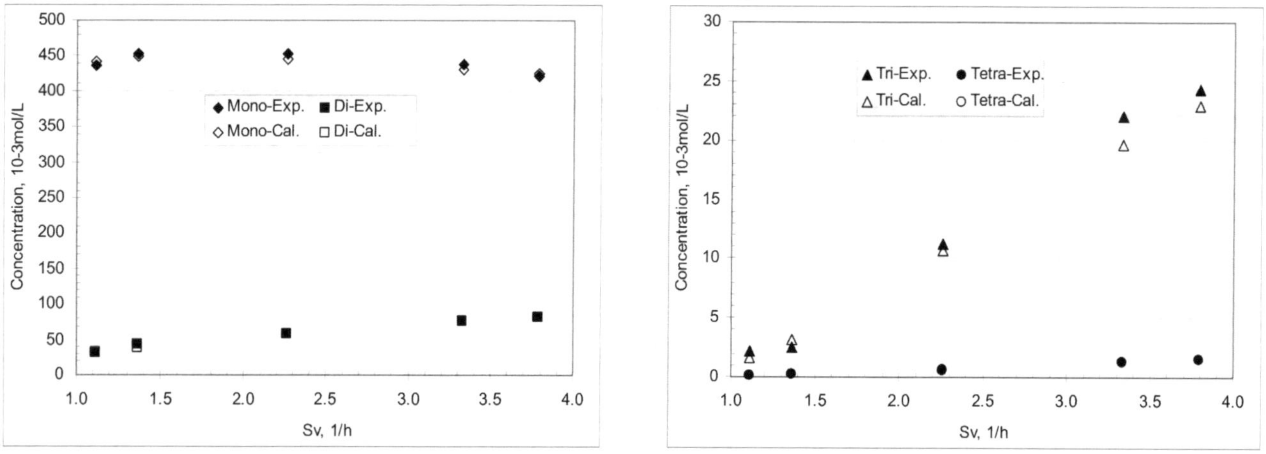

(a) Mono and Di (b) Tri and Tetra

Figure 4. Aromatics concentrations in the effluent-model fitting vs. experimental data

hindrance effect of the two methyl groups at the 4 and 6 positions. As observed in this study, since the HDS of both DBT and 4,6-DMDBT were inhibited by aromatics, it can be concluded that the inhibition of aromatics affected both hydrogenolysis and hydrogenation reactions. This is consistent with the conclusions obtained by other researchers (Lo, 1981; Farag et al. 1999; Egorova and Prins, 2004)

The kinetics parameters above also illustrate that the adsorption equilibrium constant decreases with the increase in the number of aromatic rings. These adsorption equilibrium constants represent the lumped adsorption ability of a number of compounds with the same number of aromatic rings. Although such data have not been reported in the literature, Koltai et al. (2002) reported a similar trend that fluorene had a higher adsorption equilibrium coefficient than that of phenanthrene. From the experimental findings in this study it was also concluded that aromatics with the greater number of aromatic rings were more reactive than those with fewer aromatic rings in the order of: Tetra > Tri > Di > Mono. Koltai et al. (2002) also reported this observation in their study. It is those aromatic compounds with higher adsorption coefficients that are the major competing species for hydrogenation sites on catalyst surface with sulphur compounds.

5. CONCLUDING REMARKS

The HDS kinetics of DBT and 4,6-DMDBT in the presence of aromatics was studied over a commercial $NiMo/Al_2O_3$ hydrotreating catalyst with designer feeds containing up to 40wt% aromatics. The aromatic compounds in these feeds were 1,3,5-thimethylbenzene, tetralin, 2-methylnaphthalene, fluorene, phenanthrene and pyrene. Langmuir-Hinshelwood type of kinetics was assumed for HDS, with the aromatic compounds being the competitive adsorption species. The aromatic compounds in the system were lumped into 5 different groups according to the number of aromatic rings in the compounds, namely naphthenes, mono-aromatics, di-aromatics, tri-aromatics and tetra-aromatics. The hydrogenation kinetics of these aromatic groups was obtained by assuming power law kinetics. It was found that the hydrogenation reactivities of the aromatics were in the order of Tetra > Tri > Di > Mono while the adsorption equilibrium constants were in the order of Mono > Di >Tri > Tetra. The intrinsic reactivity of DBT was much higher than that of 4,6-DMDBT.

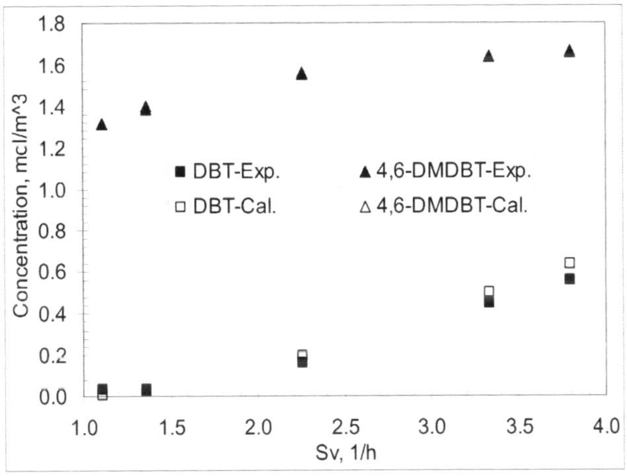

Figure 5. DBT and 4,6-DMDBT concentrations in the effluent-model fitting vs. experimental data

6. ACKNOWLEDGEMENTS

The authors are grateful to the following individuals for their help and support: Dennis Carson for conducting the pilot plant experiments, Yevgenia Briker for sample analysis, and Norm Sacuta for paper editing. Partial funding for NCUT has been provided by the Canadian Program for Energy Research and Development (PERD), the Alberta Research Council (ARC) and the Alberta Energy Research Institute (AERI).

7. REFERENCES

Chen, J., Te, M., Yang, H., and Ring, R. "Hydrodesulphurization of Dibenzothiophenic Compounds in a Light Cycle Oil", Petrol. Sci. Tech., 21(5&6), 911-935, 2003.

Chen, J., Yang, H., and Ring, Z., "HDS Kinetics Study of Dibenzothiophenic Compounds in LCO" Catal. Today, 98, 227-233, 2004.

Chen, J. and Ring, Z., "HDS Reactivities of Dibenzothiophenic Compounds in a LC-Finer LGO and H_2S/NH_3 Inhibition Effects", Fuel, 83(3), 305-313, 2004.

Egorova, M. and Prins, R., "Competitive Hydrodesulphurization of 4,6-Dimethyldibenzothiophene, Hydrodenitrogenation of 2-methylpyridine, and Hydrogenation of Naphthalene over Sulphided NiMo/ -Al_2O_3", J. Catal., 224, 278-287, 2004.

Farag, H., Sakanishi, K., Mochida, I., and Whitehurst, D. D., "Kinetics Analyses and Inhibition by Naphthalene and H2S in Hydrodesulphurization 4,6-Dimethyldibenzothiophene (4,6-DMDBT) over CoMo-Based Carbon Catalyst", Energy & Fuels, 13, 449-453, 1999.

Ho, T. C., "Inhibiting Effects in Hydrodesulphurization of 4,6-Dimethyldibenzothiophene", J. Catal., 219, 442-451, 2003.

Isosa, T., Nagao, S., Ma, X., Korai, Y., and Mochida, I., "Hydrodesulphurization of Refractory Sulphur Species. 1. Selective Hydrodesulphurization of 4,6-Dimethyldibenzothiophene in the major Presence of Naphthalene over CoMo/Al_2O_3 and Ru/Al_2O_3 Blend Catalysts", Energy & Fuels, 10, 482-486, 1996.

Koltai, T., Macaud, M., Guevara, A., Schulz, E., Lemaire M., Bacaud, R., and Vrinat, M., "Comparative Inhibiting Effect of Polycondensed Aromatics and Nitrogen Compounds on the Hydrodesulphurization of Alkyldibenzothiophenes", Applied Catalysis A: General, 231, 253-261, 2002.

Lecrenay, E., Sakanishi, K., and Mochida, I., "Catalytic Hydrodesulphurization of Gas Oil and Model Sulphur Compounds over Commercial and Laboratory-made CoMo and NiMo Catalysts: Activity and Reaction Scheme", Catalysis Today, 39, 13-20, 1997.

Lo, H., S., "Kinetic Modeling of Hydrotreating", Ph. D. thesis, University of Delaware, Newark, 1981.

Nagai, M. and Kabe, T., "Selectivity of Molybdenum Catalyst in Hydrodesulphurization, Hydrodenitrogenation, and Hydrodeoxygenation: Effect of Additives on Dibenzothiophene Hydrodesulphurization", J. Catal. 81, 440-449, 1983.

Shin, S., Yang, H., Sakanishi, K., Mochida, I., Grudoski, A. A., Shinn, J. H., "Inhibition and Deactivation in Staged Hydrodenitrogenation and Hydrodesulphurization of Medium Cycle Oil over NiMoS/Al_2O_3 Catalyst", Applied Catalysis A: 205, 101-108, 2001.

Song, C., "An Overview of New Approaches to Deep Desulfurization for Ultra-Clean Gasoline, Diesel Fuel and Jet Fuel", Catal. Today, 86, 211-263, 2003.

Song, T., Zhang, Z., Chen, J., Ring, Z., Yang, H. and Y. Zheng, "Deep Hydrodesulfurization of Dibenzothiophene and 4,6-Dimethyldibenzothiophene, Kinetics and Effect of Aromatics, Proceedings of the International Symposium on Natural Resources Processing, Tianjin, China, page 297-310, 2003.

Paper No. [61d]

Equilibrium Limitations in Distillate Hydrotreating

Laura Jones and Peter Kokayeff
UOP LLC, Des Plaines, Illinois, USA

Prepared for Presentation at the AIChE 2005 Spring National Meeting, April 10-14, 2005
[61] - Advances in Hydrotreating and Ultra-Clean Fuel Production

Equilibrium Limitations in Distillate Hydrotreating

Laura Jones and Peter Kokayeff

Abstract

As the sulfur requirements for diesel fuel are decreased to lower and lower levels the existence of a "sulfur floor", referring to equilibrium limitations for hydrodesulfurization, has become an important issue to address. The concern is that thermodynamic equilibrium limitations will make it impossible to reach necessary sulfur concentrations at certain operating conditions; e.g., low pressures and high temperatures regardless of space velocity. There are two known pathways for the hydrogenolysis of sulfur in hydrotreating, direct desulfurization and partial hydrogenation prior to sulfur extraction. The hydrogenation mechanism requires the partial saturation of the aromatic rings of the dibenzothiophenes prior to hydrogenolysis. Saturation of the aromatic rings decreases the streric hindrance to hydrogenolysis for the more "difficult" sulfur species. Aromatic saturation is exothermic and reversible, and therefore has equilibrium limitations at high temperatures. In order to better understand and verify the existence of the "sulfur floor" effect data have been compiled from both pilot plant and numerical simulation experiments. The data available indicated that the significance of a sulfur floor effect is dependent on the feedstock properties as well as the operating pressure. Interestingly, a similar phenomenon was observed for the total nitrogen in the product suggesting that there may also be a "nitrogen floor". The presence of a "nitrogen floor" is a concern because nitrogen is a known inhibitor of desulfurization and the presence of a "nitrogen floor" may contribute to the presence of a "sulfur floor".

Introduction

As the sulfur requirements for diesel fuel are decreased to lower and lower levels the existence of a "sulfur floor", referring to equilibrium limitations for hydrodesulfurization, has become an important issue to address. The concern is that thermodynamic equilibrium limitations will make it impossible to reach necessary sulfur concentrations at certain operating conditions; e.g., low pressures and high temperatures regardless of space velocity. Hydrodesulfurization (HDS) for non-sterically hindered sulfur species is reasonably represented by an irreversible reaction inhibited by H_2S which competes for catalyst sites. It is widely accepted that there are two primary desulfurization paths for "difficult" or sterically hindered sulfur species, direct desulfurization (also known as sulfur extraction) and hydrogenation.[1] Direct desulfurization follows the same reaction chemistry as non-sterically hindered sulfur species. Unfortunately the sulfur of substituted dibenzothiophenes (DBTs) is not readily accessible to catalyst sites due to the steric hindrance and rigidity of the fused aromatic rings. The hydrogenation mechanism requires the partial saturation of the at least one of the aromatic rings of the dibenzothiophenes prior to hydrogenolysis. Saturation of the aromatic rings decreases the streric hindrance to hydrogenolysis. Aromatic saturation is exothermic and reversible, and is therefore subject to equilibrium limitations that could potentially limit the extent of sulfur removal under certain conditions.

The potential for thermodynamic limitations in deep desulfurization at temperatures approaching end-of-run is discussed in the literature by a number of researchers. Hu et al describe a sulfur "wall effect" or "sulfur floor" that limits the lowest sulfur level that can be achieved in a diesel stream by raising the temperature.[2] According to Hu et al, the actual concentration limit depends on the catalyst type and hydrogen partial pressure albeit the floor may be as high as 20 wppm under realistic hydrotreater conditions. In a study of the desulfurization kinetics of dibenzothiophenes in light cycle oil (LCO) Chen et al observed that the HDS rate was less sensitive to temperature changes at high temperatures.[3] The lower temperature sensitivity at high temperatures was attributed to the

---

[1] Farag, H., K. Sakanishi, I. Mochida, and D. D. Whitehurst, Kinetic Analyses and Inhibition by Naphthalene and H_2S in Hydrodesulfurizaiont of 4,6-Dimethyldibenzothiophene (4,6-DMDBT) over CoMo-Based Carbon Catalyst, *Energy & Fuels*, 13, 449-453, 1999.

[2] Hu, Michael C., Zbigniew Ring, Jenny Briker, Mure Te. *Rigorous Hydrotreater Simulation*, *PTQ*, 85-91, Spring 2002.

[3] Chen, Jinwen, Hong Yang, Zbigniew Ring. HDS kinetics study of dibenzothiophenic compounds in LCO, Catalysis Today, 98, 227-233, 2004.

hydrogenation/dehydrogenation equilibrium effect at high temperature. Similarly, Shafi observed a decrease in 4,6-dimethyldibenzothiophene conversion at high temperatures (>311 °C, 592 °F) and low pressure (30 bar, 435 psig).[4] This maximum sulfur conversion was not observed at higher pressure (45 bar, 650 psig) indicating the importance of both thermodynamics and kinetics in the system.

Interestingly, a sulfur floor is not observed in model component studies conducted with sterically hindered dibenzothiophenes in hydrocarbon solvents. Ho discusses potential thermodynamic limitations at low pressure and high temperature; however, in an experiment conducted with 4,6-diethyl dibenzothiophene in dodecane solvent he did not observe an equilibrium limitation to the extent of desulfurization at pressures as low as 0.79 MPa (115 psig).[5] Processing 4,6-diethyl dibenzothiophene (4,6-DEDBT) in a hydrocarbon solvent over a commercial catalyst at 2.17 MPa (315 psig) and temperatures ranging from 265 °C to 370 °C (509-698 °F) revealed that temperature inhibited the hydrogenation of the 4,6-DEDBT but not the sulfur removal.[6] Ho suggested that the sulfur removal at test conditions was driven by the sulfur extraction rate rather than the hydrogenation rate. Ho also mentions the possibility of hydrogenation sites converting to hydrogenolysis sites at high temperature. The interconversion of hydrogenation sites (τ-sites) and hydrogenolysis sites (σ-sites) is also discussed by Froment who was the first to introduce a sophisticated and experimentally support mathematical model of τ- and σ-sites.[7] Froment postulated that the catalyst activity and selectivity are impacted by interconversion of the τ- and σ-sites under the action of hydrogen and H_2S and that this phenomenon ought to be accounted for in HDS kinetic models.

Experimental

The experimental data compiled for this study were obtained by processing distillate range feedstocks over Ni/Mo and Co/Mo commercial hydrotreating catalysts in a once through hydroprocessing pilot plant. The oil and hydrogen are passed over an essentially isothermal catalyst bed. The reactor effluent is directed to a separation system to recover the gas and liquid products for analyses. The plant is mass balanced for each test period. The feedstock properties for the relevant experiments are shown in Table 1. Both catalysts were tested at a variety of temperatures and pressures with both feedstocks. Selected data have been extracted from these experiments to illustrate the general trends observed.

Table 1: Feedstock properties

	Feed 1	***Feed 2***
	Straight Run Diesel	***Light Coker Gas Oil***
Density, g/cc	0.8801	0.8733
Gravity, API	29.28	30.69
Sulfur, wt %	1.4	3.21
Nitrogen, wppm	550	1027
IP-391 Aromatics, wt%		
Mono Cyclic, wt%	20.2	19.9
Di Cyclic, wt%	14.2	14.7
Tri+ Cyclic, wt%	2.8	2.4
D2887 Distillation, °F		
IBP / 5wt%	312 / 416	300 / 374
10 / 30	457 / 550	401 / 467
50 / 70	598 / 631	517 /570
90 / 95	678/697	631/655
EBP	749	718

[4] An exploratory study into the Hydrodesulphurization of Hindered Dibenzothiophenes, by Raheel Shafi, Liverpool University, UK, June 2000. (private correspondence, unpublished at this time)

[5] Teh Ho. Deep HDS of diesel fuel: chemistry and catalysis, Catalysis Today, 98, 3-18, 2004.

[6] Teh Ho. Inhibiting effects in hydrodesulfurization of 4,6-diethyldibenzothiophene, Journal of Catalysis, 219, 442-451, 2003.

[7] Froment, G.F. Modeling in the development of hydrotreatment processes, Catalysis Today, 92, 43-54, 2004.

Aromatic Saturation

The reversible step of the hydrodesulfurization (HDS) and hydrodenitrogenation (HDN) reaction routes is the saturation of at least one of the aromatic rings of the sulfur or nitrogen containing heterocyclic aromatic compounds. Therefore, aromatic saturation is the most logical starting point for this discussion of the equilibrium limitations in distillate hydrotreating. Aromatic saturation is a significant reaction in distillate hydrotreating. The aromatic content of the feedstock varies widely depending on the source of the material; however, in all cases the aromatic compounds undergo saturation during hydrotreating. For modeling purposes the mechanism that was adopted was a series of reversible reactions

$$A + H_2 \xleftrightarrow{k1/k2} B + H_2 \xleftrightarrow{k3/k4} C + H_2 \xleftrightarrow{k5/k6} D \qquad (1)$$

where *A*, *B*, and *C* represent tri+, di-, and mono- aromatic compounds respectively, and *D* represents cyclic compounds (saturated aromatics). Due in part to aromatics saturation, the hydrogen consumption increases far beyond the stoichiometric requirement for desulfurization as ULSD sulfur levels are targeted.

For the present discussion the di- and tri+ aromatic compounds have been combined and labeled as poly-cyclic aromatics. The saturation of one of the rings of a poly-cyclic aromatic compound is similar to the partial hydrogenation prior to hydrogenolysis of sulfur and nitrogen containing heterocyclic aromatic compounds. The reversible nature of saturation may be observed in Figure 1 and Figure 2. Two distillate range feedstocks were processed at 700 psig in a once through isothermal pilot plant at temperatures ranging from 580 °F to 700 °F for Feed 2, Figure 1, and from 660 °F to 760 °F for Feed 1, Figure 2. The total poly aromatic concentration decreased as the temperature was increased from 580 °F to 640 °F; above ~640 °F the total poly aromatic concentration began to increase with increasing temperature, Figure 1 and Figure 2. Note that the trend continues from Figure 1 to Figure 2 despite the feedstock difference. Additional data not shown here have confirmed that the extent of poly aromatic conversion is limited by thermodynamics at high temperature and moderate to low pressure. The temperature at which poly-cyclic aromatic conversion via hydrogenation begins to decline varies with pressure. The equilibrium limitations are only of concern when the pressure is such that the maximum poly aromatic conversion occurs at a temperature within the feasible operating range for a distillate hydrotreater, as seen at 700 psig. Although this data does not directly prove that thermodynamic limitations affect the extent of HDS or HDN, they do suggest the potential for such limitations.

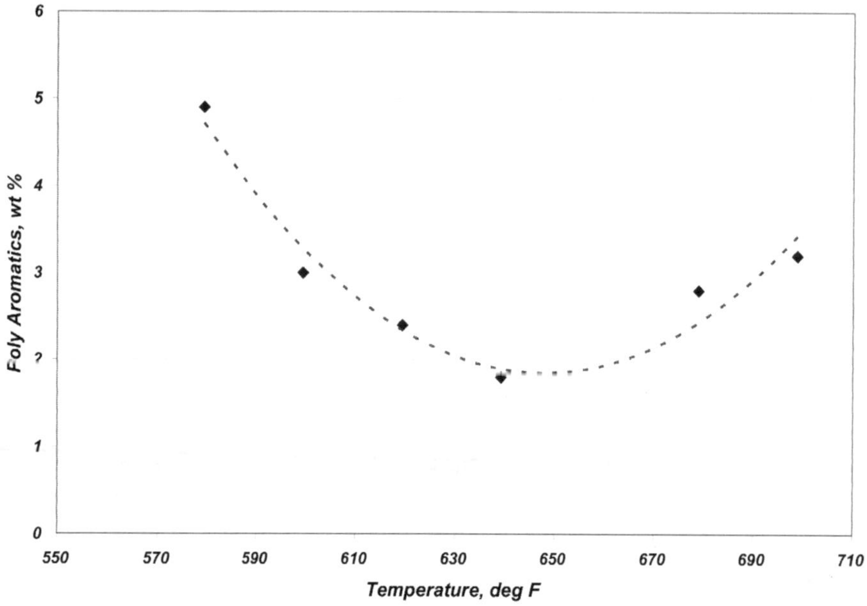

Figure 1: Poly-cyclic aromatics content versus temperature for Feed 2 (Light Coker Gas Oil) processed over a NiMo commercial hydrotreating catalyst at 700 psig.

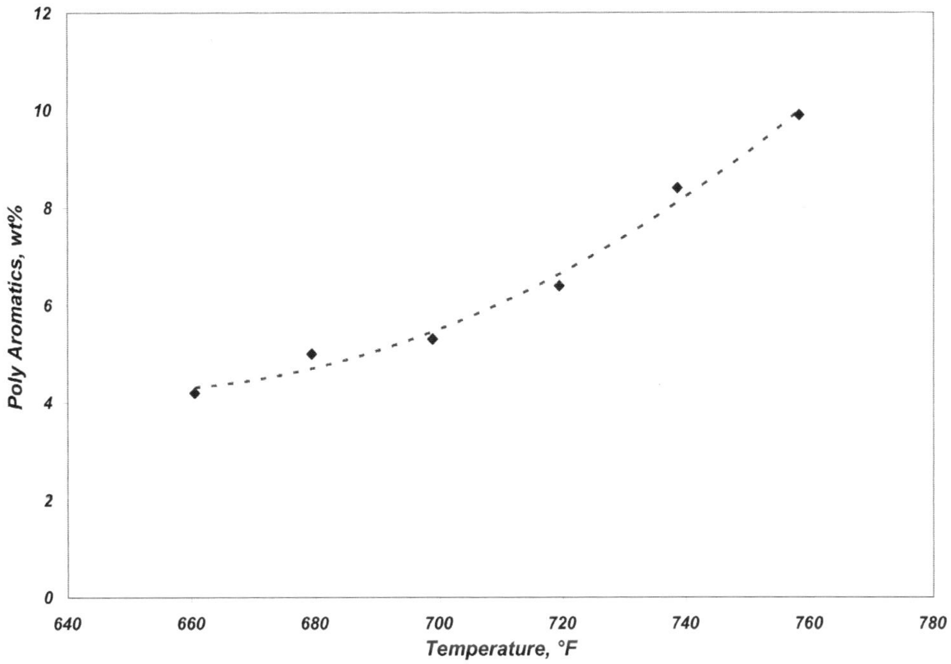

Figure 2: Poly-cyclic aromatics content versus temperature for Feed 1 (Straight Run Diesel) processed over a NiMo commercial hydrotreating catalyst at 700 psig.

Sulfur Floor

In an experiment specifically designed to investigate the "sulfur floor" Feed 1 was processed at 350 psig and 700 psig over a 660-760 °F temperature range, over both Co/Mo and Ni/Mo commercial hydrotreating (HDT) catalysts. If the sulfur floor exists, the rate of change for the product sulfur concentration with temperature would be expected to approach zero as the temperature was increased causing the product sulfur to plateau at a concentration greater than zero. As shown in Figure 3, the temperature response did slow down as the temperature was increased at 350 psig suggesting an equilibrium effect may in fact dictate the lowest achievable sulfur concentration. The product sulfur remained greater than 20 wppm at 760 °F. When operating at 700 psig the product sulfur concentration dropped to less than 1 wppm as the temperature was increased, Figure 4. This indicates that the pressure has a significant effect on the position of the equilibrium and elevating the pressure may alleviate concerns of the "sulfur floor". The increased pressure also allows the desired sulfur concentration to be achieved at more reasonable temperatures. The start of run temperature at 350 psig based on this data set would be greater than 740 °F to meet a 10 wppm sulfur specification. The lack of a sulfur floor effect at 700 psig may also be the result of the acceleration of the so called drainage reaction, the hydrogenolysis of the hydrogenated DBTs.[6] Hence, the acceleration of the hydrogenolysis reaction may nullify or at least minimize the effect of the thermodynamic equilibrium limitations.

The impact of the catalyst type may also be observed from these data. At 350 psig, the Ni/Mo HDT catalyst stabilized at a higher sulfur concentration than the Co/Mo HDT catalyst while at 700 psig this trend was reversed. Co/Mo catalysts are known to have superior direct desulfurization (DDS) activity to the Ni/Mo catalysts, while Ni/Mo catalysts typically have superior hydrogenation (HYD) activity. As a result, the hydrogenation/dehydrogenation equilibrium appears to have a greater impact on the Ni/Mo catalyst; however, due to the pressure dependence of the equilibrium the Ni/Mo catalyst has an advantage at higher pressures that favor aromatic saturation. The Co/Mo catalyst is better suited to overcome the equilibrium limitations of the hydrogenation reaction due to its superior sulfur extraction capability. As previously mentioned, as the sulfur extraction rate increases the thermodynamic equilibrium limitations may become less significant.

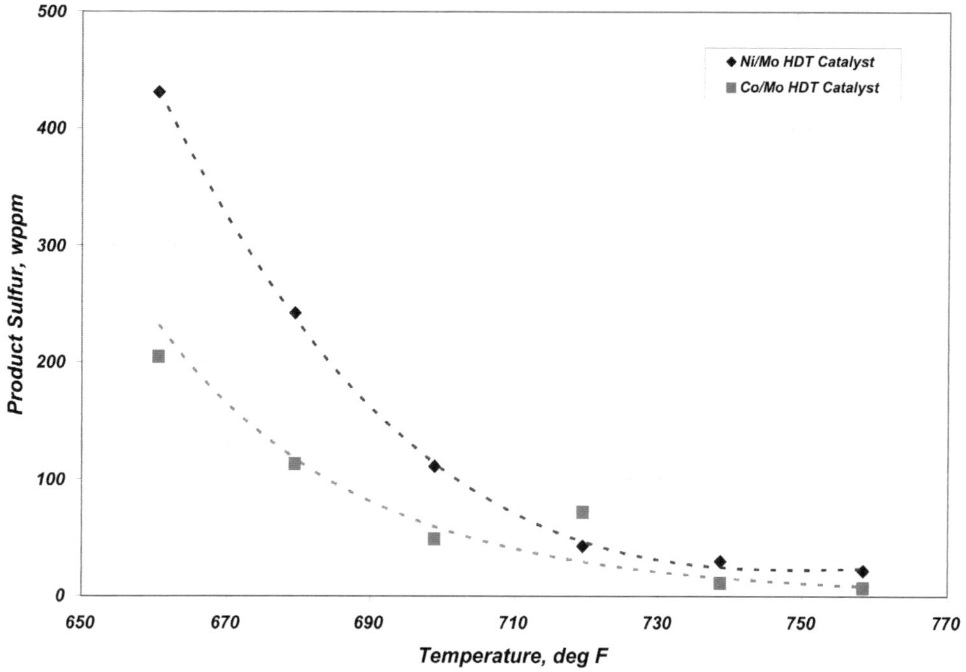

Figure 3: Hydrotreated product sulfur versus temperature; Feed 1 processed at 350 psig over commercial HDT catalysts.

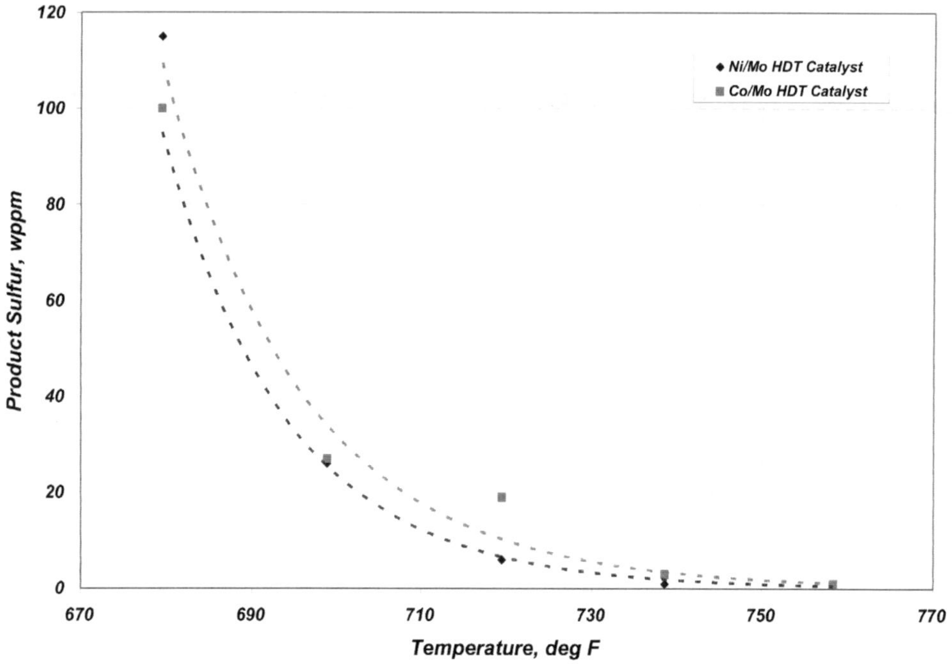

Figure 4: Hydrotreated product sulfur versus temperature; Feed 1 processed at 700 psig over commercial HDT catalysts.

In addition to the bulk sulfur concentration the sulfur components in the product were further broken down into specific species and groupings by boiling ranges. While none of the sulfur species or lumps exhibited a "floor" all exhibited a very "sluggish" temperature response at the low pressure, suggesting the possible existence of an equilibrium limitation and decline in the rate of desulfurization with temperature at low pressures. The behavior of

4,6-dimethyldibenzylthiophene (4,6-DMDBT) at 350 psig and 700 psig is shown in Figure 5 and Figure 6 as a representative compound for the sterically hindered species. Note the strikingly similar trend for the 4,6-DMDBT and the bulk sulfur suggesting that the "difficult", i.e. sterically hindered, sulfur species drive the HDS performance.

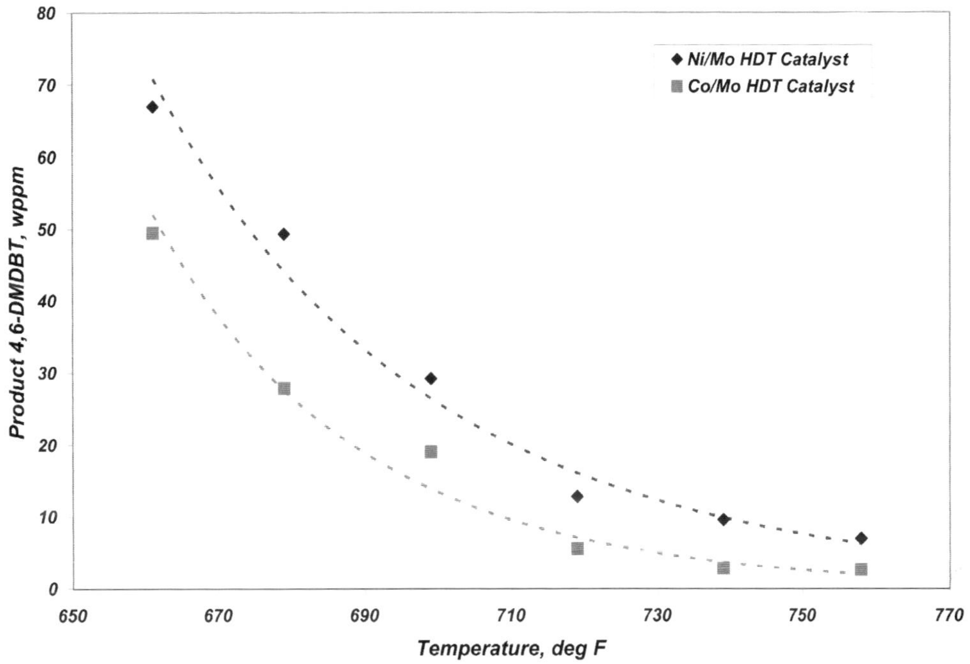

Figure 5: Concentration of 4,6-dimethyl dibenzothiophene (4,6-DMDBT) versus temperature at 350 psig, Feed 1.

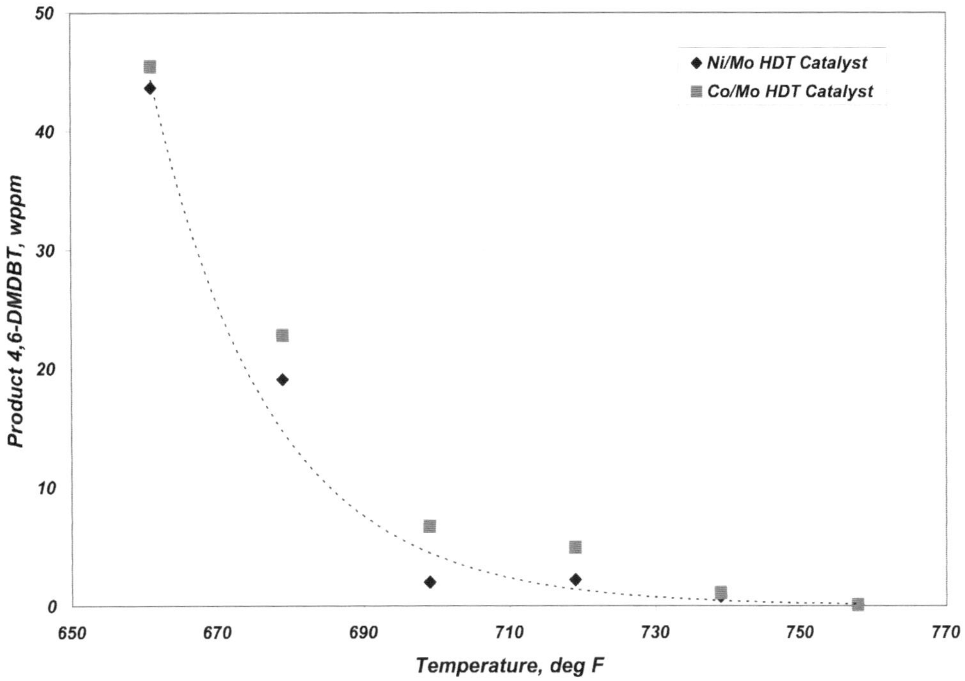

Figure 6: Concentration of 4,6-dimethyl dibenzothiophene (4,6-DMDBT) versus temperature at 700 psig, Feed 1.

Nitrogen Floor

Hydrodenitrogenation (HDN) plays a significant role in distillate hydrotreating due to the known inhibition of HDS by organo-nitrogen. The nitrogen compounds of primary concern are nitrogen containing heterocyclic aromatic compounds.[8] These compounds are classified as basic or non-basic nitrogen compounds based on the nature of the lone-pair electrons of the nitrogen atom. A basic nitrogen compound such as quinoline or acridine may be converted to a non-basic nitrogen by saturating the heteroring, hence delocalizing the lone-pair electrons. Hydrogenation of the nitrogen containing aromatic ring is required before nitrogen extraction is possible. As such the rate of nitrogen removal may be impacted by the hydrogenation equilibrium, decreasing under conditions where hydrogenation is not favorable. The potential thermodynamic limitation imposed on the HDN reactions may lead to the existence of a "nitrogen floor" similar to the potential sulfur floor. The presence of a "nitrogen floor" is a concern because nitrogen is a known inhibitor of desulfurization and the presence of a "nitrogen floor" may contribute to the presence of a "sulfur floor".

In contrast to the case for HDS, where the activities of the two catalysts depended on pressure, the Ni/Mo HDT catalyst demonstrated superior HDN performance at both 350 psig and 700 psig, Figure 7 and Figure 8. The reversal in HDS activities of the two catalysts was linked to the direct desulfurization and hydrogenation activities of the catalysts. Due to its superior direct desulfurization activity the Co/Mo catalyst out performed the Ni/Mo catalyst at conditions that are unfavorable for the hydrogenation route. The direct desulfurization route, an irreversible reaction, is not hindered by the equilibrium limitations. For organo-nitrogen compounds there are not dual routes of denitrogenation, a reversible hydrogenation step must occur before the nitrogen extraction. Therefore, even at conditions where the saturation of the nitrogen containing aromatic ring is unfavorable, the Ni/Mo catalyst out performs the Co/Mo catalyst due to its higher saturation activity.

Interestingly, the "nitrogen floor" effect is more pronounced than the "sulfur floor" effect. A comparison of the sulfur and nitrogen data from the same experiment, Figure 4 and Figure 8, reveals that while the sulfur continued to decrease to <1 wppm at high temperature the nitrogen plateaued at values greater than 4 wppm. In fact, a slight up turn in the measured nitrogen was observed with the Co/Mo catalyst. The mandatory partial saturation step in the HDN mechanism offers a reasonable explanation of why the equilibrium effect is more pronounced for HDN than HDS.

The "nitrogen floor" hypothesis is further supported by the basic and non-basic nitrogen data. The total nitrogen, basic nitrogen, and non-basic nitrogen concentrations are plotted as a function of the reaction temperature for the Co/Mo catalyst at 350 psig in Figure 9. Recall that the basic nitrogen is reversibly converted to non-basic nitrogen via saturation of the nitrogen containing aromatic ring. The nitrogen may then be irreversibly extracted from the non-basic nitrogen compound. The equilibrium limitation is most clearly observed by examining the basic nitrogen temperature response. The basic nitrogen concentration appears to have reached a minimum at ~740 °F and then began increasing. This impacted the overall denitrogenation; the decrease in the total nitrogen concentration was insignificant at temperatures greater than 720 °F despite the steady decline in the non-basic nitrogen concentration.

---

[8] Grigis, M.J., B.C. Gates, Reactivities, Reaction Networks, and Kinetics in High-Pressure Catalytic Hydroprocessing, Ind. Eng. Chem. Res., 30, 2021-2058, 1991.

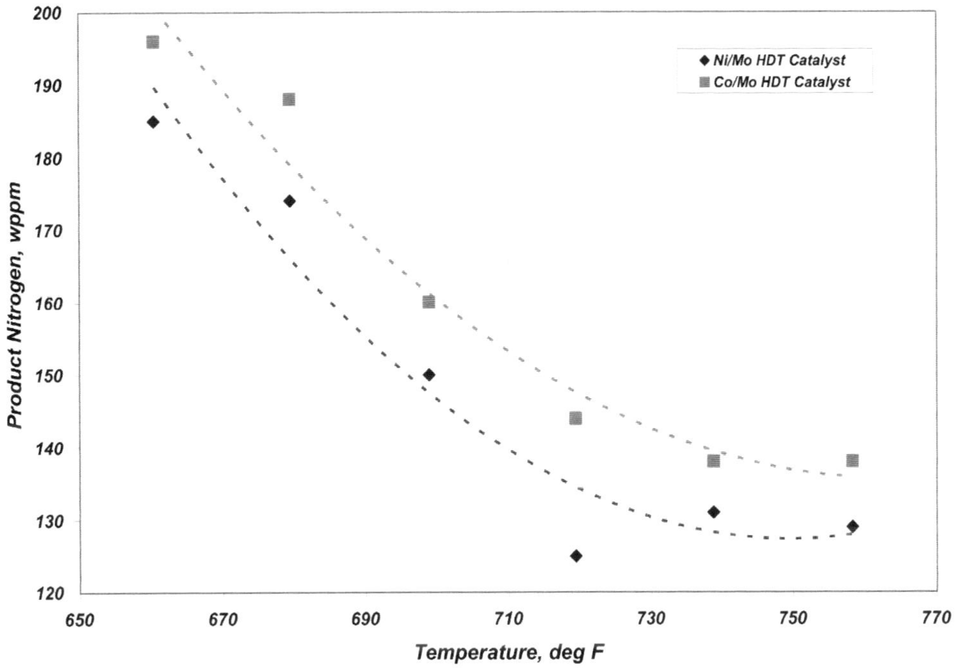

Figure 7: Total product nitrogen versus temperature for Feed 1 processed at 350 psig.

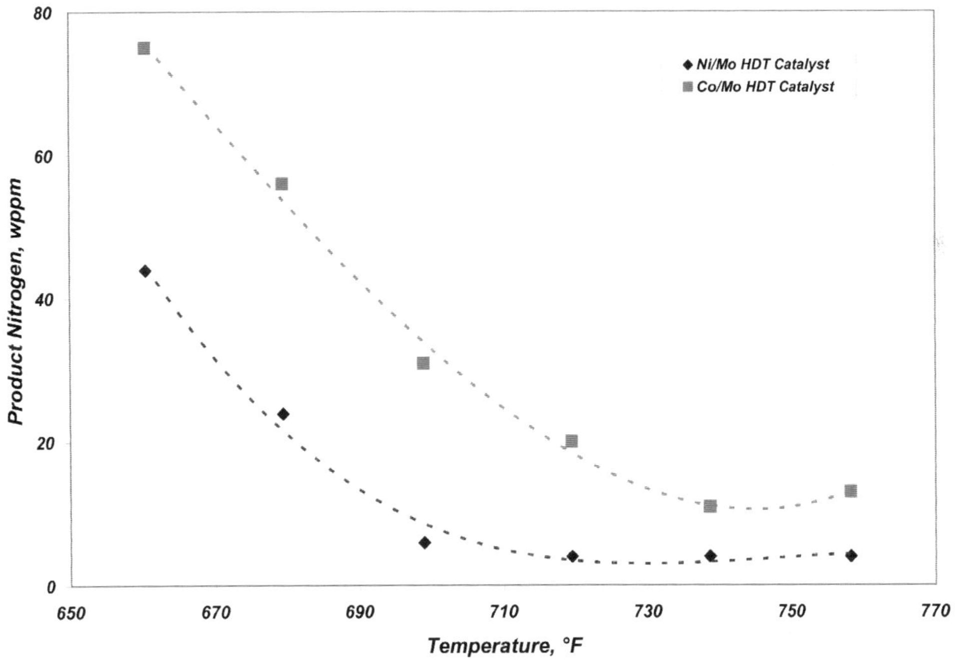

Figure 8: Total product nitrogen versus temperature for Feed 1 processed at 700 psig.

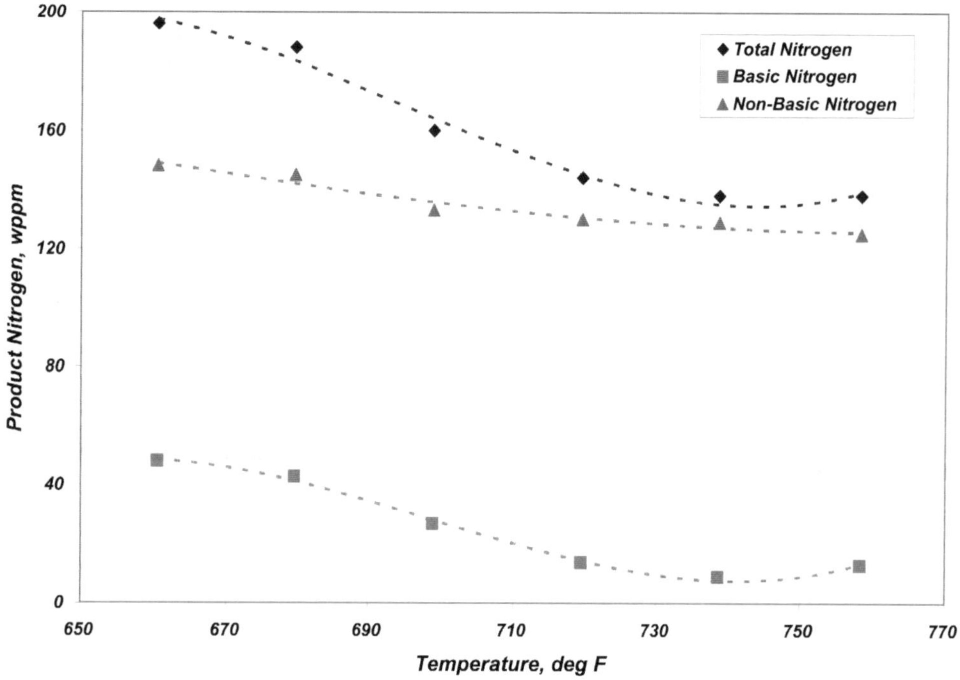

Figure 9: Total, basic, and non-basic product nitrogen versus temperature with Co/Mo HDT catalyst at 350 psig, Feed 1.

Factors Influencing the Sulfur Floor

Although the "sulfur floor" was not clearly observed in this set of experiments, the potential for significant equilibrium limitations in distillate hydrotreating was not eliminated. The existence and significance of a "sulfur floor" is a complicated issue due to the numerous factors that impact the maximum extent of desulfurization. Mayo et al concluded that the level of the sulfur floor is difficult to predict due to the numerous factors influencing it.[9] Influential factors mentioned by Mayo et al include the following: concentration of sterically hindered dibenzothiophenes, nitrogen content, hydrogen partial pressure, space velocity, and catalyst type. There may be reason to believe that the aromatic and hydrocarbon content of the feedstock ought to be added to the list of factors influencing the level and existence of the sulfur floor.

Perhaps these numerous influential factors offer an explanation for the diversity of conclusions made by researchers regarding the existence and significance of the sulfur floor when studying model components compared to commercial feedstocks. Ho did not observe any equilibrium limitations to the extent of conversion of 4,6-DEDBT in dodecane solvent at 0.79 MPa (~115 psig)[5], while others report equilibrium limitations impacting ULSD production at realistic commercial conditions.[2,3,4] The reaction environments in the liquid filled catalyst pores are very different in an experiment conducted with model compounds, such as 4,6-DEDBT in dodecane, and a commercial feedstock, particularly a highly aromatic feedstock such as LCO. As a result, the seemingly contradictory findings may in reality be quite easily reconciled.

As hydro-desulfurization (HDS), hydro-denitrogenation (HDN), and aromatics saturation (HDA) reactions occur in parallel in a shared environment they compete for the necessary resources, namely hydrogen and catalyst sites.[10] There is significant evidence that the presence of poly-cyclic aromatic compounds affects the rate of HDS.

---

[9] Mayo, S. E. Brevoord, F. Plantenga, N. Gudde, G. Hoekstra. *AM-02-38, ULSD in Real Life: Commercial Performance of STARS and Nebula Technology*, 2002 NPRA Annual Meeting, San Antonio, Texas.

[10] Laura Jones, Peter Kokayeff. *Distillate Hydrotreating to ULSD: The Impact of Aromatics*, in: Proceedings of 2004 AIChE Spring National Meeting (7th International Conference on Refinery Processing), New Orleans, LA, USA, April 25-29, 2004, Paper No. 90a.

The impact of aromatics may be the result of competitive adsorption onto the active sites similar to nitrogen or it may be the result of a competition for the co-reactant, hydrogen. The rate of aromatic saturation is ~100 times faster than HDS rates for sterically hindered sulfur species. The fast rate of aromatic saturation relative to HDS and the significant hydrogen consumption result in the diffusion of hydrogen into the liquid filled pores of the catalyst being potentially rate limiting. The solubility of hydrogen is influenced by temperature, pressure, and the bulk liquid or solvent composition. At low hydrogen partial pressure the hydrogen solubility will be lower than at high pressure decreasing the hydrogen readily available in the catalyst pores. Furthermore, hydrogen is less soluble in aromatic solvents than in saturated hydrocarbons, particularly normal paraffins.[5] Therefore, when the solvent is paraffinic the probability of observing a sulfur floor is lower than for the case of an aromatic solvent.

The significance of the existence of the "nitrogen floor" is magnified by the inhibition of HDS by organo-nitrogen compounds. If the nitrogen concentration ceases to decrease with increasing temperature, the HDS activity will be adversely affected. In a separate series of experiments than previously discussed a commercial feedstock was processed over commercial Co/Mo and Ni/Mo HDT catalysts at 450 psig, LHSV ranging from 0.7 to 2.0 hr^{-1}, and temperatures ranging from 630 to 730 °F. The product sulfur and nitrogen trends with temperature were not strikingly different from those observed in the experiments discussed previously. However, a "floor" effect is visible when the sulfur and nitrogen concentrations are plotted as a function of the residence time, Figure 10 and Figure 11. For the Ni/Mo catalyst, Figure 10, there is a clear correlation between the nitrogen and sulfur concentrations; in fact, at the exact point that the nitrogen concentration plateaus the change in the sulfur concentration becomes very sluggish. The relationship is not as clear for the Co/Mo catalyst which is not unexpected as Ni/Mo catalysts tend to be more sensitive to nitrogen than Co/Mo catalysts. Therefore, it is possible that the so called "sulfur floor" effects that have been observed experimentally may truly be the result of a "nitrogen floor" limiting the degree of denitrogenation and hence inhibiting HDS. This is consistent with Ho's hypothesis that although the rate of DBT hydrogenation is inhibited by thermodynamics at high temperature, the overall sulfur removal rate is not significantly impacted.[6] Therefore, the absence of nitrogen compounds in the dodecane solvent offers a valid explanation of Ho's observation that the extent of desulfurization of 4,6-DEDBT in such a solvent does not exhibit a "floor" effect.[5]

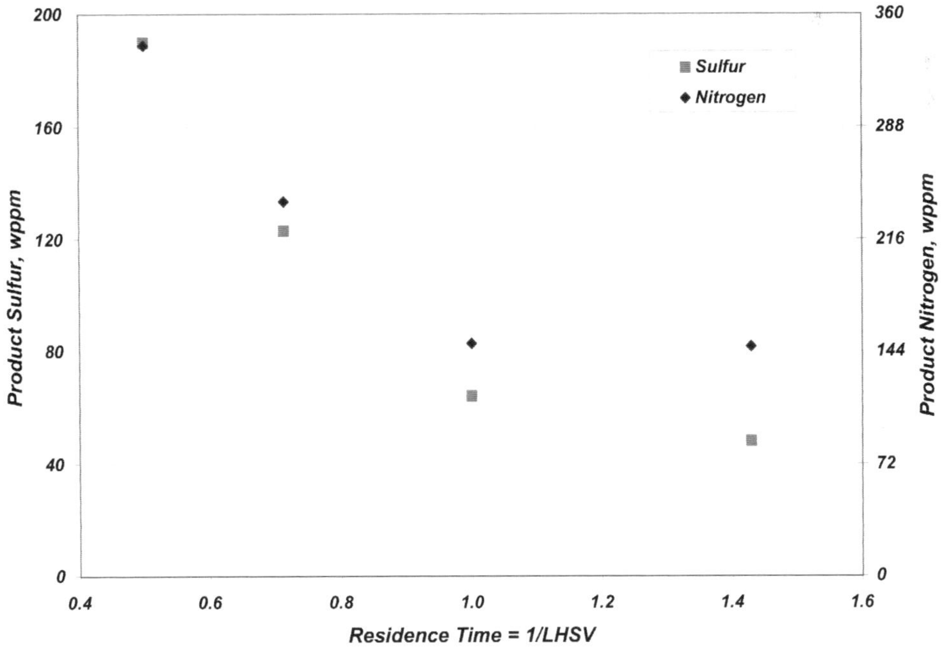

Figure 10: Product sulfur and nitrogen concentration for commercial feedstock processed over a commercial Ni/Mo HDT catalyst at 730 °F and 450 psig.

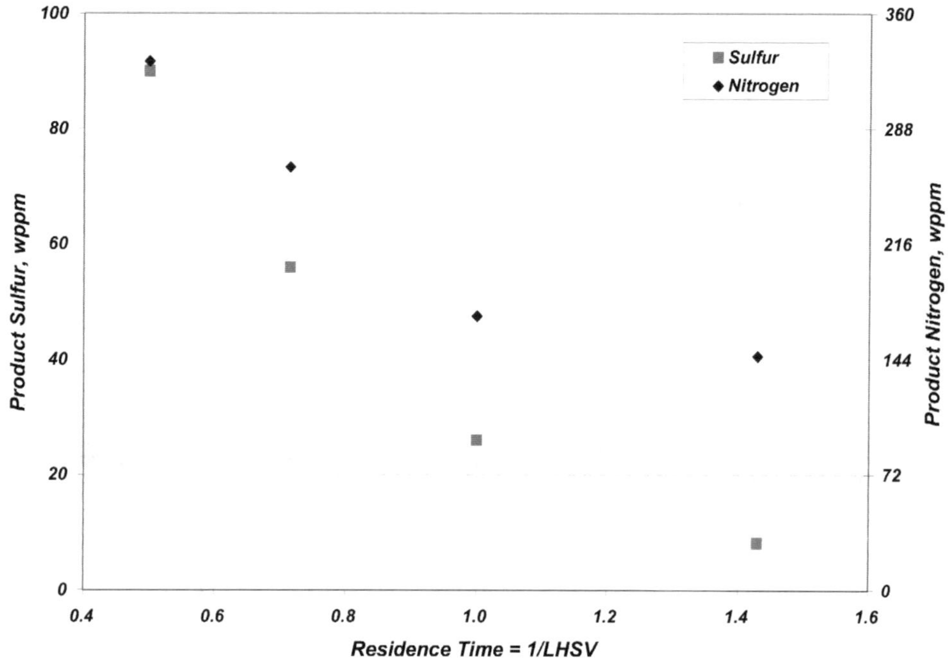

Figure 11: Product sulfur and nitrogen concentration for commercial feedstock processed over a commercial Co/Mo HDT catalyst at 730 °F and 450 psig.

Conclusions

Although a "sulfur floor" was not definitively observed in the pilot plant experiments conducted, the data do clearly show that the HDS temperature response decreases dramatically at high temperatures particularly at lower pressure. Therefore, it may be concluded that the extent of sulfur removed is impacted by thermodynamic limitations under certain conditions. However, the more significant question is whether the thermodynamic limitations will impact the commercial production of ULSD. It is conceivable that thermodynamic effects will result in a "sulfur floor" higher than the 10 wppm sulfur specification for ULSD. The minimum sulfur level achievable by increasing temperature will be influenced by the feedstock composition, operating conditions, and the catalyst.

On the other hand, the existence of a "nitrogen floor" was more clearly observed than the "sulfur floor". The equilibrium limitations are more significant for denitrogenation due to the lack of a direct extraction reaction path. The nitrogen compounds must undergo saturation prior to nitrogen extraction. In the presence of nitrogen there are apparent "thermodynamic" limitations to the extent of desulfurization which may be more accurately linked to the thermodynamic limitations of the HDN reactions and the consequent inhibition of HDS by nitrogen. As a result, the "nitrogen floor" is of great significant due to the inhibiting effect of nitrogen on HDS.

The possibility does exist for the hydrogenation/dehydrogenation equilibrium to impact ULSD production; however, in many cases the actual floor will not be observed until the temperature is increased beyond realistic end-of-run temperatures for a commercial unit. Furthermore, a unit designed for ULSD production is unlikely to operate at low pressures where the "thermodynamic" limitations would be of greatest concern. Having said that, the potential for "thermodynamic" limitations highlights the importance of careful consideration of the feedstock composition and operating conditions at SOR and EOR when selecting a catalyst or designing a new distillate hydrotreater to produce ULSD.

Effect of Hydrotreating Conditions on Maya Asphaltenes Composition and Structural Parameters

Fernando Trejo [a,b] *and Jorge Ancheyta,* [*, b]
[a] *Facultad de Química. UNAM. Ciudad Universitaria, México, D.F. 04510*
[b] *Instituto Mexicano del Petróleo, Eje Central Lázaro Cádenas 152, Col. San Bartolo Atepehuacan. México, D.F. 07730.*
[*] *Author to whom correspondence should be addressed. Tel.: (+52 55) 9175 8443; Fax (+52 55) 9175 8429; E-mail: jancheyt@imp.mx*

Abstract

Maya crude was hydrotreated in a pilot plant. The pressure was varied in the range of 70-100 kg/cm^2, temperature between 380-420°C, and space-velocity was varied between 0.33-1.5 h^{-1}. The hydrogen-oil ratio was kept constant at 5000 scft/bbl. Asphaltenes were precipitated from the hydrotreated products by using a batch reactor with *n*-heptane as solvent. The effects of reaction conditions on asphaltenes properties during hydrotreating are discussed in terms of changes in heteroatoms contents and structural parameters by using elemental analysis, metals content, VPO aggregate molecular weight, XRD and NMR measurements.

Keywords: Asphaltenes hydrocracking, structural parameters

1. Introduction

It is well-known that heavy oil and residues can contain significant amounts of asphaltene which are responsible for the coke deposition that causes catalytic deactivation in the thermal and catalytic processing of petroleum residues. The main problems caused by asphaltenes during catalytic hydroprocessing of heavy oils are: (1) reduction of the overall rate of hydrodesulfurization (HDS) reaction, (2) precipitation on the catalyst surface and blockage of the pore mouth, and (3) they act as coke precursors. [1]

Vacuum residua are the most difficult feeds to convert catalytically because of their heavy molecular structures and impurity contents such as sulfur, nitrogen and metals (mainly vanadium and nickel) concentrated in asphaltene fractions. [2]

The molecular nature of petroleum asphaltenes has been the subject of numerous investigations. They appear to be very complex aromatic molecules surrounded and linked by aliphatic chains and heteroatoms. A great variety of analytical techniques have been employed to investigate asphaltene molecular structure. Particularly, ^{1}H and ^{13}C NMR provide reliable average molecular parameters such as aromatic carbon fraction (f_a), average number of carbon per alkyl side chains (n), average percent of substitution of aromatic carbon (A_S), and aromatic ring number (R_a) [3].

X-ray diffraction (XRD) and small-angle X-ray scattering (SAXS) can also be used to obtain structural information of asphaltene aggregates [4]. XRD provides internal structural information and the crystallite parameters of the molecules associated in the aggregates. SAXS gives a topological view of the aggregate systems in which the X-rays are scattered by the electron density differences between the aggregate and the medium [5].

Due to self-association propensity, molecular weight using vapor pressure osmometry (VPO) corresponds to the molecular weight of asphaltene aggregates [6]. For this reason, measurements made by VPO will be considered in this work as aggregate molecular weight (AMW) [7]. Molecular weight of asphaltenes decreases whereas HDT temperature is increased and asphaltenes tend to increase its aromaticity due to the shortening of alkyl side chains. [8]

The shift of the molecular weight distributions toward lower masses can be explained as a consequence of the conversion of higher mass molecules to lighter ones due to the dealkylation of side alkyl chains during the hydrogenation process. [2] At higher temperature, nitrogen and metals contents increase and sulfur content decreases whereas asphaltenes aromaticity, which was measured as aromaticy factor (f_a), increases. [9] As the reaction temperature is increased, the H/C ratio of the asphaltenes is reduced. The sulfur remaining in the product is predominantly found in the asphaltenic fraction [10].

The study of the changes of asphaltenes properties during hydrotreating of heavy oil is an important concern nowadays. In the present paper we precipitated asphaltenes from different hydrotreated products and the effects of reaction conditions on asphaltene properties are examined.

2. Experimental

2.1. Hydrotreating Experiments

Maya crude oil was employed for all experiments and their main properties are shown in **Table 1**. Maya crude oil was hydrotreated in a pilot plant described in detail elsewhere [11]. The reactor was verified for isothermal behavior and the greatest deviation from the desired temperature values was about 4-5°C.

The catalyst employed for hydrotreating reactions was a commercial NiMo/Al_2O_3 sample (175 m^2/g specific surface area, 0.56 cm3/g pore volume, and 127 Å mean pore diameter). Different catalyst sizes were tested by crushing pellets and it was found that with 0.25 mm diameter particles, internal diffusion limitations can be neglected. 100 mL of crushed catalyst was employed as catalytic bed. The catalyst is loaded to the reactor in the oxide form and in situ sulfided at the following conditions: pressure of 28 kg/cm2, hydrogen-to-oil ratio of 2000 ft3/bbl, LHSV of 2 h-1, and temperatures of 260°C (3 h) and 320°C (8 h), with a mixture of straight-run gas oil and dimethyl disulfide (1.0 wt% sulfur).

The HDT experiments were carried out at the following operating conditions: pressure of 70, 85, and 100 kg/cm^2, H_2-to-oil ratio of 5000 ft^3/bbl, liquid hourly space-velocity of 0.33, 0.5, 0.75, 1.0, 1.25, and 1.5 h^{-1}, and reaction temperature of 380, 400 and 420°C. Product samples were collected at 7-30 h intervals, after allowing an 8 h stabilization period under each set of conditions.

2.2. Precipitation of asphaltenes

Asphaltenes were precipitated by using n-heptane in a Parr batch reactor model 4842 equipped with stirring system, temperature control, pressure control, and vent. Hydrotreated products and n-heptane were put into the reactor. Pressure was adjusted at 25 kg/cm^2 by using nitrogen as inert, temperature at 60°C and the stirring rate was fixed at 750 rpm once temperature was reached. The solvent-to-hydrotreated products ratio was 5:1 (volume/weight). More details of these experiments were reported elsewhere [12].

2.3. Characterization of oils and asphaltenes

Nickel and vanadium contents of asphaltenes in Maya crude and in hydrotreated products were measured by atomic absorption using a SOLAAR AA Series Spectrometer analyzer. Carbon, hydrogen, nitrogen, and sulfur contents in asphaltenes were determined by combustion in a Perkin Elmer 2400 analyzer.

Aggregate molecular weight was measured by vapor pressure osmometry in a Corona Wescan 232A equipment. Structural parameters were determined by liquid state ^{1}H and ^{13}C NMR in a JEOL Eclipse 300 spectrometer operating at ^{1}H resonance frequency of 300 MHz and ^{13}C resonance frequency of 75 MHz [3].

XRD measurements were carried out in a model Siemens diffractometer equipped with a Cu Kα_1 radiation (λ=1.5416 Å), operating at 40 kV and 30 mA, in the 4°-40° scan range of 2θ with a scan rate of 0.02°/s. 0.1 g sample of asphaltenes was mounted on an aluminum plate sample. To determine the crystallite parameters, the layer distance between aromatic sheets (d_m) was first estimated from the maximum of the graphene band, using Bragg equation [13]:

$$d_m = \frac{\lambda}{2\sin\theta} \qquad (1)$$

The average height of the stack of aromatic sheets perpendicular to the plane of the sheet (L_c) was calculated as follows [5]:

$$L_c = \frac{\lambda}{\omega\cos\theta} \qquad (2)$$

where ω is the full width at half-maximum, obtained by using the graphene band. The number of aromatic sheets in a stacked cluster (M) is then calculated with:

$$M = \frac{L_c}{d_m} + 1 \quad (3)$$

3. Results and discussion

3.1. HDT Experiments

Some properties of the hydrotreating feed (Maya heavy crude) and properties of asphaltenes precipitated from this heavy oil are presented in **Table 1**. It is evident that Maya is heavy in nature and hence high percentage of heavy molecules is concentrated in its asphaltenes. Nickel, vanadium, and asphaltenes contents were substantially reduced when Maya crude was hydrotreated.

Table 1. Properties of Maya crude and of their asphaltenes

	Maya crude	Asphaltenes
API gravity	20.9	---
C, wt%	83.96	82.54
H, wt%	11.80	8.46
N, wt%	0.37	1.11
S, wt%	3.4	7.10
H/C atomic ratio	1.686	1.230
Insolubles in n-C_7, wt%	12.4	---
V, wppm	299	1509
Ni, wppm	55	320
AMW	550	5190
R_a	---	62.0
A_s	---	38.9
n	---	6.80
f_a	---	0.52

3.2. Characterization of Asphaltenes

The elemental analyses (C, H, N, S) of asphaltenes from selected hydrotreated products are shown in **Table 2**. As can be seen, carbon, hydrogen, and sulfur contents progressively diminished as the severity of the reaction was increased. Nitrogen content remains almost constant since this heteroatom forms stable molecules when it is bounded in aromatic rings due to its location inside the asphaltene molecule. [14] The H/C atomic ratio of asphaltenes decreases as the temperature and pressure are increased and LHSV is reduced, which indicates that more aromatic molecules are formed as a consequence of the breakage of side alkyl chains.

Ni and V contents in asphaltenes from hydrotreated products are also presented in **Table 2**. Both metal contents are increased as temperature and pressure are incremented. This behavior has been previously reported [9] and has been attributed to the reduction of alkyl side chains because of removal of aliphatic carbons whereas metals tend to remain unchanged and concentrate on the asphaltenes.

Table 2. Elemental analysis, metals content and structural parameters of asphaltenes

	400°C and LHSV=1 h^{-1}		
	70 kg/cm^2	85 kg/cm^2	100 kg/cm^2
C, %wt	84.00	83.97	83.82
H, %wt	7.02	6.98	6.96
N, %wt	1.50	1.47	1.44
S, %wt	3.52	3.39	3.41
H/C	1.003	0.997	0.996
Ni, wppm	311	325	385
V, wppm	1601	1630	1682
Structural parameters			
R_a	28.32	27.86	25.90
A_s	33.89	34.29	32.70
	100 kg/cm^2 and 400°C		
	1.50 h^{-1}	1.00 h^{-1}	0.33 h^{-1}
C, %wt	84.10	83.82	83.35
H, %wt	7.01	6.96	6.86
N, %wt	1.47	1.44	1.34
S, %wt	3.45	3.41	3.32
H/C	1.000	0.996	0.988
Ni, wppm	354	385	428
V, wppm	1645	1682	1748
Structural parameters			
R_a	46.97	25.90	28.10
A_s	46.54	32.70	21.53
	100 kg/cm^2 and LHSV=1		
	380°C	400°C	420°C
C, %wt	83.85	83.82	83.71
H, %wt	6.98	6.96	6.94
N, %wt	1.44	1.44	1.43
S, %wt	3.47	3.41	3.32
H/C	0.999	0.996	0.995
Ni, wppm	311	385	433
V, wppm	1637	1682	1698
Structural parameters			
R_a	40.18	25.90	31.35
A_s	36.61	32.70	26.95

Figure 1 shows the results obtained by XRD and in **Table 3** some crystallite parameters are summarized. As can be seen, with increasing temperature from 380°C up to 420°C and keeping constant the pressure at 100 kg/cm^2 and LHSV at 1 h^{-1}, the average number of aromatic sheets tends to diminish as a consequence of the saturation of some aromatic molecules. Also, the average height of the stack of the molecules is reduced due to the same. The distance between aromatic sheets is slightly reduced. When pressure is varied keeping constant the temperature at 400°C and LHSV at 1 h^{-1}, the average number of aromatic sheets diminishes drastically from ~20 up to ~10.

The major change is observed when pressure is increased from 70 up to 85 kg/cm^2. The same behavior can be seen in the average height of the stack and the distance between aromatic sheets. At 70 kg/cm^2, the peak obtained is sharp and corresponds to an ordered arrangement similar to graphite.

At the most severe reaction conditions, the average height of the stack is the lowest; the aromatic sheets are removed as the LHSV is reduced because the contact time is higher and conversion increases.

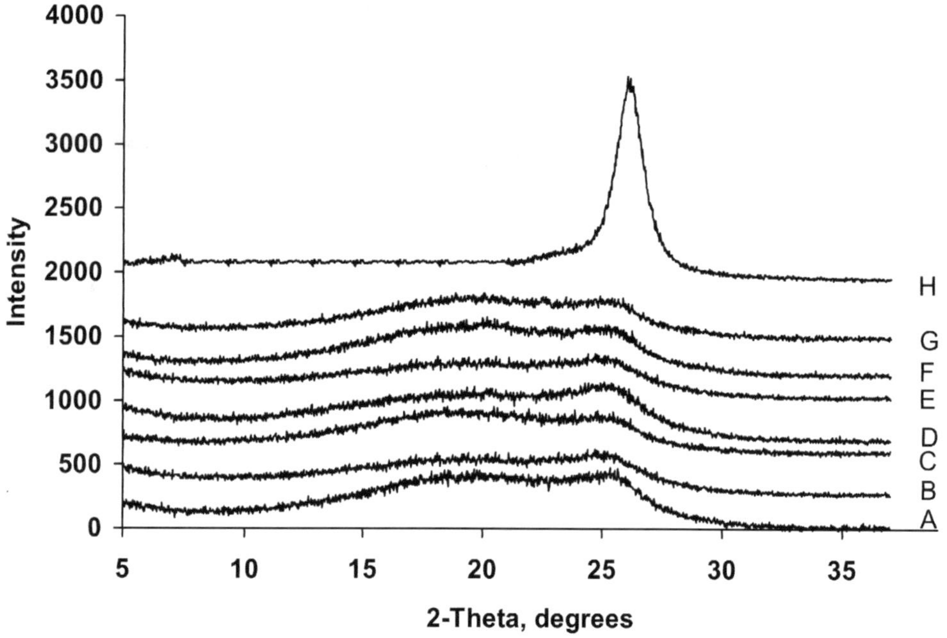

Figure 1. X-ray diffraction patterns of asphaltenes: (A) Maya crude asphaltenes; (B) T=400°C, P=100kg/cm^2, LHSV=0.33; (C) T=400°C, P=100kg/cm^2, LHSV=1.0; (D) T=400°C, P=100kg/cm^2, LHSV=1.5; (E) T=380°C, P=100kg/cm^2, LHSV=1.0; (F) T=400°C, P=100kg/cm^2, LHSV=1.0; (G) T=400°C, P=85kg/cm^2, LHSV=1.0; (H) T=400°C, P=70kg/cm^2, LHSV=1.0;

Table 3. Crystallite parameters of asphaltenes obtained from Maya crude and hydrotreated products.

Maya crude asphaltenes			
	d_m, Å	L_c, Å	M
---	3.60	32.53	10.1
400°C and LHSV=1 h^{-1}			
Pressure, kg/cm^2	d_m, Å	L_c, Å	M
70	3.43	67.7	20.7
85	3.52	33.3	10.5
100	3.53	33.5	10.5
100 kg/cm^2 and 400°C			
LHSV,h^{-1}	d_m, Å	L_c, Å	M
1.50	3.48	35.2	10.7
1.00	3.53	33.5	10.5
0.33	3.60	26.3	8.3
100 kg/cm^2 and LHSV=1 h^{-1}			
Temperature, °C	d_m, Å	L_c, Å	M
380	3.58	35.6	10.9
400	3.53	33.5	10.5
420	3.51	31.4	10.0

Figure 2 shows the ^{13}C NMR spectra of asphaltenes precipitated from Maya crude and from hydrotreated products obtained at different reaction conditions. The differences between aliphatic and aromatic peaks are notorious. Some structural parameters obtained by NMR were shown in **Table 2**.

Aromatic rings number (R_a) tends to decrease generally when pressure and temperature are increased and space-velocity is reduced. Percent of substitution of aromatic rings (A_s) is also reduced as severity conditions are increased. These observations could lead to figure out that polycondensed rings are turned into other lighter fractions.

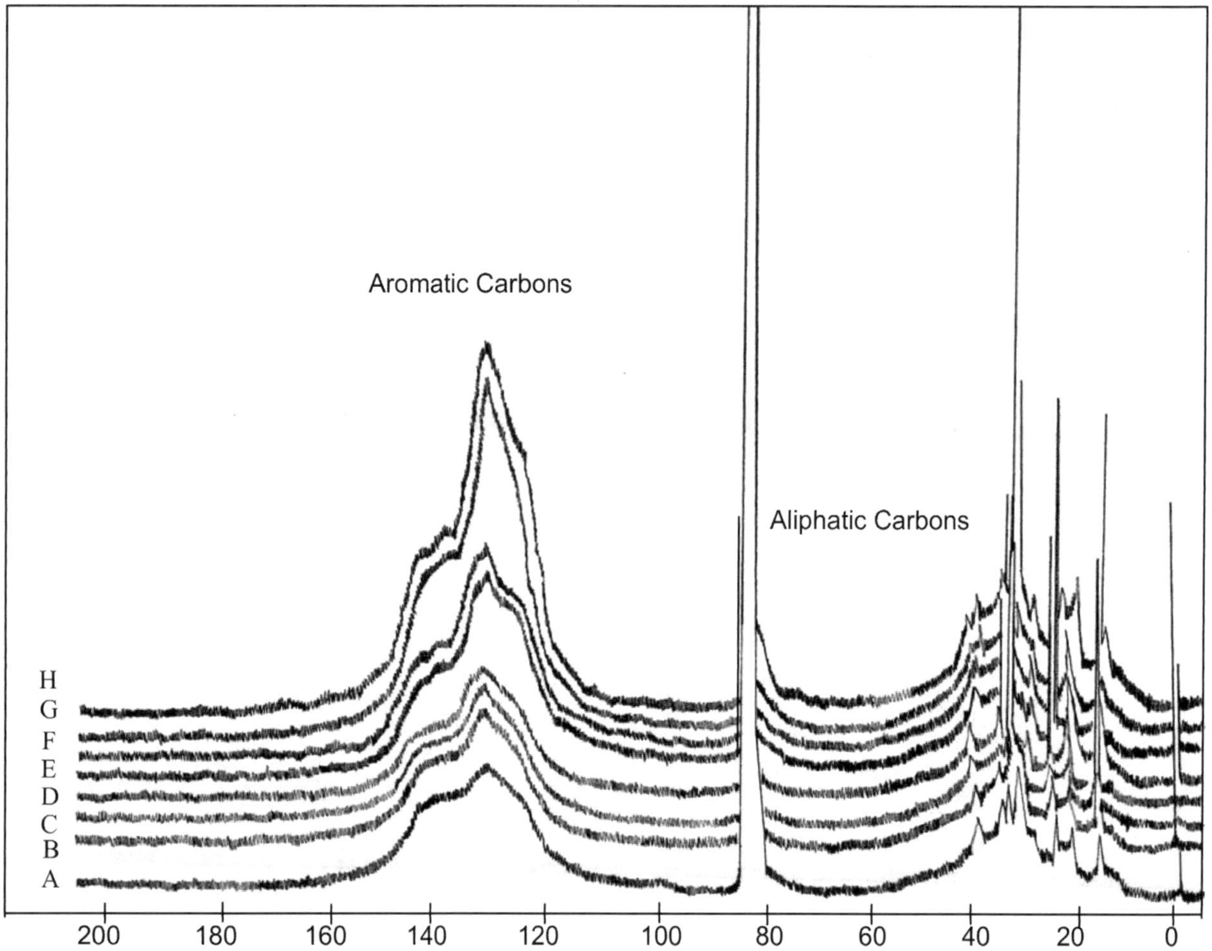

Figure 2. NMR spectra of asphaltenes. (A) Maya crude asphaltenes; (B) T=400°C, P=70kg/cm^2, LHSV=1.0; (C) T=400°C, P=85kg/cm^2, LHSV=1.0; (D) T=400°C, P=100kg/cm^2, LHSV=1.5; (E) T=380°C, P=100kg/cm^2, LHSV=1.0; (F) T=400°C, P=100kg/cm^2, LHSV=1.0; (G) T=420°C, P=100kg/cm^2, LHSV=1.0; (H) T=400°C, P=100kg/cm^2, LHSV=0.33

3.3. Effect of reaction conditions

Figure 3 shows the changes in asphaltenes as function of temperature when pressure and space-velocity are constants. When temperature increases, a shortening of the alkyl side chains occurs and the average number of carbons gradually reduces. For this reason, there is less aliphatic carbons in asphaltenes and aromatic carbons become predominant so that H/C atomic ratio tends to reduce. The aggregate molecular weight was reduced as reaction temperature was increased which indicates that the size of asphaltene molecule is also decreased because the heteroatoms located at the external part of the asphaltene are released and in consequence they are easily removed.

At high pressure, aromaticity factor increases and H/C atomic ratio reduces, which means that breakage of the aliphatic carbon in side alkyl chains has occurred. **Figure 3** shows the effect of pressure on the properties of asphaltenes keeping constant the space-velocity and the temperature. Aggregate molecular weight is also diminished as pressure is increased.

The number of carbons in alkyl side chains and aggregate molecular weight are reduced as space-velocity is also diminished. When LHSV is low the conversion is high which yields more selectivity toward hydrocracking. **Figure 3** shows how space-velocity affects some structural parameters of asphaltenes. AMW decreases as LHSV diminishes because the temperature produces the most important change in the aromaticity factor due to the shortening of the alkyl side chains.

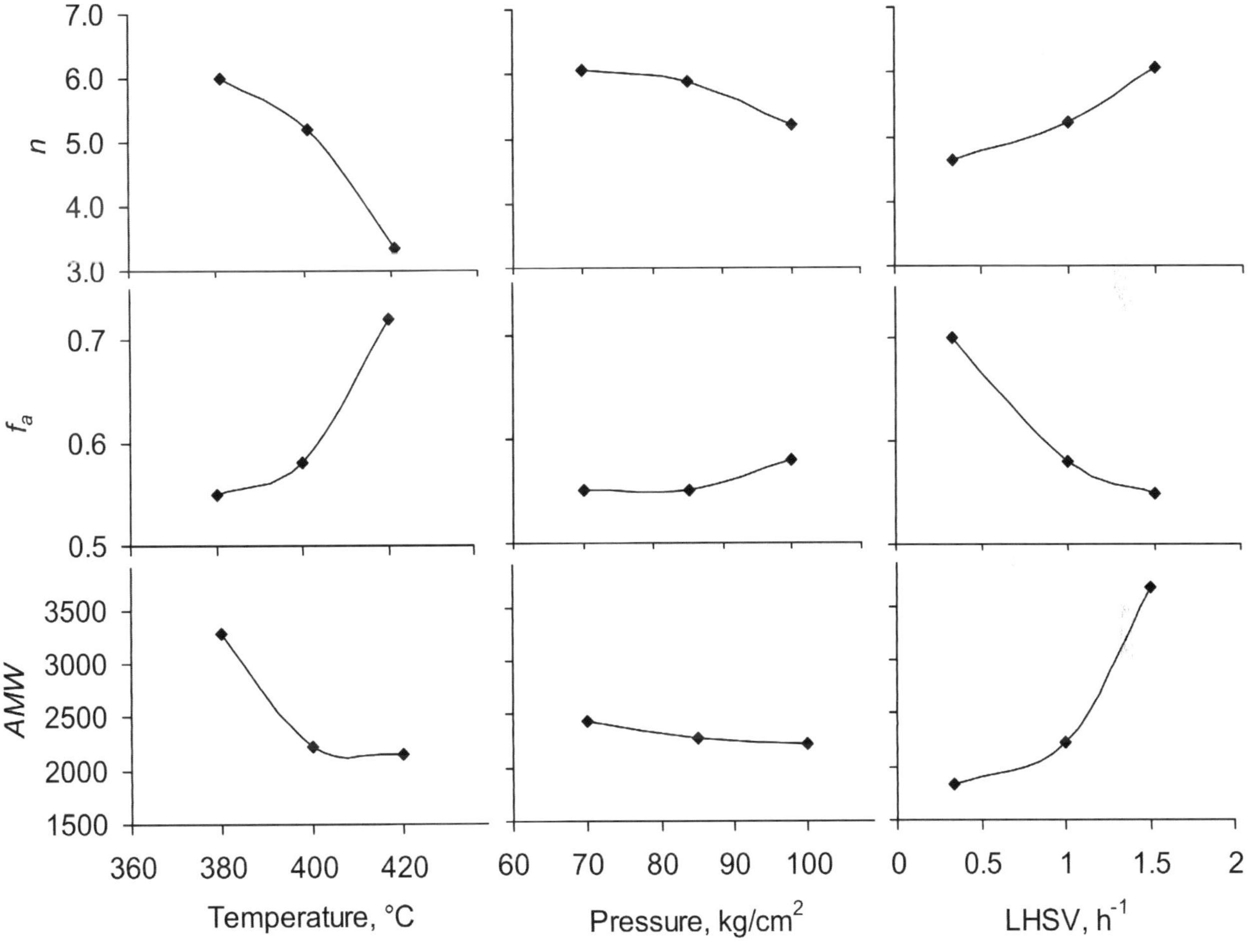

Figure 3. Effect of temperature, pressure, and space-velocity on the average number of carbons per alkyl side chain (*n*), aromaticity factor (f_a), and aggregate MW

From all these observations, it can be established that hydrocracking of asphaltenes is more prominent at temperature above 400°C, pressure greater than 85 kg/cm^2 and space velocity less than 1 h^{-1}. In all cases, the effect of pressure is less significant than the effect of temperature and space-velocity on shortening of alkyl side chains, and on changes in aromaticity factor and reduction of AMW.

4. Conclusions

On the basis of our hydrocracking experiments and characterization results, we can conclude that precipitated asphaltenes exhibit different changes in composition and in structural parameters depending on the severity of the reaction.

Changes in structural parameters of asphaltenes supported the fact that metals concentrate in asphaltenes as the temperature is increased, since they are located inside the asphaltene structure and remain without change while the average number of carbons per alkyl side chain decreases.

Asphaltenes aromaticity increased as temperature and pressure were also incremented and space-velocity was reduced. The H/C atomic ratio was slightly reduced and confirmed that asphaltenes became more aromatic at the most severe reaction conditions.

As reaction conditions are more severe, there is a reduction of the stack of the aromatic sheets, confirming the tendency toward hydrocracking of asphaltenes at high temperature, and pressure, and low space-velocity.

Acknowledgement

The authors thank Instituto Mexicano del Petróleo for its financial support. F. Trejo also thanks CONACyT and UNAM for financial support.

References

[1]. B. Avid, S. Sato, T. Takanohashi, I. Saito, Energy Fuels 18 (2004) 1792-1797.
[2]. Merdrignac, C. Truchy, E. Robert, I. Guibard, S. Kressmann, Petr. Sci. Techn. 22 (2004) 1003-1022 .
[3]. V. Calemma, P. Iwanski, M. Nali, R. Scotti, L. Montanari, Energy Fuels 9 (1995) 225-230.
[4]. J. W. Shirokoff , M. N. Siddiqui, M. F. Ali, Energy Fuels 11 (1997) 561-565.
[5]. R. Tanaka, E. Sato, J. E. Hunt, R. E. Winans, S. Sato, T. Takanohashi, Energy Fuels 18 (2004) 1118-1125.
[6]. E. Y. Sheu, Energy Fuels 16 (2002) 74-82.
[7]. F. Trejo, G. Centeno, J. Ancheyta, Fuel 83 (2004) 2169-2175.
[8]. H. Seki, F. Kumata F, Energy Fuels 14 (2000) 980-985.

[9]. J. Ancheyta, G. Centeno, F. Trejo, G. Marroquín, Energy Fuels 17 (2003) 1233-1238.
[10]. J. Bartholdy, S. I. Andersen, Energy Fuels 14 (2000) 52-55.
[11]. J. Ancheyta, S. K. Maity, G. Betancourt, G. Centeno, P. Rayo, M. T. Gómez, Appl. Catal. A 216 (2001) 195-208.
[12]. G. Centeno, F. Trejo, J. Ancheyta, A. Carlos, Rev. Soc. Quím. Méx. 48 (2004) 186-195.
[13]. M. N. Siddiqui, M. F. Ali, J. Shirokoff. Fuel. 81 (2002) 51-58
[14]. S. Mitra-Kirtley, O. C. Mullins, J. V. Elp, S. J. George, J. Chen, S. P. Cramer, J. Am. Chem. Soc. 115 (1993) 252-258.

Gas-liquid Distribution in Trickle Flow Regime through a Fixed Catalyst Bed Reactor Used in Hydrodesulfurization Process

A. Koudil, I. Harter, Th. Chapus
Institut Français du Pétrole. BP 3. 69390 Vernaison. France

Abstract

Transportation fuel quality specifications are undergoing profound changes that are largely driven by environmental constraints. Most current regulatory efforts are directed towards limiting sulfur content in order to improve catalytic converter performance and to reduce motor vehicle SOx emissions.

The trend to low sulfur fuels particularly affects gasoline and diesel production. Refiners must adapt their production schemes and in some cases, revamp or build dedicated units to keep pace with the new specifications.

Conventional processes for sulfur removal use hydrodesulfurization (HDS) technologies. These processes can be modified through improvements in catalysis and technology to increase their efficiency. The main challenge for diesel hydrotreatment processes is to increase HDS activity. Many efforts are currently done in order to improve the global efficiency of the process. Catalysts' activity has of course to be increased, but it is also essential to have a near-ideal gas-liquid distribution, in order to ensure optimum catalyst utilization. Recent advances in the design of distributors for trickle-bed reactors are discussed in this paper. One of the major concerns for this kind of reactors is the phase distribution. In order to increase the efficiency of a two-phase gas-liquid reactor, it is important to achieve a near uniform distribution of the two phases flowing through the catalyst bed. An uneven distribution causes an inefficient use of the catalyst volume, resulting in a net loss of yield and conversion. Another effect of poor distribution is for exothermic processes the occurrence of hot spots inside the catalyst bed, which harms the catalyst and thereby the global reactor efficiency.

A cold mock-up consisting in a 600 mm inner diameter column is used for many studies regarding the phase distribution in a two-phase gas-liquid flow through a fixed bed. The liquid distribution is obtained using a set of 60 liquid collectors each connected to a compartment inside the experimental column. These compartments are of equal section and cover the entire cross-section of the column. The liquid level inside each collector is converted into a liquid flow rate. Images of the distribution are reconstructed using the values of the normalized differences between the ideal and the actual flow rates. A gamma-ray tomographic system is used to compare some images obtained with the collector system.

1. Introduction

Increased environmental awareness has resulted in a worldwide trend for transportation fuel reformulation. This requirement originates from the inhibitive effect of sulfur on catalytic converter performance.

Among the changes in diesel fuel specifications in recent years, only the sulfur specification has been constantly tightened, as shown in Table 1. Sulfur specification in Europe has been lowered from 350 ppm down to 50 ppm in 2005, then to 10 ppm in 2009. A gradual shift has already taken place, favored by tax incentives, in countries such as Germany. The US will follow the same trend; sulfur content in automotive diesel will be reduced from the current 500 ppm to 15 ppm in 2006. The suppression of sulfur in diesel fuel is a strong trend worldwide as other countries adopt severe sulfur specifications : 30 ppm in Korea in 2006 and 10 ppm in Japan in 2005.

Table 1. Sulfur Motor Fuel Specifications.

Country	European Union		United States		Korea	Japan
Year	2005	2009	2002	2006	2006	2005
Sulfur, wt ppm	50	10	500	15	30	10

1.1. Diesel ultra deep desulfurization processes

The main challenge to desulfurizing diesel streams is to convert dibenzothiophenes, particularly the refractory dibenzothiophene compounds containing alkyl groups in either or both the 4- or 6- positions relative to sulfur atom [1, 2, 3, 4]. Desulfurizing diesel streams is still more difficult when treating diesel from FCC units (LCO) or from coking units (CGO) because of the presence of aromatics and nitrogen compounds which strongly inhibit catalyst desulfurization activity [5].

As a result of the new ULSD specifications, conventional HDS technology has been required to adapt and account for very low quantities of very difficult to treat sulfur compounds. Significant effort has been undertaken to conquer two major ULSD technology hurdles :

- significantly increasing catalyst activity,
- ensuring near-ideal flow dynamics in the hydrotreater.

Figure 1. Scheme of a simplified diesel hydrodesulfurization process.

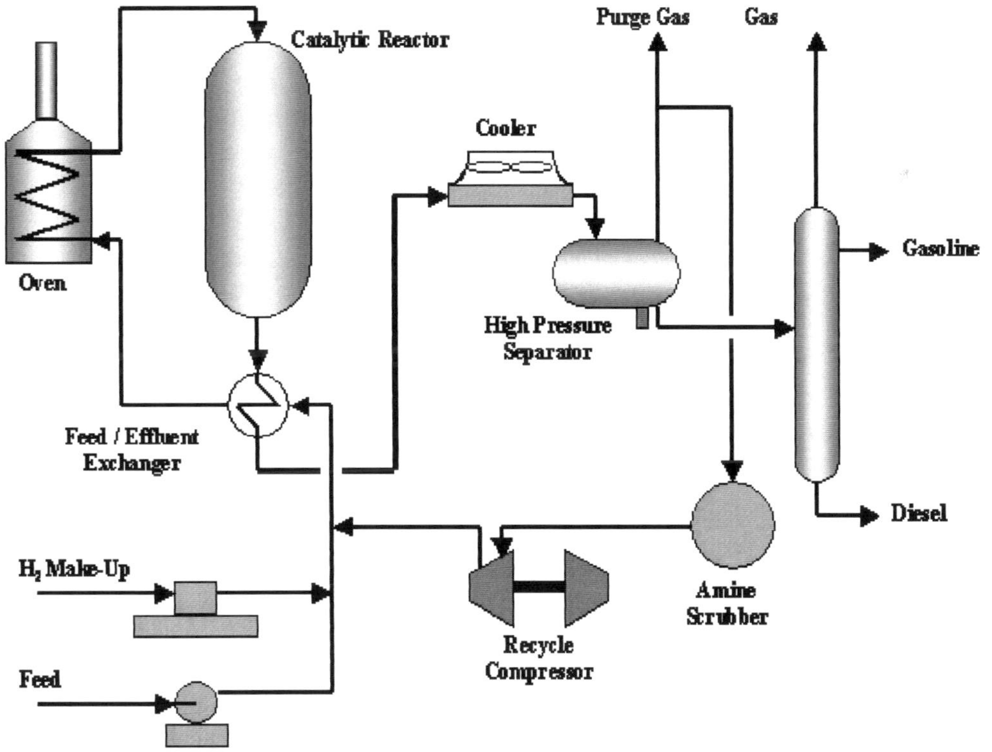

Multiple options are available for refiners in order to attain ULSD specifications through revamping of existing units, including :

- adding catalyst volume : using dense loading technology or adding additional reactor volume,

- employing the most recent high activity catalysts,

- reducing the H_2S content in the recycle gas stream by adding an amine scrubber. An operating temperature reduction of several degrees can easily be obtained with this implementation,

- adding a hydrogen purification unit to increase make-up hydrogen purity,

- installing optimized reactor internals.

1.2. Increasing catalyst activity

This is of course of major importance. New catalyst generations are based on an Advanced Catalytic Engineering (ACE) platform. These catalysts are tailored for ULSD and other high activity usages and are manufactured in such a way as to enable nano-scale construction of the active sites. These catalysts are ideally suitable as drop-in catalyst replacements in any unit. Depending on the particular application, catalysts used are generally containing CoMo or NiMo active phases on an alumina support. ACE technology allowed a significant activity improvement for deep hydrodesulfurization, coupled with a much higher tolerance to organic nitrogen.

1.3. Gas-liquid distribution

Even the most active catalysts can fail to meet ULSD specifications if the design of the reactor internals does not provide near-ideal gas/liquid distribution. As sulfur targets approaches zero, even the smallest amounts of liquid or vapor bypassing the catalyst would result in a very critical decline in performance. Based on LHSV impact on HDS performances, it can be estimated that bypassing 10% of the catalyst volume implies a loss of desulfurization activity of about 3°C, and as much as 6°C if there is a 20% loss of catalyst utilization. Those values are even underestimated, since a maldistribution in the reactor also implies an increase of catalyst deactivation, due to hot spots / high radial temperature profiles.

Inside gasoil hydrodesulfurization reactor, liquid trickles on catalyst whereas gas flows in remaining space. In order to maximize the performances, a gas-liquid distributor is put at the top of the reactor. It must provide a homogeneous distribution. The improvement of the contact between the three phases (gas, liquid and catalyst) enhances both the reaction kinetics and the transport phenomena.

In this study, two-phase gas-liquid flows through a fixed bed column are investigated in a cold flow mock-up representative of the industrial units at a smaller scale (1/10 - 1/6). It is mainly focused on trickle flow conditions.

2. Experimental Setup

Industrial reactors have an inner diameter of several meters and are about ten meters long. A pilot plant reactor is often designed as small as possible to reduce its cost [6]. In order to be representative of industrial reactors in terms of catalyst loading and flow configuration or distribution, it is important however to use a sufficiently large experimental column. The column used in the present study to test the new distributor is of 600mm ID. The experimental setup is shown in Figure 2.

Figure 2. Experimental setup.

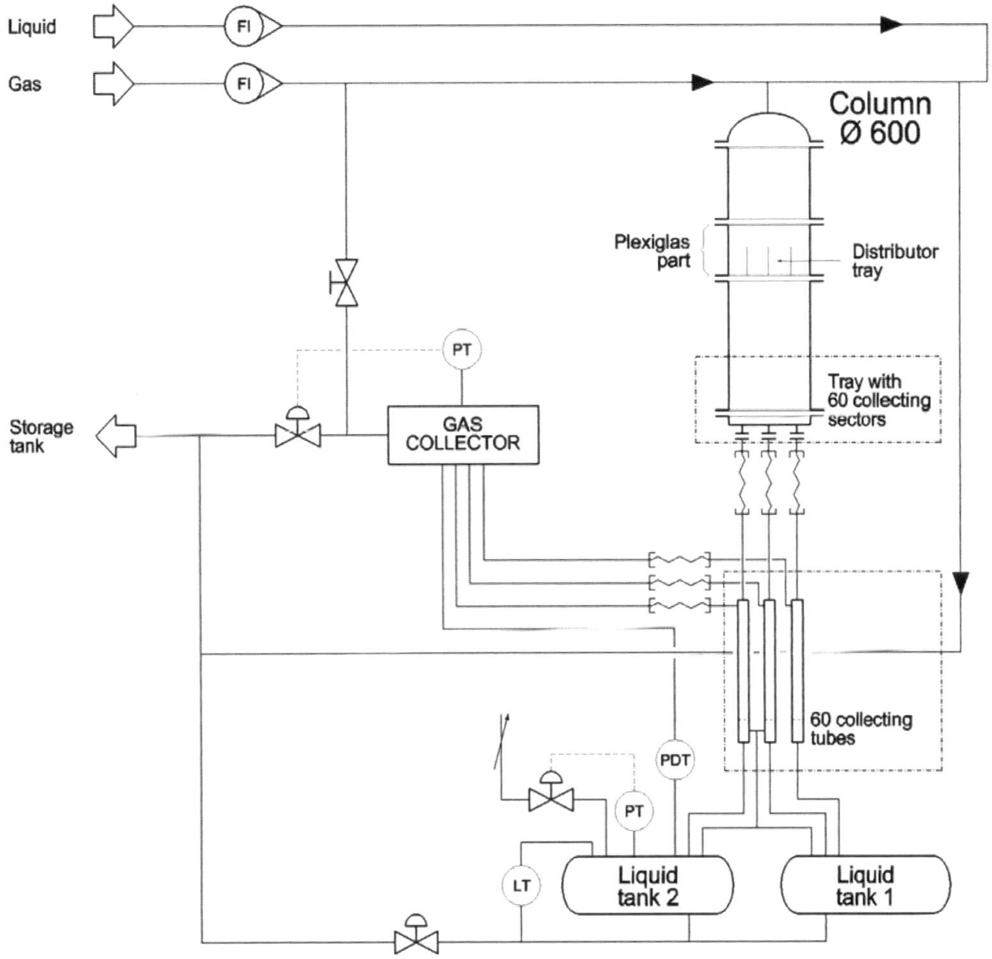

2.1. Description of the recirculating loop

The main loop element consists of a stainless steel column. A portion of the column at the inlet is made of Plexiglas in order to visualize the liquid level and the flow at the inlet and outlet of the distributor. The liquid enters the column via a diffusing system, and it then impinges on the liquid surface at the distributor. After crossing the distributor tray, it reaches a collecting system that divides the total cross section in 60 sectors of equal section. Each sector is linked to a collector, a vertical Plexiglas tube (Figure 3). The gas disengages at the top of the collector and enters an upper gas collector. The liquid falls down and is evacuated to two liquid tanks before being recirculated. Heptane and nitrogen are used at ambient temperature and near atmospheric pressure. Heptane is actually representative of the industrial liquids in terms of flow properties as the density which is approximately 0.7 and the surface tension of the order of 20mN/m. Its viscosity is lower than the industrial one, 0.4cP instead of few centipoises. Industrial reactors work at different pressures depending on the process being involved. Using nitrogen near atmospheric pressure is sometimes an ambitious extrapolation. Nevertheless, the effects of the fluid properties are investigated during experiments carried out on a smaller scale using a higher viscosity liquid (iso-hexadecane) and a higher density gas (sulfur hexafluoride, SF_6) [7]. These measurements confirmed some of the results obtained with the experimental setup shown in Figure 2 and permitted the selection of the relevant flow conditions to meet each industrial application.

Figure 3. Details of the collecting system.

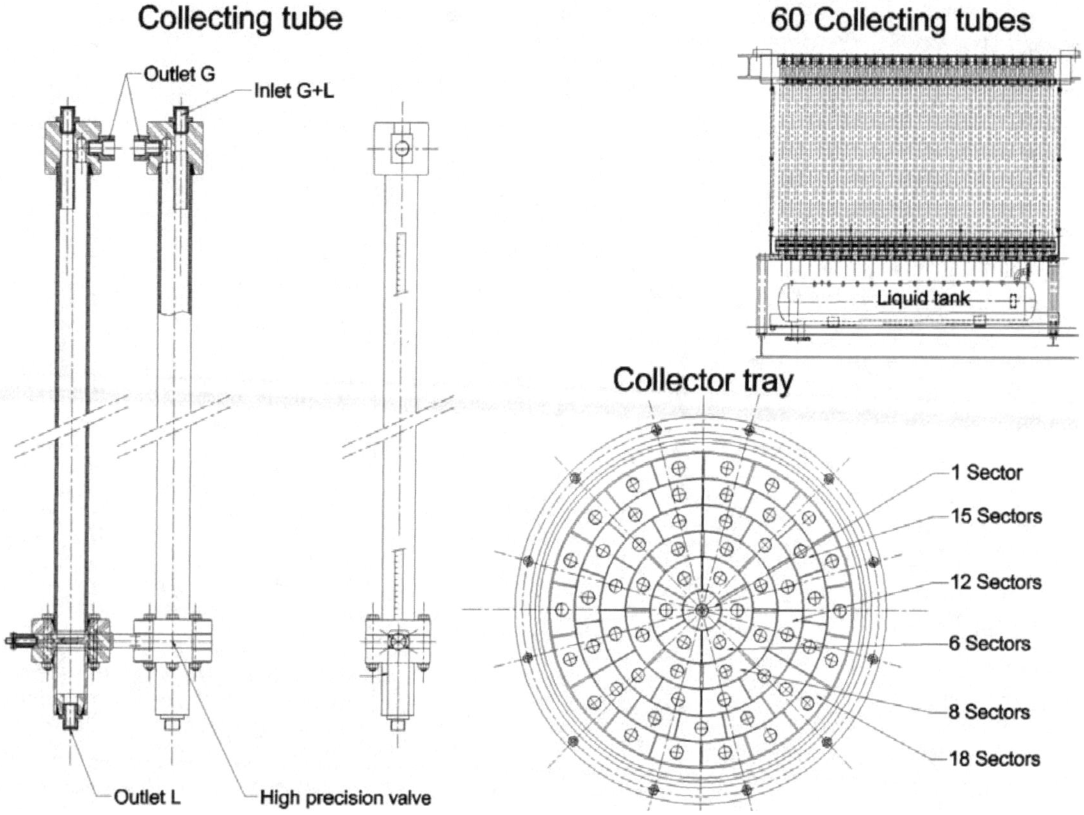

2.2. Measurement

The measurements procedure starts once steady state conditions are reached which is obtained quickly. After setting the relevant apertures of the valves at the outlet of each collecting tube, all the liquid levels are recorded. Each level h_i is transformed into a flow rate q_i using the following relation applied for each collector i :

$$q_i = C_i \sqrt{2gh_i}\, s_i \tag{1}$$

where s_i is the section of aperture of the valve, h_i is the liquid level and g is the gravitational acceleration. The discharge coefficient C_i is obtained after a calibration procedure [8].

The measurements are validated only when the sum of the flow rates in the collectors is equal to the total inlet flow rate within 10%. During the experiments, it was found that equation (1) does not work when the gas flow rate is relatively high. In this case, the sum of the calculated flow rates q_i does not match the total inlet flow rate. This is due to the fact that a pressure difference exists between the free surface of the liquid and the discharging tank. The measurement of this pressure difference is not feasible for all 60 tubes. A new procedure is therefore adopted which consists in recording for each collector two different liquid levels at two different apertures of the valve, so the liquid flow rate is expressed in two ways

$$q_i = C_i \sqrt{2gh_{i1} + \frac{2\Delta p_{i1}}{\rho_l}}\, s_{i1}$$
$$q_i = C_i \sqrt{2gh_{i2} + \frac{2\Delta p_{i2}}{\rho_l}}\, s_{i2} \tag{2}$$

where ρ_l is the liquid density. By visualizing the flow at the outlet of the distribution tray, it is made sure that the pressure differences do not influence the measurements. In equation (2), Δp_{i1} and Δp_{i2} are equated and the flow rate crossing the collector is then given by

$$q_i = C_i \sqrt{2g\left(h_{i1} + \frac{s_{i2}^2 h_{i2} - s_{i1}^2 h_{i1}}{s_{i1}^2 - s_{i2}^2} \right)}\, s_{i1} \tag{3}$$

Using equation (3) instead of equation (1) allows obtaining the total liquid flow rate by summing the partial flow rate in each collector, even at high gas flow rates.
The liquid flow rates in (3) are used to reconstruct images of the flow rate distribution in a section where the collecting system is positioned. For that purpose, the values of the normalized flow rates q_{ni} are used:

$$q_{ni} = \frac{(q_i\text{-}Q_l/60)}{Q_l/60} \tag{4}$$

where Q_l is the total liquid flow rate. The values of q_{ni} could be positive or negative. A zero value for q_{ni} means that the collector carries a flow rate equal to the 1/60th of Q_l. It is obvious that a good distribution implies values of q_{ni} close to zero.
To discriminate between two similar distributions, it is judicious to use statistical tools. The values of q_{ni} are considered as a discrete distribution of 60 values of the random variable q_{ni}. The distribution has a mean value and a standard deviation defined as follows

$$\overline{q}_n = \sum_{i=1}^{60} \frac{q_{ni}}{60}$$
$$\sigma = \sum_{i=1}^{60} \sqrt{\frac{q_{ni}^2}{60}} \tag{5}$$

Replacing the expression of q_{ni} (4) in equation (5) gives a zero value for the mean flow rate $\overline{q}_n$. This is actually the condition of mass conservation for the liquid. The mean value is not a discriminating factor. The comparison between several distributions can be done using the value of the standard deviation. The best distribution is obtained with the lower value of the standard deviation σ.

3. Gamma-ray tomographic system

In addition to the measurements with the collectors, some measurements have been carried out using a gamma-ray tomographic system. The flows investigated are generally steady state so that a photon attenuation technique is adapted to investigate the phase distribution at various levels of stainless steel experimental column. When the photon flux crosses the column filled with a real catalyst, the attenuation is quite high. Hence, a suitable gamma-ray source with high-energy photon is necessary. A 300 mCi activity Cs137 source collimated to produce a fan beam photon flux is used. This source is coupled with a series of 32 BGO photo-scintillator transducers. These detectors are separated by spaces of the same thickness as their detection width, 13mm, so that with one lateral movement the photon fluxes are sampled over 64 chords. The whole system, gamma-ray source and the detectors, is rotated around the fixed bed column and the photon fluxes are counted on 64, 128 or 256 angular positions. An algebraic algorithm is used to reconstruct a 2D map of the phase distribution across a reactor cross-section. In order to account for cases with high attenuation gradients inside the bed, a new algebraic reconstruction algorithm is used [9]. A full description of the system and its calibration can be found in [10].

Figure 4. Geometry of the tomographic system.

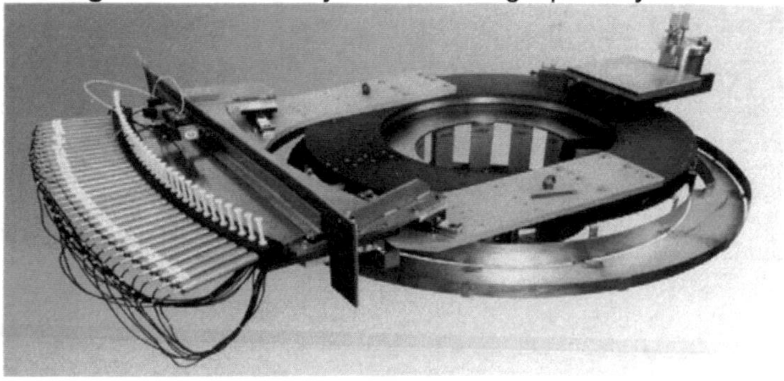

The evaluation of the performances of the tomographic system is a fundamental step toward a good use and interpretation of the reconstructed images. The system has therefore been tested with physical phantoms of known attenuations inserted inside the column in order to simulate, inside the bed, gas-liquid mixtures with various gas proportions. Nine samples have been tested simulating from 0% to 100% gas volume fraction. The values obtained after reconstruction and the values measured during calibration do not differ by more than 3%.

4. Limitations of the collecting system

The collecting device has some limitations. The first limitation concerns the reconstructed images and is inherent to the fact that the spatial resolution of the measurement device is limited to $\pi/4\,(\mathrm{ID})^2/60 \approx 50\mathrm{cm}^2$. Maldistributions cannot be detected at a level lower than 7cm. Nevertheless, this resolution is sufficient for industrial distributors. To highlight the spatial resolution, two experiments have been carried out to compare the results obtained with the system of 60 collectors and the tomographic system even if the measured parameters are not the same. For the collector tray, flow rates are measured whereas phase distribution is measured with the gamma-tomography. For the reason that the gas-liquid interaction is weak compared to the liquid-solid interaction, the images obtained with the two methods can be compared.

Figure 5 shows, for two particular configurations, that the collector system can grasp the liquid distributions. In the first configuration (upper images), the liquid is injected centrally, while the second one (lower images) involves seven injection points. The images obtained using the collectors whose spatial resolution is 50cm^2 are in good agreement with the far better gamma-ray tomographic image whose resolution is 1cm^2 [11].

Figure 5. Comparison of liquid collection system to gamma ray tomography.

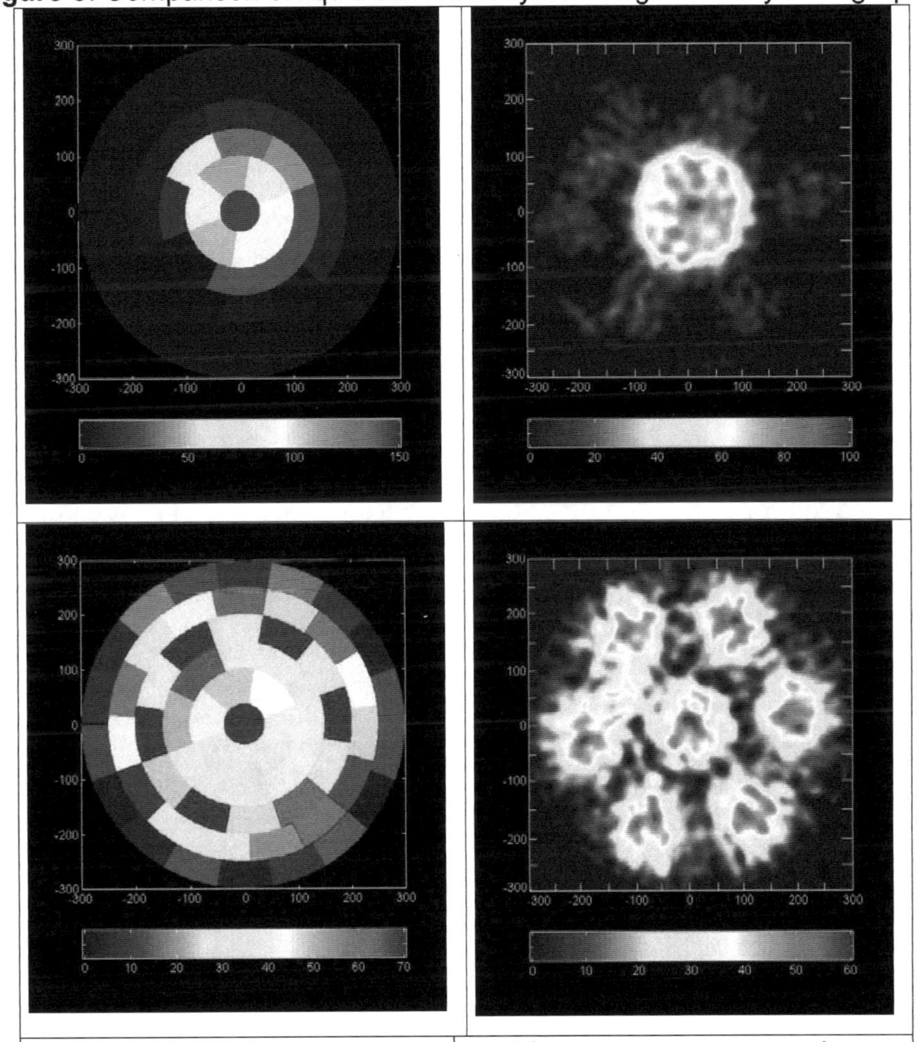

a) liquid collectors	b) gamma-ray tomography

There are also limits concerning the liquid flow rate. The upper limit corresponds approximately to a superficial velocity of V_{sl} =1cm/s. This is due to the fact that when a maldistribution occurs at this upper limit, it is expressed by a blocking of one or more collectors. At the other extreme, there is no real lower limit for the liquid flow rate. When a distribution results in very low liquid flow rate in one or several collectors, the valve is closed and the time t to reach a certain level h is recorded so that the flow rate can be estimated using the relation :

$$q = \frac{h}{t} s_c \qquad (10)$$

where s_c is the section of the tube collector. Nevertheless, the lowest liquid velocity considered is 0.05cm/s. The resolution of the collecting system in terms of liquid flow rate is close to this limit.

For the gas phase, there is no lower limit; some experiments have even been done without gas flux. The upper limit of $800Nm^3/h$ is dictated by the capacity of the two compressors used to deliver the gas flow.

5. Results

The measurement device described above is dedicated to investigate the influence on the distribution of many parameters or to characterize different distribution trays. To make use of the measurements, images are reconstructed or curves giving the evolution of the standard deviation as a function of the parameter under investigation are plotted.

5.1. Extension of the operating range of a distributor tray

A distributor tray was designed with the help of the collecting system. The first measurements were carried out within the following limits [13] :

- superficial liquid velocity V_{sl} was varied from 0.2 up to 0.7cm/s,
- superficial gas velocity V_{sg} was varied from 0 up to 30cm/s.

In this study, the velocity intervals are extended. Both lower and higher values of the liquid velocity are tested. The limits achieved are 0.05cm/s as the lowest value and 1cm/s as the highest one. For the gas velocity, higher values of the order of 70cm/s are reached.

Figure 6 shows the evolution of the standard deviation as a function of the liquid flow rate for a fixed gas flow rate at the highest value of the velocity, i.e. V_{sg} =70cm/s. Apart from the lowest liquid velocity, the distributor tray shows an equal and satisfying performance. It is confirmed by the figure 7 which shows the evolution of the standard deviation as a function of the gas flow rate for a fixed liquid flow rate, i.e. V_{sl} =0.5cm/s. The distributor works well whatever the value of the superficial gas velocity (V_{sgmax} =75cm/s).

The worst distribution is obtained when both liquid and gas velocity are low. For these conditions, it is very hard to achieve a good distribution. Nevertheless the capability of the tray to distribute the liquid is confirmed for more severe conditions. The new distributor tray design is therefore extended to cover larger liquid and gas flow rate intervals.

Figure 6. Standard deviation as a function of the liquid velocity (V_{sg} =70cm/s).

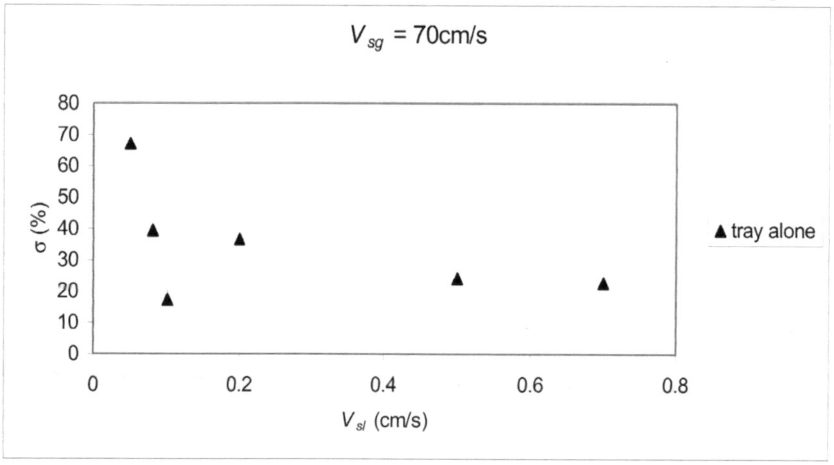

Figure 7. Standard deviation as a function of the gas velocity (V_{sl}=0.5cm/s).

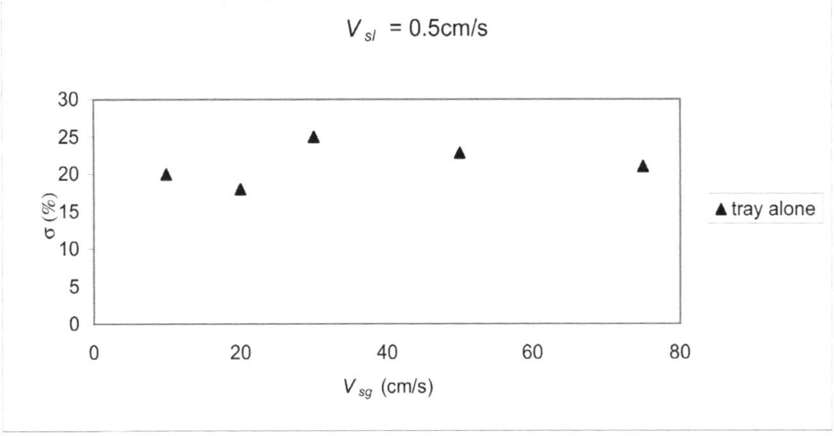

5.2. Effect of catalyst loading

The volume of the catalyst bed is an important parameter. The design of a reactor is done such that the ratio of the flow rate to the bed volume is as high as possible. High density loading permits high conversion because more catalyst is available for the chemical reaction. However, high density loading is not always feasible. The limitation is due to the fact that high density loading means higher pressure drop because of the reduction of the cross section of the flow. Therefore, it is sometimes preferable to perform lower loading density of the bed in order to limit the pressure drop and thereby the investment costs of the flow devices such as pumps and compressors.

Two different loading densities have been tested : dense loading and sock loading. The procedures to obtain each loading can be found in reference [12].

Figure 8 shows the curves of the standard deviation (%) as a function of the superficial gas velocity for a fixed superficial liquid velocity (V_{sl}=0.5cm/s) and for dense and sock loadings in comparison to the tray alone, i.e. without a downstream bed. The dependence of the distribution on the loading density is moderate. A limited impact is found at low gas velocity at which the sock loading experiences a better distribution. It could be explained by the fact that the radial transport is more pronounced for the sock loading which in fact induces lower resistance to radial flow. No significant effect is found above 10cm/s.

When the distribution upstream is good enough, the catalyst bed provided it is homogeneous insures that this good distribution remains far away in the bed.

Figure 8. Effect of loading on the distribution (2mm spherical particle -V_{sl}=0.5cm/s) .

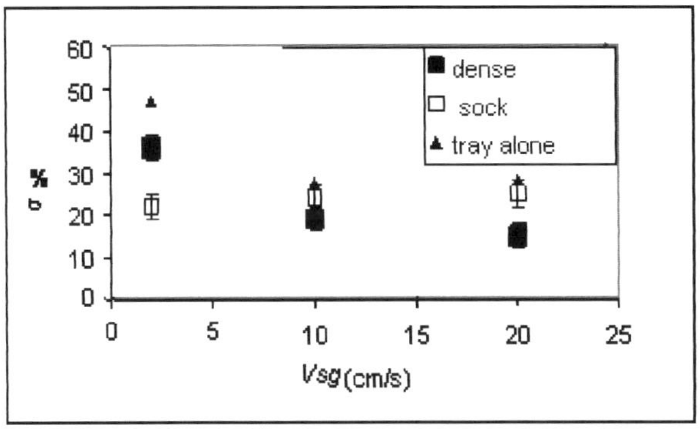

Figure 9. Liquid distribution along the bed for a 1.2mm diameter trilobe catalyst dense loaded.

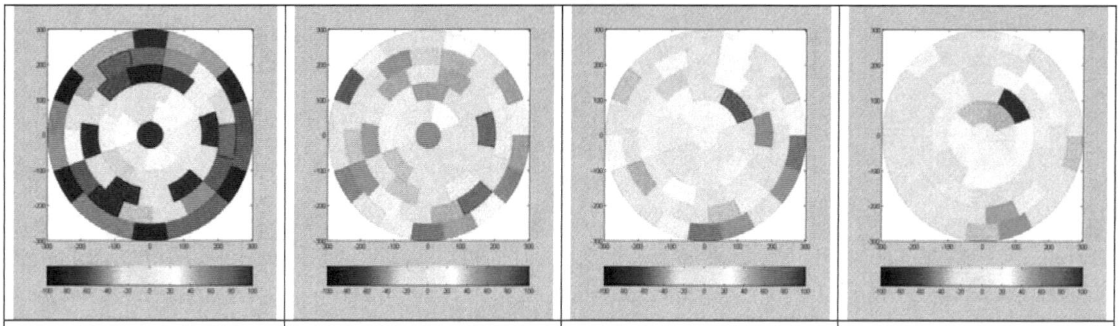

1 m , σ = 64%	2m , σ = 31%	3m , σ = 25%	5m , σ = 24%

Figure 9 shows the axial evolution of the distribution along an extrudate catalyst bed densely loaded. A non optimized distributor tray is used in this case. After one meter downstream of the distributor, the liquid is only present in certain areas corresponding to the injection points. The distribution improves progressively. After 3m, the standard deviation reaches its asymptotic value which corresponds to a satisfactory distribution.

5.3. Effect of catalyst pre-wetting

The effect of the catalyst pre-wetting is not negligible. Experiments have been carried out with both pre-wetting and no pre-wetting of a spherical catalyst bed. Figure 10 shows that when a pre-wetting is performed, the distribution is far better. It can be explained by the fact that when the bed is drained after flooding, liquid bridges remain between adjacent catalyst grains and therefore insure the connection of the liquid films which helps the spread of the flowing liquid across the entire section of the bed.

Figure 10. Effect of the pre-wetting conditions.

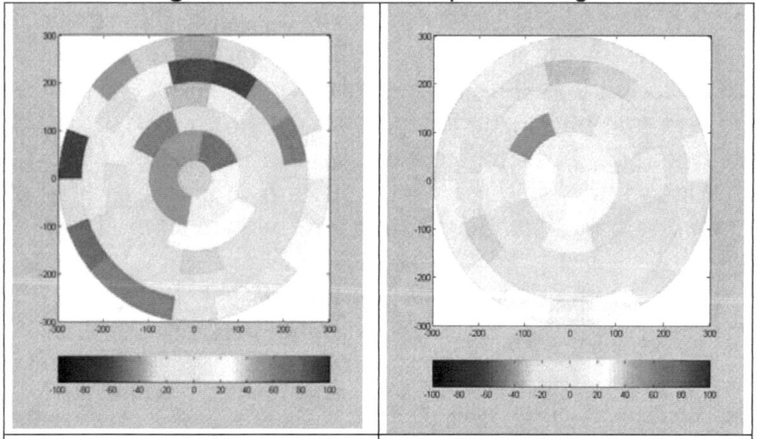

(a) : no prewetting σ = 37 %	(b) : prewetting σ = 24 %

Conclusion

During the latest years, important R&D work has been done with the aim to achieve ULSD. Many aspects were investigated such as :

- crude oil and feed analysis based on crude oil origin and characteristics. In order to predict accurately the performances at very low sulfur level, it is essential to take into account the composition of diesel streams entering the reactor,

- knowledge of feed transformation under hydrotreating conditions based on a predictive kinetic model,

- development of high performance hydrotreating catalysts,

- development of new reactor internals (distributors and mixing trays) designed to afford near-ideal gas-liquid distribution for optimum catalyst utilization.

The work completed in this study relates to this last aspect. Collector system is used to characterize experimentally liquid distribution in hydrodesulfurization reactor. Measurements obtained with this system are in good agreement with those obtained with the gamma-ray tomography. This collector system is devoted to investigate the effects on the distribution of several parameters. Numerous distributions are compared using statistical parameters, mainly the standard deviation.

In a former study, a distributor tray was designed with the help of measurements obtained with the described system. In this study, new measurements allow the extension of the use of the distributor tray to more severe conditions, mainly higher gas velocities.

In addition to the characterization of the capabilities of the tray, the collecting system is used to investigate the effect of the loading density. A moderate effect of the loading density is shown especially at sufficiently high gas flow rates. At the lower gas flow rates, sock loading seems to provide better liquid spreading. Measurements were also carried out to assess the effect of the wetting conditions. The catalyst pre-wetting has been proved to be important in order to achieve a good distribution.

Further effects are investigated such as the fluid properties and the catalyst particle form but are not presented in this study. The present collector system is also used to construct charts giving the value of design parameter such as distributor porosity.

REFERENCES

[1] Houalla, M., Broderick, D.H., Sapre, A.V., Nag, N.K., De Beer, V.H.J., Gates, B.C., Kwart, H. : "Hydrodesulfurization of methyl-substituted dibenzothiophenes catalyzed by sulfided Co---Mo/-Al2O3". Journal of Catalysis, 61, p.523, 1980.

[2] Whitehurst, D.D., Isoda, T., Mochida, I. : "Present State of the Art and Future Challenges in the Hydrodesulfurization of Polyaromatic Sulfur Compounds". Advances in Catalysis, 42, 345, 471, 1998.

[3] Meille, V., Schultz, E., Lemaire, M., Vrinat, M. : "A new route towards deep desulfurization: selective charge transfer complex formation" Applied Catalysis A General, 1999, 187, 179-186

[4] Lopez Garcia, C., Roy-Auberger, M., Chapus, Th. : "Analysis and Kinetic Modeling in ULSD Hydrotreating". ACS New York 2003

[5] Schulz, H., Böhringer, W., Waller, P., Ousmanov, F. : "Gas oil deep hydrodesulfurization: refractory compounds and retarded kinetics". Catalysis Today, 49, 87, 1999

[6] Burkhardt, T., Verstraete, J, Galtier, P., Kraume, M. : "Residence time distributions with a radiotracer in a hydrotreating pilot plant : Upflow versus downflow operation". Chem. Eng. Sci., 57, p.1859, 2002

[7] Raynal, L., Harter, I. : "Influence des propriétés physiques sur le fonctionnement d'un élément de plateau Equiflow". IFP Confidential Report n° 57786, 2003

[8] Koudil, A., Harter, I., Raynal, L., Boyer, C. : " Experimental characterisation of the gas-liquid distribution through a fixed catalyst bed reactor". Proceedings of the 9th International Conference on Multiphase Flow in Industrial Plant, Rome, 2004

[9] Koudil, A., Boyer, C., Bentolila, Y. : "Measurement of liquid and gas flow distribution in a fixed bed reactor using a gamma-ray tomographic system". Proceedings of the 3th World Congress on Industrial Process Tomography, Banff, p.445, 2003

[10] Boyer, C., Fanget, B. : "Measurement of liquid flow distribution in trickle bed reactor of large diameter with a new gamma-ray tomographic system". Chem. Eng. Sci., 57, p.1079, 2002

[11] Harter, I., Boyer, C. , Raynal, L, Ferschneider G., Gauthier Th. : "Studies of gas-liquid flow through reactors internals using VOF simulations". Ind. & Eng. Chem. Res., 40, N° 23, p.5262, 2001

[12] Koyama H., Nagai E., Tori H., Kumagai H. : "Japanese refiner solves problems in resid desulfurization unit". Oil & Gas Journal. 93 (46) p.82 Nov 13, 1995

NONLINEAR ANALYSIS OF CATASTROPHIC FAILURE EFFECTS ON MULTI OBJECTIVE INTRA REFINERY OPTIMIZATION

Ahmed Sirag Khogeer, Dept of Chemical Engineering, Colorado State University. Fort Collins, Colorado and M. Nazmul Karim, Chair, Department of Chemical Engineering, Texas Tech University, Lubbock, Texas

Abstract

Petroleum refining is a capital-intensive business. With the stringent environmental regulation on the process industry, political instability and declining refining margins, higher pressures are exerted on refineries to optimize performance. Most refinery optimization methods do not cover the multiple objectives such as catastrophic failure effects. With the increased risk of war and terrorist attacks, refineries and fuel transportations grids are targeted since they supply fuels to essential nodes such as power plants and war efforts. This work covers this case by analyzing the catastrophic failure effects and dividing the effects into two categories: A. Catastrophic failure having either on or off operations, and B: Increased risk of operations such as dangers from terrorist hijacking tankers or trucks. The first category was modeled as binary constraints and alternatives routes were examined in order to secure supplies. For the second category, penalties for increased risk were added to all streams or facilities undergoing high-risk operations. Case studies were done on several refineries serving different markets with different products needs. Simulation runs were done using ASPEN PIMS utilizing nonlinear techniques to solve the optimization problem. The results addressed the need to debottleneck some refineries or transportation media to be able to meet the demand for essential products under partial or total failure scenarios. It also addressed how importing some high spec products can help recover some of the losses and what needed to be done in order to accomplish that. In addition, the results showed nonlinear relations among local and global objectives for some refineries.

Introduction

Petroleum refining is a capital intensive business and thus optimization can lead to saving of millions of dollars. However, with the increased risk of war and terrorist attacks, refineries and fuel transportations grids are targeted since they supply fuels to essential nodes such as power plants and war efforts and thus they need to secure supply in case of a catastrophic failure event. This is done using multiple objectives optimization, where catastrophic failure effects were quantified and incorporated into the objective function. This way, catastrophic failure effects on objective could be accounted for in the economic objective. In literature, not much attention has been paid to this field until recently. Most work in literature is concentrated on the effects of process accidents on refinery emissions and safety. For example, for some refiners, the scope might be instability and phase equilibria research rather than economic consideration [2, 3, 4, 5].

The idea behind having an optimization scheme under a catastrophic failure is to study the effects, and explore alternatives of the refinery operations and product supply to the markets when a catastrophic failure resulting from a process accident or a deliberate action such as terrorism causes partial or total failure of the system. This type of condition should be divided into two groups:

a) Catastrophic failure having on or off (0/1) operations. 0/1 indicates that a particular equipment or process unit is either working or not. For example, in the case of pipeline rapture, the pipeline cannot usually be used and should be shutdown. For this group, constraints were treated as binary (0/1) constraints in the model.

b) Relative increased risk of operations due to process or non process reasons. This includes war risks and increased risk of terrorist acts such as hijacking of tankers or trucks. For example, in case of war, tankers are penalized $1.0 per barrel for shipment from the Arabian Gulf but not from the West Coast of the Arabian Peninsula.

Supply-Demand Chain

A Supply–Demand chain was constructed which included the major refinery feed, processing, storage and consumptions nodes. To have a good model of a refining system Supply –Demand chain, the following should be identified:

a. Different crude types.
b. Different feeds other than crude, including fuels (condensate, fuel gas, reduced crude, MTBE, etc.).
c. Different refineries with each refinery having different process units.
d. Process limitations and capacities. For example, CCR maximum reactor temperature in refinery A is 500C. Feed capacity at this reactor temperature cannot exceed 7 KBPD.
e. Different transportation media such as tankers, pipelines and trucks.
f. Major consumer nodes such as desalination plants, power plants, etc.
g. Markets and products requirements for that market such as gasoline, diesel, fuel oil, etc. for summer and winter modes.
h. Supply/demand growth per year.

Figure 1 shows a schematic example of a simplified supply demand chain which gives an example of feed pools available to refineries, refineries as blocks, internal units in refineries, major consumption nodes and interconnections among refineries and within the chain.

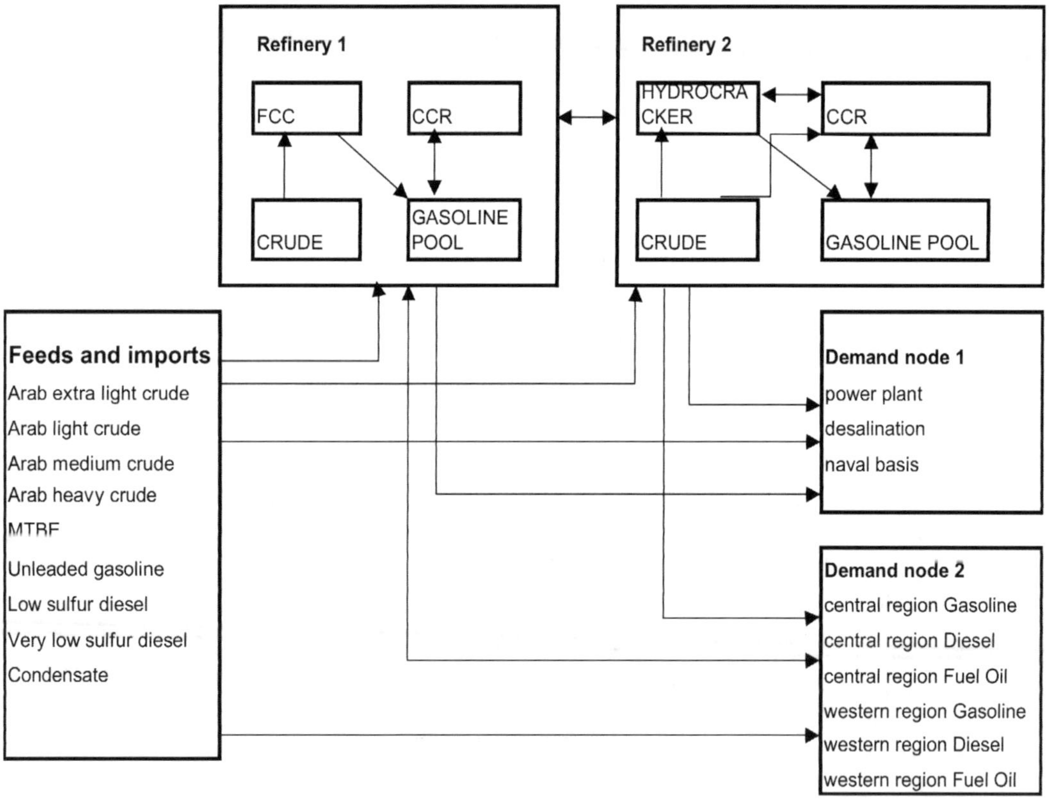

Figure 1: Refining-Supply / Demand Chain

Objective Function

The objective function can be expressed mathematically as follows [1]:

$$\text{Minimize } \Phi = -(P_1 - P_2 - P_3 - P_4) \text{ ---(1)}$$

Where

- Φ is the profit-based objective function,
- P_1 is the selling price of the products which is the major income of a refinery
- P_2 is the price of the raw materials,
- P_3 is the operating costs,
- P_4 is the product-property related parameter and is usually set to zero.

P_1 is given by the following equation:

$$P_1 = \sum (\text{Prod}_j \,.\, \text{Price}_j \,.\, \text{wt}_j) \text{ --- (2)}$$

Where

- j = 1, ..., 7 is the product for the refinery such as LPG, Gasoline, etc.
- Prod_j is the production rate of product $_j$, (BBL/Day)
- Price_j is the price of a unit quantity of product j, \$/BBL
- wt_j is the weighting on the jth term of the objective function

P_2 is given by the following equation:

$$P_2 = \sum (\text{Feed}_k .\, \text{Price}_k \,.\, \text{wt}_f) \text{ --- (3)}$$

Where

- F is the feed or imported stream flow rate (BBL/Day). This can be crude oil, intermediate product (such as vacuum gas oil VGO for FCC feed) or finished product (such as ultra low sulfur diesel) imported from other source.
- Price k is the cost of the feed, \$/BBL
- wt_k is the weighting on the term related to the cost of raw materials.

P_3 is given by the following equation:

$$P_3 = \sum (\text{utility}_n \,.\, \text{UC}) \text{ ---(4)}$$

Where

- Utility_n, (m^3/Day) is the utility consumed in process unit n;
- UC is the cost of the utility (\$ /m^3).

Termination Criterion

The termination criterion is

$$\Phi_{n+1} - \Phi_n < \varepsilon \text{ ---(5)}$$

Where Φ denotes the value of the objective function and the subscript indicates the iteration number. ε represents the absolute difference in magnitude of the objective function in two successive iterations and is defined by the user.

Optimization Algorithm

Simulation runs were done using ASPEN PIMS utilizing nonlinear techniques to solve the optimization problem. The nonlinear algorithm is Successive Linear Programming SLP which has a wide acceptance and success in solving mildly nonlinear term in large refining system linear programming models [6]. SLP uses a Taylor series expansion to approximate the nonlinear terms in the constraints and the objective function around a feasible point.

Description of the Refining System

The model consisted of seven refineries; five of them are domestic refineries with variable capability of exporting excess products to other refineries or international markets. These refineries where built with the vision of supplying local markets so there are limitations on the type and quality of products they can produce. The other two are export refineries and are state of the art refineries with capabilities of meeting stringent products specifications.

Cases

Different scenarios were done to simulate the consequences of catastrophic events on the products supply and profit margin. Different refineries had different scenarios depending on the location, proximity to oil supply channels (such as seaport or pipelines), complexity factor of the refinery (which is a factor of presence of treating and conversion units) and availability of alternatives. All of the cases presented are based on real refining situation with some figures and constraints modified for confidentiality purposes. Following are the major cases in the model:

1. Increased Risk from Targeting Crude Oil Vessels Supplying a Refinery with a High Turndown Ratio but Single Feed Source

In this case, the aim was to simulate the increased risk of targeting crude oil supplies to a refinery with high turndown ratio but single feed source that is vessels (ship crude carriers). The simulation started with Refinery # 3 which is a small refinery having several crude units and is relatively far from the tension zones. The main constraint for this refinery was crude supply since it receives crude oil through vessels which can be targeted. A penalty for the increased risk (such as insurance cost or delay in delivery) was gradually added to the objective function and embedded in the crude cost. Another constraint was that this refinery can only process light crudes in the range of 32-35 API due to both crude units' overhead condensers limitations and distillation tower internal design. Cooling problems are serious issues in countries in the Middle East especially at summer time where it can limit not only the fractionator's output but the whole refinery throughput.

2. Increased Risk from Targeting Crude Oil Pipelines Supplying a Refinery with a Single Feed Source

This case was done on Refinery # 4, which is a large topping refinery with only one source of feed and having the ability to process only light crudes. The refinery is supplied by crude through a pipeline portion that is connected to a main header feeding several other refineries. Increased penalties were imposed on the transfer cost of the portion of the pipeline feeding that refinery only, even though other refineries share the main pipeline header. This simulates the case of damage to the pipeline portion feeding this refinery only. In this case, it was assumed that crude was transferred by pipeline to an oil terminal and from there it was distributed to other refineries either by pipelines or by small tankers.

3. Total & Partial Capacity Loss as a Result of a Catastrophic Failure

These cases were simulated on Refinery # 1, which is a large refinery with capability of processing different crudes and condensate. This refinery is considered as the work horse of the supply for distillate and jet fuels which is particularly important at war times. However, the turndown ratio for this refinery is lower than other refineries since it is a modern refinery assumed to be built after 1980, where the economics favors having a single large crude fractionator. Another condensate splitter was added. Even though this configuration gives the refinery very low operating costs, and much less parts inventory and maintenance problems, it does not give much flexibility in processing different crude types simultaneously or having low turndown ratios. Note that the products from each refinery are derived from the following equation:

$$\text{Product}_i = \Sigma\, \text{Product}_{ij}\, Tr_k, \qquad Tr_k \in [0,1] \text{ binary}$$
$$= \text{Product}_{i1}\ (0 \text{ or } 1) + \text{Product}_{i2}\ (0 \text{ or } 1) + \ldots \text{Product}_{in}\ (0 \text{ or } 1)$$

Where:

- Product is any petroleum product such as low sulfur diesel, JET-A1, etc.
- i is product number.
- J stands for the refinery number.
- Tr_K stands for the unit or train abbreviation and K is a binary variable to show that the selected unit or train in a unit is either on or off.

Three scenarios were done for this refinery:

1. Partial loss in the condensate splitting capacity, which can be due to one or more of the following considered reasons:
   a. Logistics of supplying the refinery with condensate such as an attack on the pipeline or constraint on the flow due to pumping limitation as a result of an attack, mechanical failure or cooling limitations.
   b. Loss of processing capacity in the condensate splitter column, products coolers, overhead condenser or downstream product treatment of the condensate section of the refinery.
2. Total loss in the condensate capacity along with decreasing crude capacity to the minimum running capacity. This can be a result of a partial outage in the crude unit such as loss of one or more crude trains.
3. Hydrocracker unit flexibility loss. In this scenario, no constraint was imposed on the refinery capacity but a constraint was put on the Hydrocracker unit decreasing its capacity of running in diesel mode. This is important in producing high quality distillates which is needed not only for production of diesel but also for blending with straight run high sulfur diesel from the crude unit to make it salable. A loss in the hydrocracked diesel of 5000 BPD may cause more than double this amount of loss in total diesel available for sale. In addition, very low sulfur diesels are usually produced from hydrocrackers and this constraint can affect the supply of low sulfur diesel to the whole country as discussed in the results.

Results

For the first case of Refinery # 3, the refinery had to gradually drop production due to the penalty imposed. However, a minimum production has to be kept to meet demand that need to be met from this particular refinery and cannot be switched to other refineries.

For Refinery # 3, an interesting result was found that the local objective of the refinery increased with the increased oil prices until the crude price reached $26/BBL then it decreased. However, global objective decreased as the crude prices increased. This is shown in Figure 2. The local optimum for this refinery is clearly not the global optimum and the refinery shows a strange phenomenon in the objective with increasing feed prices. This is due to complicated price structure for this refinery since it makes money by selling intermediate products to a nearby lube refinery and these intermediates prices are based on crude prices. Thus an increase in the crude prices will cause a temporary gain in the price of intermediate heavy fuels sold to the lube refinery but to a certain point. After that point, the effect of the increasing crude prices will affect the refinery economics since most of the crude is consumed by the refinery.

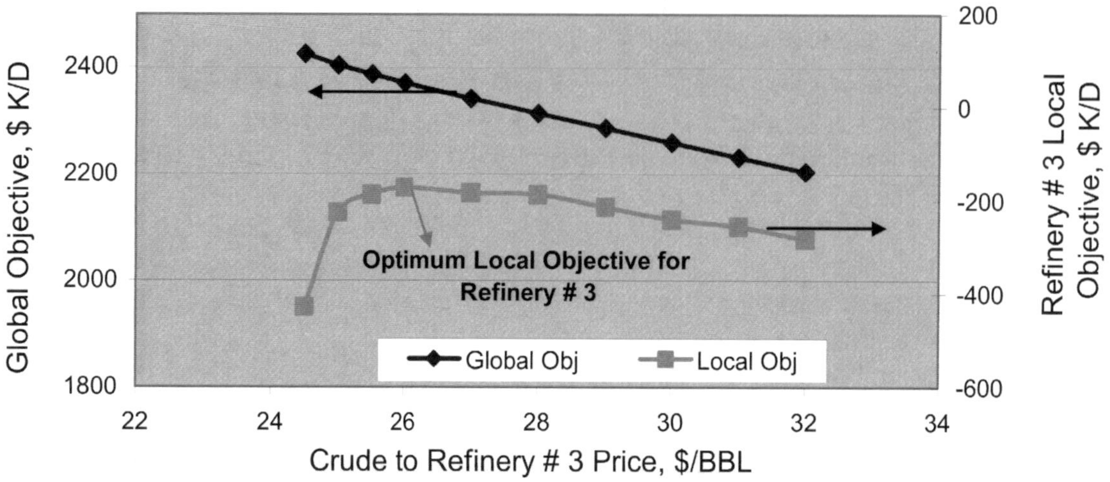

Figure 2: Effects of Increased Crude Price (due to increased risk) on Local and Global Objectives

For Refinery # 4, the local objective did not trail the global objective but was mainly a function of the refinery capacity. This is another example of the nonlinear relation between both objectives as can be seen in Figure 3. The figure also shows that penalties did not drive the refinery capacity in a continuous but rather in a discrete manner.

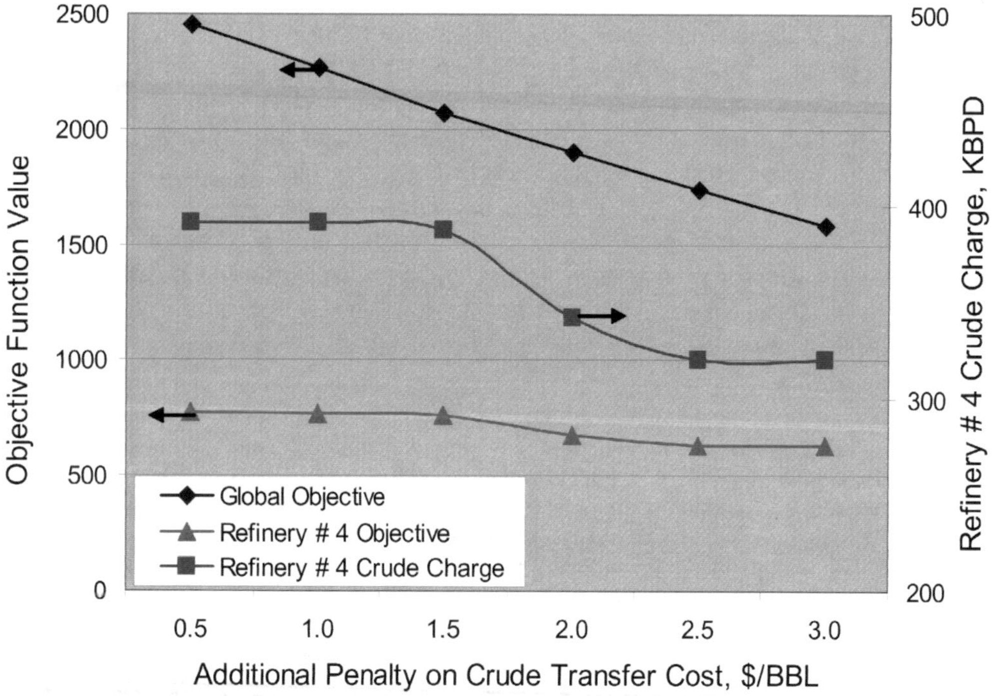

Figure 3: Effect of Transportation Cost Penalties on Capacities and Objectives

For Refinery # 1, the local and global objective were more correlated than the other two refineries for all scenarios as shown in Figure 4. This is due to the large size of the refinery, where it consist about one third of the total refining capacity of the country.

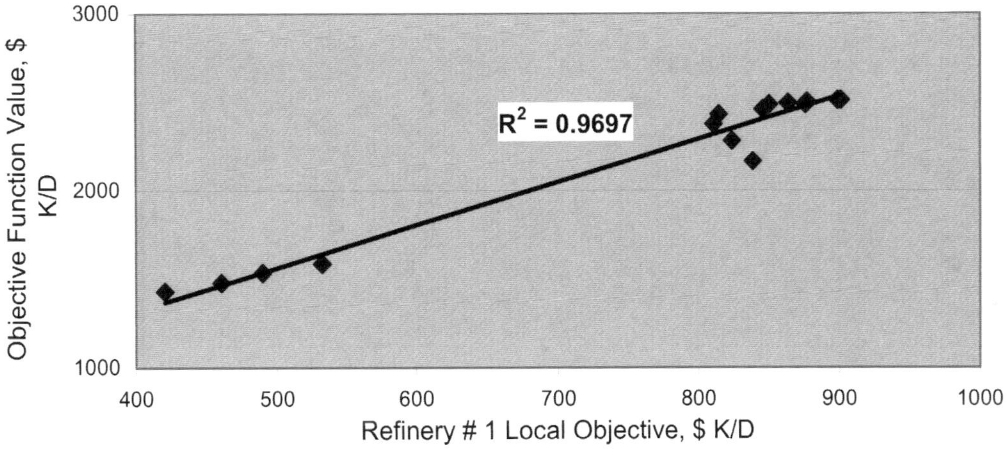

Figure 4: Relation Between Global & Refinery # 1 Objectives for All Cases

Figure 5 shows the effects of condensate capacity loss on margins & products yield for different products in scenario # 1. This is due to the availability of alternatives and the composition of the condensate. The effect was most clear on the production of unleaded gasoline (regular). The second product was low sulfur diesel. In both cases the effect was mainly observed after the capacity loss exceeded 40 %.

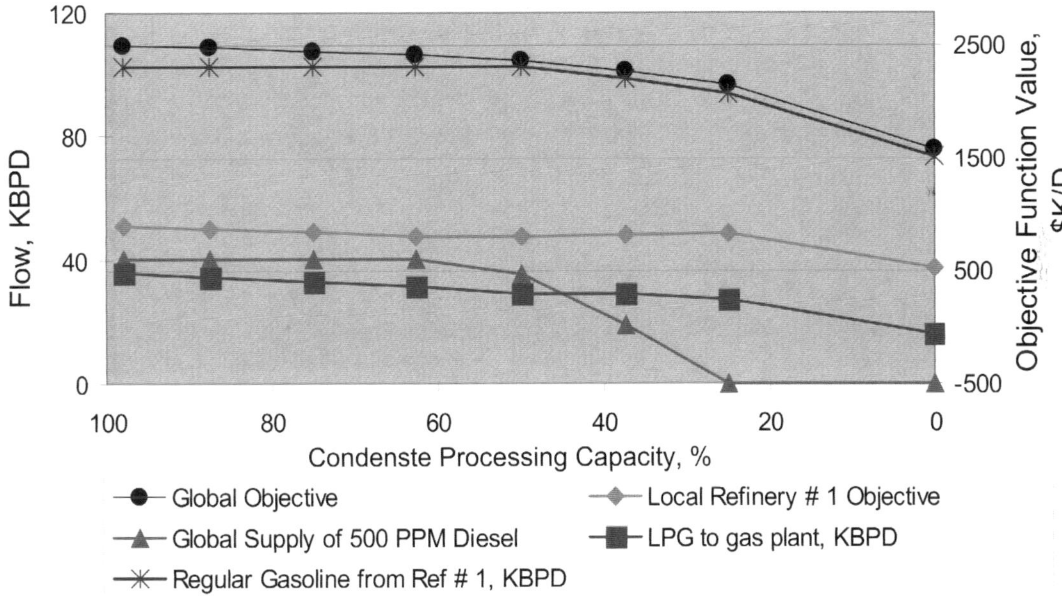

Figure 5: Effect of Condensate Capacity Loss on Objectives & Products Flow

For scenario # 2, the effect was clearer on the objective function for both local and global objectives. Figure 6 shows the effect of simultaneous crude and condensate capacities loss on selected light products yields.

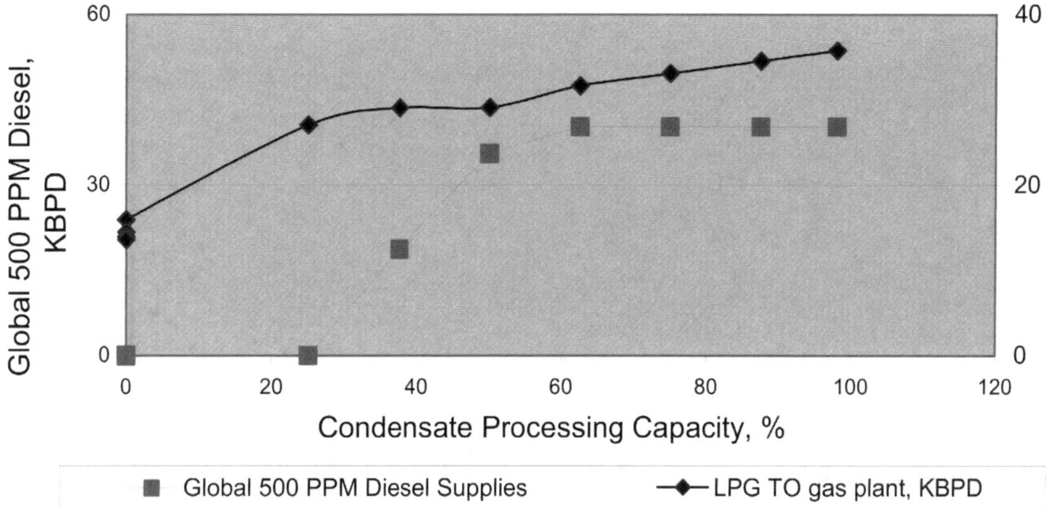

Figure 6: Effects of Condensate Processing Capacity Loss Along with Partial Loss in Crude Capacity on Selected light Products

Scenario # 3 showed interesting results as shown in Figure 7 that both global and local objectives were not significantly affected by the loss of 50 % of Hydrocracker flexibility even thought this had a significant effect on the overall refinery low sulfur diesel (LSD) production. This is mainly due to the fact that the total Hydrocracker output is not affected but it is the quality and composition of products which are affected. That is, the Hydrocracker charge in all case is about 100 % of the base case, but due to limitations on the diesel mode capacity, the unit can not run towards maximum distillate but rather can go to the maximum constrained limit and have a point between maximum naphtha and maximum distillate modes. The main issue is that the amount of LSD available to the country is critically affected even thought it does not appear to effect neither the local nor the global objective functions.

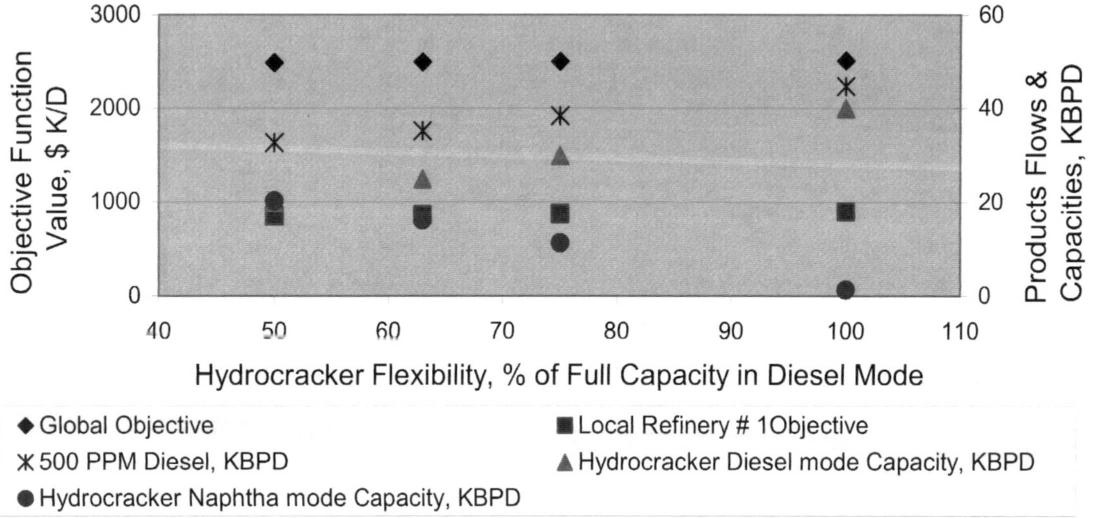

Figure 7: Effect of Hydrocracker Flexibility Loss on Objectives and Capacities

Conclusion

1. The results show that refineries can have a local multi objective optimum that does not follow the same trend as the global objective, or the local single objective optimum (discussed below).
2. In addition, the results show that the relationship between the global and local optimum may not always be linear. Moreover, the two objectives may not necessarily move in the same direction or at the same rate. This is a factor of refineries internal complexities and the type of imposed constraints.
3. For refineries, the multi objective local optimum may be different than the local optimum for the economic objective case. The deviation from the economic optimum depends on the degree of flexibility of the refinery in both selecting different feed / product scheme and in having different operating modes. These flexibilities act as a relief on the imposed multi-objective constraints by providing more alternatives to the model which increases the feasible region and thus may lead to a better optimum solution. Transportation costs of feeds and other intermediate products is also a strong factor.
4. For most refineries especially the large ones, the change in the capacity or the local objective as a result of imposed constraints usually follows a discrete behavior in contrast to the global objective which normally follows a smooth curve. This discrete behavior is due to the fact that refinery operations are hard to change and there are minimum turndown ratios that cannot be exceeded. In addition, the operating cost of refinery unit is partially fixed. That is, if a unit's capacity is reduced by 20 %, the operating expenses such as cooling costs and manpower requirements will not drop by the same percentage. In actual life, changing the operating conditions requires time to stabilize the unit in order to be able to produce on spec products.
5. For catastrophic failure effects, crude processing capacity is the main factor affecting the objectives. The loss of feed processing capacity such as crude or condensate has more effect on the objectives than the loss of upgrading and treatment capacities such as gasoline reforming. In addition, the results show the benefit of having more than one crude unit, which gives higher flexibility in running the refinery in less than the 60 % minimum turn down ratio in case of a catastrophic failure event. That is, for two refineries of the same size, if a refinery has one crude unit, it can usually run at 60 % of the maximum capacity. However, if the same refinery has 'm' crude units, it can run as low as 0.6/m of the total capacity.
6. The results show the importance of examining the effects of loss of internal processing flexibility on the product supplies which may not be clear in either local or global objective functions but it will appear at a point where the supplies cannot be met from all local refineries and thus imports are needed. Proper import tiers penalties should be imposed on imports in order to clearly address the shortage of some products under the event of partial or total loss.
7. Refinery # 3 price structure creates a conflict with the global and objectives at some points. This is due to many constraints imposed on the refinery to meet the demand for both the nearby lube refinery and the local urban market. The refinery is not well credited for these demand responsibilities which creates a negative margin for the refinery. Therefore, optimization on a local level may suggest cutting back on most of the processing operations.

Nomenclature

ABBREVIATION	Description
LP	Linear Programming (also known in industry as linear planning)
SLP	Sequential Linear Programming
NLP	Nonlinear Programming
PIMS	Process Industry Modeling System
HYSYS	Steady State or Dynamic Process simulator
RVP	Raid Vapor Pressure
RON	Research Octane Number
MON	Motor Octane Number
ON	Octane Number
BPD	Barrels per day
KBPD	Thousand barrels per day
MBPD	Million barrels per day
FCC	Fluidized Catalytic Cracking unit
CCR	Continuous Catalyst Regeneration (Type of Catalytic reforming)

References

1. Basak, K, Abhilash, K, Ganguly, S and Saraf, D. On-Line optimization of a crude distillation unit with constraints on product properties. Industrial and engineering chemistry research, v 41, 2002.
2. Bloss, K. Dynamic process optimization through adjoint formulations and constraint aggregation; PhD dissertation. Lehigh University, 2001.
3. Chaskunchuensaken, S. A combined optimization algorithm for stability, phase, and chemical equilibria (Phase separation). PhD Dissertation, Polytechnic University, 2000, ISBN 9-599-75785-X.
4. Hartmann, J. Distinguish between scheduling and planning models; Hydrocarbon Processing, Vol. 77 Issue 7, 1998.
5. Li, X. Refinery-wide optimization. Ph.D. dissertation. Texas Tech University. 2000.
6. Pike, R. Optimization for Engineering Systems. Louisiana State University 2001. http://www.mpri.lsu.edu/tcontentsindex.html.

62b Selecting "Sensors" for Safety Instrumented Systems per IEC 61511 (ISA 84.00.01 – 2004)

Dale Perry

The international standard for safety instrumented systems for the process control sector IEC 61511 was published in 2003 and details the lifecycle requirements for Safety Instrumented Systems (SIS). The ISA Standard 84.01-1996 will migrate to the IEC 61511 standard under the name ISA 84.00.01 – 2004. This presentation reviews processes and decisions designers need to make when selecting "sensors" to meet safety requirements in compliance with these new standards while reducing overall lifecycle costs. This paper will also outline a "Best Practice" approach to the selection of sensors for SIS applications that meet the requirements of IEC 61511 / ISA 84.00.01-2004 while minimizing lifecycle costs.

62c Abnormal Situation Prevention through Smart Field Devices

Douglas White, Evren Eryurek, Pete Sharpe

Detection of the faults as quickly as possible – preferably before an event actually occurs, diagnosis of the underlying problem, and providing assistance to plant operations and maintenance to correct the problem are the goals of Abnormal Situation Prevention (ASP) technology.

With the advances in computing and digital bus technologies, today's smart field devices are capable of providing more information regarding the process and its conditions in addition to their traditional process variable information. They can scan and process data at much higher rates than centralized systems which permits diagnotic solutions not previously available.

In this paper a survey of recent developments in this area is presented. Use of advanced pattern recognition technologies, multivariate signal analysis, statistical process control, as well as expert system based solution with the digital bus devices for improved process quality, abnormal event detection and prediction as part of ASP will also be discussed.

62d Use Weibull Analysis and Reliability Modeling to Improve Reciprocating Compressor Reliability

Jose Diaz, Keith Parsell, Kirtan K Trivedi

Use of Weibull techniques to characterize component failures in reciprocating compressors will be discussed. The results of Weibull analysis is used for developing a Reliability model for the reciprocating compressor to evaluate repair and maintenance strategies.

The Art of Evaluating Laboratory Fixed Bed and Fixed Fluid Bed Catalytic Cracking Data

Lawrence L. Upson and William J. Reagan
UOP LLC
Des Plaines, IL

Prepared for presentation at the 8th Topical Conference on Refinery Processing, at the 2005 Spring Meeting of the American Institute of Chemical Engineering - Advances in Fluid Catalytic Cracking Symposium
Atlanta GA, April 10 – 14, 2005

The Art of Evaluating Laboratory Fixed Bed and Fixed Fluid Bed Catalytic Cracking Data

Introduction

The Fluid Catalytic Cracking (FCC) community has used small, laboratory fixed bed units for many years to evaluate catalytic cracking results. Such units are typically used to evaluate the differences in the performance of various cracking catalysts and also are used to provide some insight into the effect of process variables and feedstock effects.

One such laboratory unit, which was used extensively in the later part of the twentieth century, was the Microactivity Test unit (MAT) (1). The test, as first described by Ciapetta et al (2) in 1967, consisted of a small bed (5 gms) of 3/32 inch pellets through which a small amount of feed (~1 cc) was injected downflow. The test was quickly modified (3) to allow evaluation of catalysts in the fluid catalyst form, to improve mass transfer and heat transfer effects.

More recently, a small fixed fluid bed unit utilizing essentially the same quantities of fluid cracking catalyst and FCC feed has been developed and is licensed by Kayser Technology Inc. The unit, designated as the Advanced Cracking Evaluation (ACE) Technology unit, has better heat and mass transfer characteristics than the MAT unit and has been gradually replacing MAT units since 1997 (4).

Since both the ACE test and the MAT test utilize a fixed bed of catalyst, there are significant differences between the cracking environment in these tests and the environment in a riser reactor, as typically used in a commercial Fluid Catalytic Cracking Unit (FCCU). Nevertheless, the laboratory test results from these units provide very useful information, and provide it in a cost effective manner. Both of these units provide data that allows for a comparison of the cracking activity of the various cracking catalysts of interest and also a comparison of the product selectivity of the various catalysts.

Such tests are usually run at a temperature chosen to simulate a particular commercial operation. The catalyst / oil weight ratio (C/O) is varied at either 1) constant on-stream time (t) by varying feed rate, or 2) by varying on-stream time at constant feed rate, hence constant weight hourly space velocity (WHSV). Either technique provides a range of conversions and hence a range of product yields. In these tests, the three variables time, C/O, and WHSV are related as:

$$t = \frac{3600}{(C/O)(WHSV)} \qquad \text{t in seconds}$$

Frequently, in these tests, only a small number of C/O variations are used (3-5) and hence only a small number of data points are available to develop the required conversion and product yield correlations. In addition, a considerable amount of experimental error is associated with each set of data, making the proper correlation even more problematic.

It is the purpose of this paper to describe practical techniques that can be applied to the handling of these types of laboratory data to obtain the maximum amount of reliable information.

Experimental

The data used in this paper to discuss these techniques were obtained via ACE unit testing. The ACE unit was operated at the following run conditions:

Reaction initial temperature	543°C (1010°F)
Reaction time	60 seconds of feed
Feed rate	1.30 grams / minute
Catalyst weight	variable to give ~ 3, 4, & 6 C/O

The feed stock was a mid-continent gas oil with the following properties:

Density	0.9064 gms/cc
UOP K	11.8
Sufur	5300 ppmw
Total nitrogen	1340 ppm
Conradson Carbon	0.415 wt%

ASTM D2887H (Sim Dist)	*°C*
IBP	236
10%	353
50%	440
90%	534
EP	572

The gas product was collected in a water displacement receiver and analyzed with an on-line gas chromatograph. The liquid product was collected in a chilled glass receiver. The boiling point distribution of the products was determined by standard simulated distillation GC procedures, ASTM 2887. Coke on catalyst was determined by in-situ air calcination of the spent catalyst and measurement of product carbon dioxide with an infrared analyzer. Mass balances were calculated and only runs with balances >95 wt% were accepted for the data correlations. Product yields were expressed on a weight percent of feed basis.

The three FCC catalysts evaluated for this paper were blends of a base equilibrium catalyst and various laboratory deactivated shape selective zeolite additives. The additives were deactivated in a cyclic deactivation unit using 50 cycles of cracking over a gas oil feed at 565°C (1050°F) and air/steam regeneration at 788°C (1450°F).

Correlating Conversion Data

In small scale laboratory FCC testing at a constant reactor initial bed temperature, tests are frequently carried out by varying C/O ratio, thus achieving a range of conversions. As in all laboratory testing, experimental error exists resulting in considerable data scatter. The question is – What is the best way to correlate the data to get the most accurate understanding of the effect of C/O ratio on conversion?

It is tempting, in this age of computer developed technology, to let the computer determine the best statistical fit to the data. This however is generally not the best approach. In many cases, only 3-4 data points are available to develop the correlation since only 3-4 C/O ratios were evaluated, and frequently replicate runs are not made. With the experimental error that exists, the statistical fit can generate nonsensical lines. If several catalysts are being tested with a single feed, the computer-generated C/O versus conversion lines will typically cross each other at curious angles, a highly unlikely situation.

It is always useful, in developing correlations, to have one point which is known to have little or no experimental error, and, through which, the correlation can be forced. In the case of the relationship between conversion and C/O, the conversion at zero C/O is fairly well defined. At zero C/O, the conversion might be expected to be 0.0 wt%. There is, however, some conversion at essentially zero C/O caused by the nearly instantaneous thermal cracking that occurs when the feed contacts the hot catalyst environment. When the feed is constant, the thermal cracking at zero C/O would be constant for all the catalysts. Thus the conversion at zero C/O represents an anchor point through which all the plots must pass. The conversion at zero C/O is in the range of 25 wt% for most feeds, when the initial reaction temperature is in the range of 525 – 550°C. This concept is particularly useful when several catalysts are being evaluated with one base case feed.

A further question is whether to directly use the measured conversion, defined as:

FCC Conversion, wt% = 100 – (wt% Light Cycle Oil + wt% Heavy Cycle Oil)

as the dependent variable, or to use some more fundamental value related to conversion.

If the traditional FCC conversion value as defined above is used, two difficulties arise:

Including the thermal conversion at zero C/O results in a very wide axis for the dependent variable, conversion. The conversions measured at the various C/O test conditions then become fairly close in value (typically in the 70 –80 wt% range). Differentiating between the catalyst correlations and how each correlation passes through the zero C/O point becomes difficult.

The relationship between the actual conversion value and C/O is strongly non-linear. Given the experimental error that exists, and the few points available to develop a correlation, a reliable characterization of the correlation that must best fit the data points, must pass through the zero C/O point, and must have the proper non-linearity is difficult.

A more useful technique is to approach conversion from a kinetic point of view. Although pure hydrocarbons have been found to crack according to a kinetic first order process (5,6), cracking of a multi-component FCC gas oil feed has been found to be well approximated by a 2nd order reaction (7). The rationale for this observation is that the cracking rates for the components in the FCC feed cover a wide range, and these rates decrease as the conversion progresses and the compounds become more difficult to crack. Combining all the varying 1st order rates results in one overall rate that approaches 2nd order. The concept of 2nd order conversion, is thus a useful tool in correlating FCC conversion data, with 2nd order conversion defined as:

$$Conversion : 2ndOrder = \frac{wt\%Conversion}{100 - wt\%Conversion}$$

The use of the 2nd order concept has several advantages.

1) Correlating 2nd order conversion against C/O provides a nearly linear fit of the data (8), particularly in the conversion range from 0 to 80-85%

.

The linear relationship readily encompasses the low conversion point at zero C/O.

2) The 2nd order concept provides a more useful spread of the conversion data, with the data points being compressed at the low conversion range, while the data points at the high conversion range spread apart. For example, on a 0-100% conversion scale, the distance between 33 and 50% conversion seems quite large, whereas on the 2nd order scale it is only ½ a number (the difference between 0.5 and 1.0). Likewise the difference between 75 and 80% conversion on the 0-100 scale seems relatively small, while on the 2nd order scale it is a whole number (the difference between 3.0 and 4.0).

3) Since the values at low conversion are compressed, the exact 2nd order intercept value at zero C/O is not critical in defining the correlation anchor point. For example, if the conversion at zero C/O is 30 wt% rather than 25 wt%, the 2nd order value only changes from 0.33 to 0.43, a trivial difference on the conversion vs. C/O plot.

Figure 1

Cat/ Oil Effect Upon Conversion

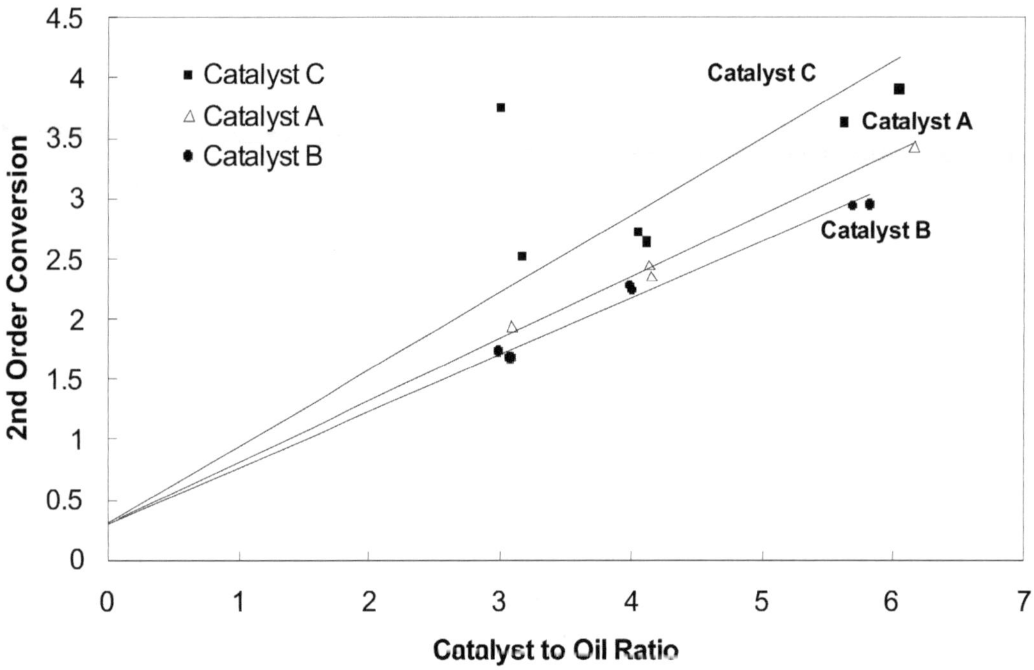

Figure 1 illustrates the technique, using data from a typical three catalyst comparison test done in UOP's ACE unit for a constant feed case. The data for Catalyst A and Catalyst B show little experimental error and provide reasonable linear fits, which define the zero Cat/Oil intercept. The data for Catalyst C have considerable experimental error.

A linear fit through the intercept determined by Catalysts A & B gives the best representation of the Catalyst C data.

Correlating Yield Data

Of considerable interest in comparing catalyst performance is the comparison of catalyst selectivity, i.e., the comparison, between catalysts, of a particular product yield at a given conversion. For many of the products produced in catalytic cracking, 2nd order conversion is a better correlating parameter than actual conversion in evaluating the effect of conversion on product yield.

The benefits of 2nd order conversion here are similar to the benefits found in the correlation of C/O ratio vs. conversion.

1) In many cases, 2nd order conversion provides a much more linear fit of the yield data.
2) 2nd conversion compresses the conversion scale at the low conversion region and allows the use of the 0% conversion point to be conveniently included in the relationship. The 0% point is the most reliable point in the set of data.

Coke Selectivity

For the more common conversion ranges (in the range of 75 wt% or lower – 2nd order conversions of 3 or less), coke yield has a nearly linear relationship, as typified by ACE unit data in Figure 2.

Figure 2

ACE Unit Data - Effect of Conversion on Coke Yield

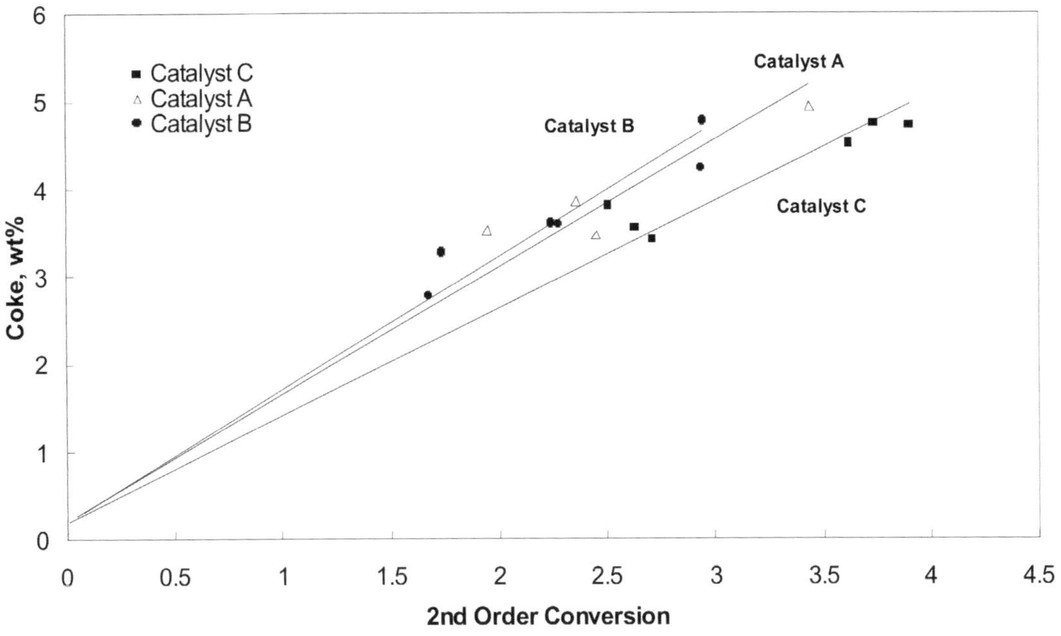

As noted in Figure 2, the coke yield at zero conversion, as determined by extrapolation, is a common value for all low metals contaminated catalysts in a constant feed study, but is not zero. The intercept appears to be related to the Conradson carbon (also designated as the carbon residue) of the feed. Typically the intercept is in the range of 50 –75% of the Conradson carbon value. It is believed that this high molecular weight feed material deposits nearly instantaneously on the catalyst as a result of the thermal shock when feed first contacts hot catalyst. This shock creates a pulse of coke yield before any other significant conversion occurs. This coke pulse provides the offset noted in the coke vs. conversion plot.

Dry Gas Selectivity (Hydrogen, Methane, and Ethane + Ethylene)

As with the coke data, the use of the 2nd order conversion values to correlate dry gas yields provides a more useful means of correlating the data. A linear correlation exists in the lower regions of conversion and the intercept at zero conversion provides a fixed point to anchor the correlation. Also, as with the coke data, the zero conversion intercept is not at 0% dry gas yields. The nearly instantaneous thermal shock at the point of contact with the hot catalyst provides a dry gas yield offset, as it does with the coke yield. This is illustrated in Figure 3, using data from the testing of Catalyst B.

Figure 3
ACE Dry Gas Data - Catalyst B

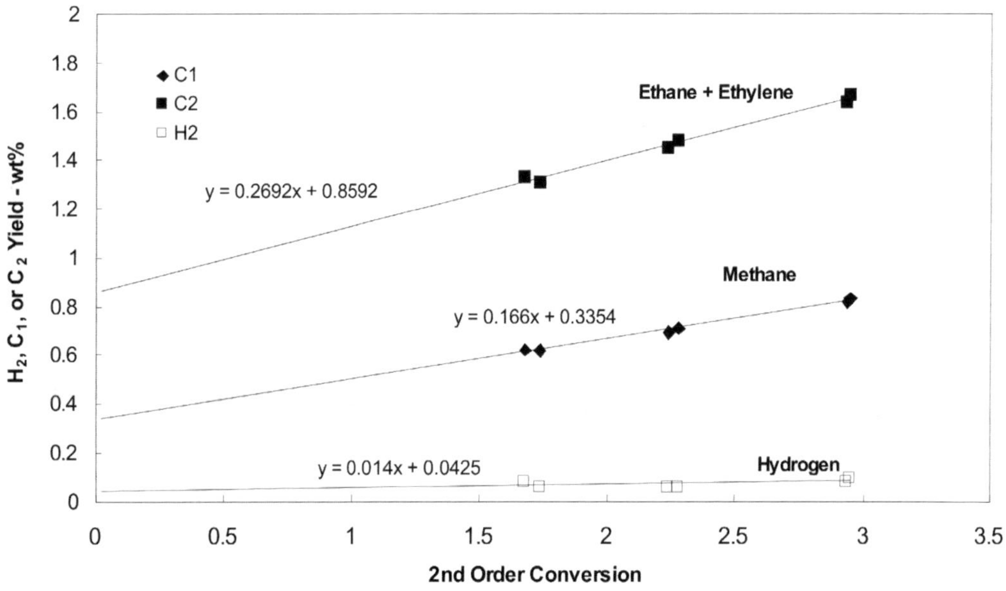

LPG Correlations

For all the LPG yields, it is reasonable to assume that at zero conversion there is zero yield of these products. Thus any correlation of the yields of these products should pass through the point: zero yield at zero conversion. As seen in the subsequent sections, the inclusion of the (0,0) point helps significantly in defining the yield/ conversion relationship.

Propane, *n*-Butane and *I*-Butane

The use of 2nd order conversion gives linear plots that pass through the origin for propane, n-butane, and i-butane yields. For propane, the yield – conversion relationship is obtained via a computer generated least squares regression passing through the data points and the origin, as seen in Figure 4.

Figure 4
Effect of Conversion on Propane Yields - ACE Data

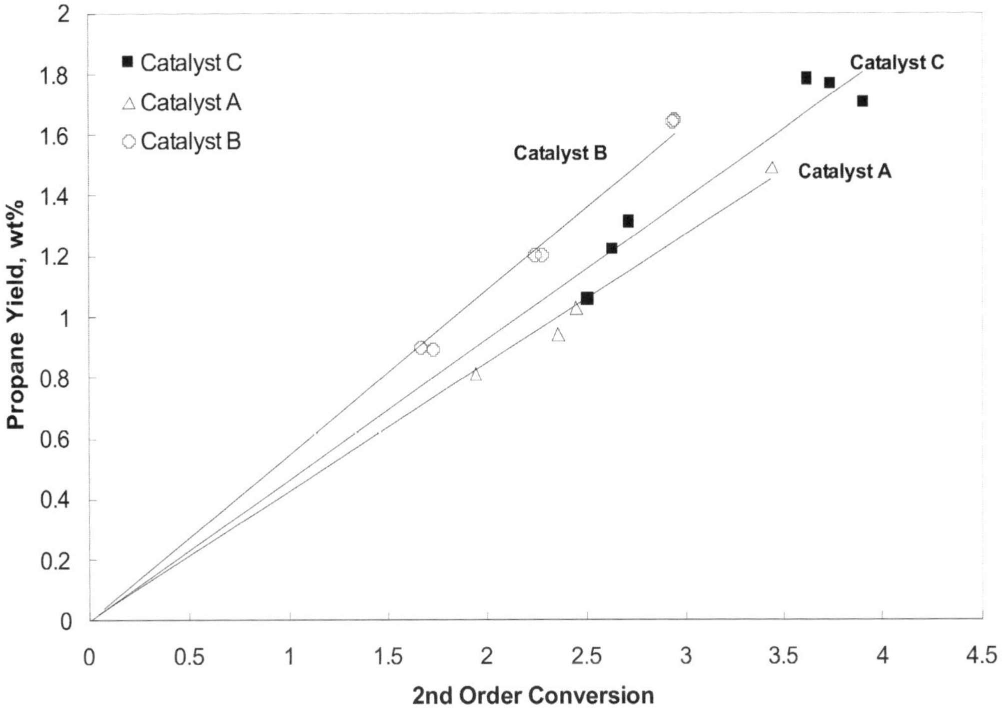

LPG Olefins – Propylene and Butylenes

In correlating the LPG olefins, zero yield at zero conversion again provides a fixed data point which helps define the relationship. The use of 2nd order conversion helps to compress the conversion scale at the low conversion range so that the zero point can be a useful anchor. However, 2nd order conversion no longer provides a linear relationship, as it did with propane and n-butane, since the olefinic products participate in secondary reactions and thus tend to go through a maximum at high conversion. The data are now nicely represented by a 2nd order polynomial. This type of least square regression going through the origin is shown in Figure 5 for propylene.

Figure 5
Effect of Conversion on Propylene Yields - ACE Data

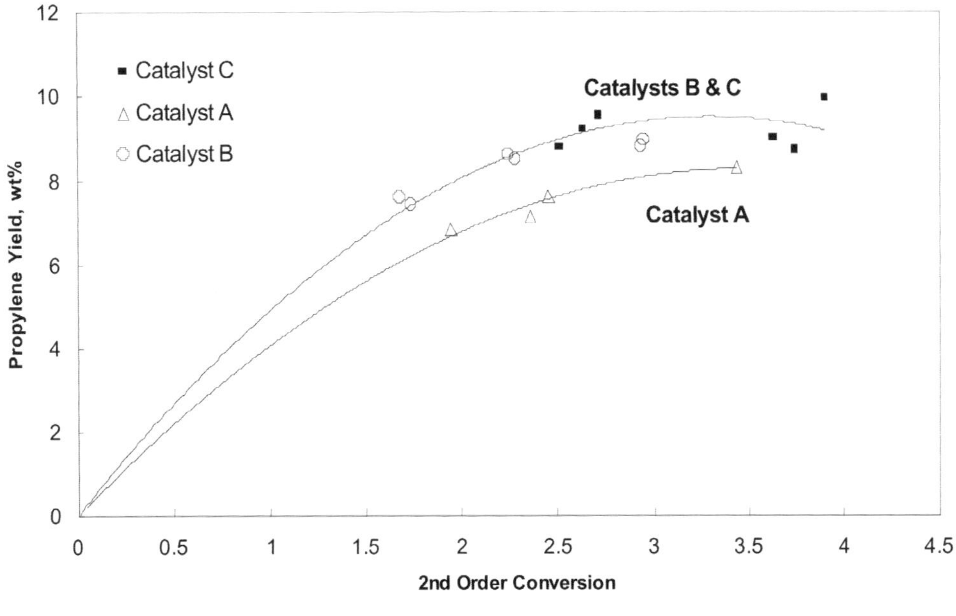

A common error in correlating these data is to ignore the requirement that the relationship must go through the (0,0) origin. Figure 6 illustrates the nonsensical plots that occur if the data are regressed without including the origin as a fixed point.

Figure 6
Effect of Conversion on Propylene Yields - ACE Data
Correlations Do Not Go Through the Origin

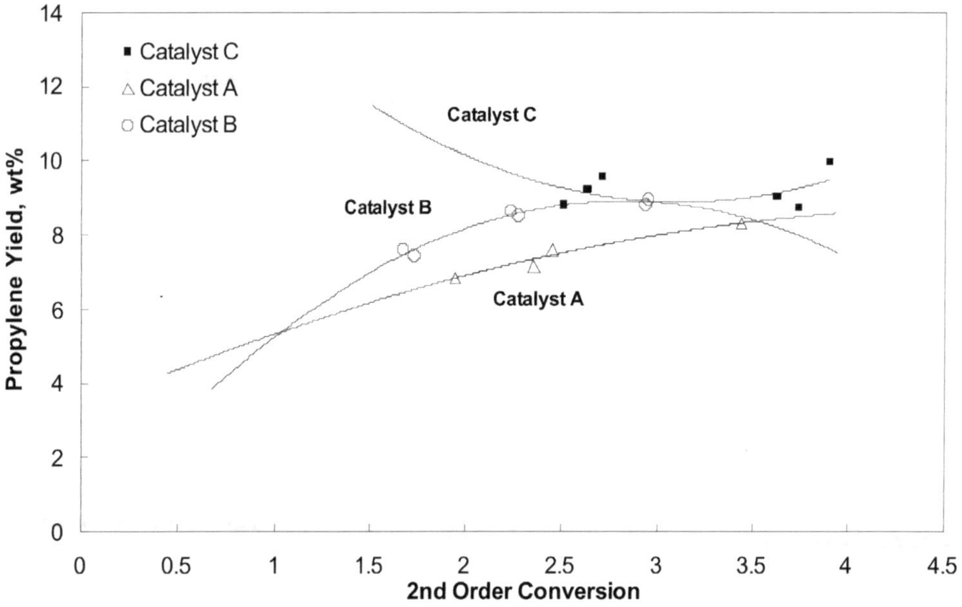

Correlating Gasoline Data

The gasoline product is typically defined as that material boiling in the n-pentane to n-dodecane range and referred to as the material in the C_5 – 221°C TBP product fraction. By definition, conversion is the sum of all the gasoline and lighter products plus coke:

$$\text{Conversion} = \text{Gasoline} + \Sigma C_4- + \text{Coke}$$

Thus the gasoline yield can be defined as:

$$\text{Gasoline} = \text{Conversion} - (\Sigma C_4- + \text{Coke})$$

Since the relationship with conversion for all the C_4- products has been established in the previous sections, those values can be used to define the correct gasoline relationship. This approach to obtain the gasoline relationship is illustrated for Catalyst A.

For Catalyst A, the following regression equations for the product yields, in wt%, were obtained as:

Component	*Yield – wt%*
Coke	= 1.4538X + 0.20
H_2	= 0.0216X + 0.0425
Methane	= 0.1399X + 0.3354
Ethane + Ethylene	= 0.1673X + 0.859
Propylene	= - 0.6864X^2 + 4.7667X
Propane	= 0.422X
n-Butane	= 0.315X
i-Butane	= 1.4567X
i-Butylene	= - 0.4825X^2 + 2.3586X
n-Butylenes	= - 0.663X^2 + 4.0906X

For all of these relationships, X = 2nd order conversion.

Using these relationships, the gasoline yield over a range of conversions can be determined by difference. Table 1 illustrates this approach, using conversions at 50, 66.7, 75, and 80 wt% (1.0, 2.0, 3.0, and 4.0 2nd order conversion). The individual yields at these conversions are shown in the table. The gasoline yield at these conversions is then obtained by subtracting the sum of these products from the related conversion.

Table 1
Gasoline Yields for Product A – ACE Catalyst Testing
Obtained by Difference from Product Yields for Catalyst A

Conversion – wt%	50.0	66.7	75.0	80.0
Conversion – 2nd Order	1.00	2.00	3.00	4.00
Product Yields – wt%				
Coke	1.65	3.11	4.56	6.02
H_2	0.064	0.086	0.107	0.129
C_1	0.48	0.62	0.76	0.89
C_2 + C_2=	1.03	1.19	1.36	1.53
Σ Dry Gas	1.57	1.90	2.23	2.55
C_3=	4.08	6.79	8.12	8.08
C_3	0.42	0.84	1.27	1.69
Σ C_3s	4.50	7.63	9.39	9.77
nC_4	0.31	0.63	0.94	1.26
iC4	1.46	2.91	4.37	5.83
iC_4=	1.88	2.79	2.73	1.71
nC_4=	3.43	5.53	6.30	5.75
Σ C_4s	7.08	11.86	14.34	14.53
	====	=====	====	====
Σ C_4- + Coke	14.80	24.50	30.52	32.89
Gasoline = Conv – (ΣC_4- + Coke)	35.20	42.17	44.48	47.11

As with all the previous yield correlations, a reliable correlation of the gasoline data must include the gasoline yield at zero conversion. Thus, in Figure 7, the gasoline vs. conversion points (small dots) are combined with the (0,0) point to develop the gasoline relationship shown in Figure 7. Also shown in Figure 7 are the Δ, •, and ■ data points representing the experimental gasoline yield data points for Catalysts A, B, and C respectively, which, as expected, closely agree with the line obtained by differences.

Since gasoline is an intermediate product, which cracks to lighter products, the gasoline yield typically goes through a maximum at high conversion.

Figure 7

Effect of Conversion on Gasoline Yield - ACE Data
Gasoline Plots Obtained by Difference

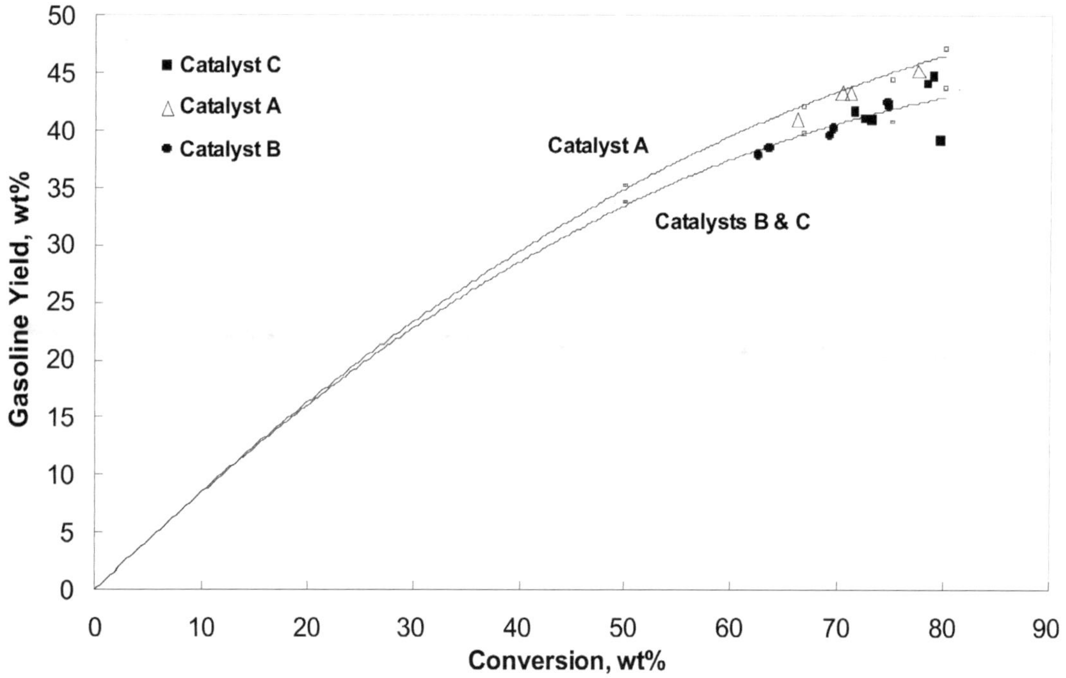

LCO Yields (430° to 650°F TBP Cut)

When correlating light cycle oil (LCO) yields versus conversion, consideration needs to be taken for the LCO yield at both zero conversion and 100% conversion. Generally, there is some LCO in the feed, i.e., that material in the feed boiling below the defined initial boiling point of the material defined as "true feed". "True Feed" is typically defined as material boiling above 650°F (343°C). The feed in this study contained 8% of material boiling below 650°F. Thus at 0% conversion, 8% of the feed was already LCO. The 0% conversion intercept for this study is thus 8%, which becomes a fixed point on the LCO plot.

The conventional definition of FCC conversion is given as:

$$\text{Conversion} = 100 - (\text{LCO} + \text{HCO})$$

At 100% conversion, the LCO yield, by definition, is therefore zero. Thus, the LCO yield versus conversion has a second fixed point of 0% at 100% conversion.

2nd order conversion is not a convenient parameter to use when plotting LCO yields. As conversion approaches 100%, 2nd order conversion approaches infinity, which makes its use impractical. The plot of LCO conversion vs. standard conversion is shown in Figure 8.

Figure 8
Effect of Conversion on Light Cycle Oil Yields

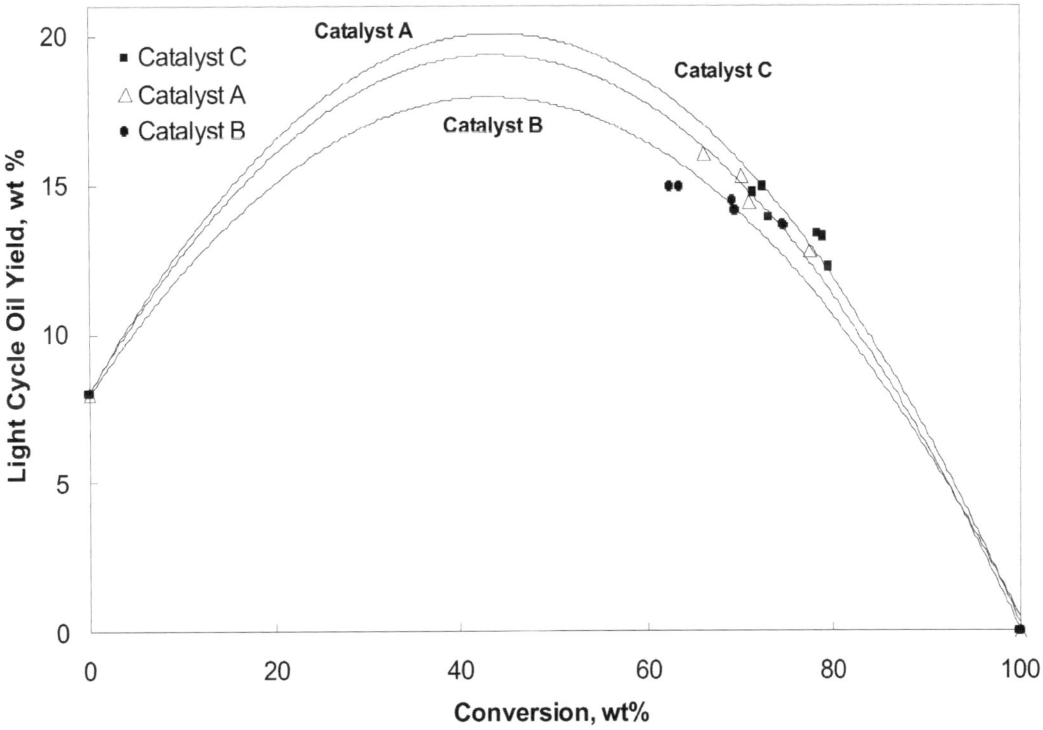

Heavy Cycle Oil Yields (650ºF + TBP Cut)

By convention, the heavy cycle oil (HCO) yield is defined as the amount of material in the liquid product boiling above 650ºF (343ºC). As mentioned above, the feed for this study contains 8% LCO (8% boiling below 650ºF) and thus at zero conversion, there is only 92% boiling above 650ºF, hence only 92% HCO. This becomes a fixed point on the HCO yield vs. conversion plot.

As with LCO, the definition of conversion requires the HCO yield to be 0% at 100% conversion. This becomes a second fixed point on the HCO yield vs. conversion plot. The resulting HCO yield plot is seen in Figure 9.

Because the HCO scale covers a much wider range (0 – 100 wt%), the differences between catalysts at any given conversion is sometimes hard to discern in these plots. For example, in Figure 9, it is difficult to see that the HCO yields for Catalyst A lie between the HCO yields for Catalyst B and the HCO yields for Catalyst C. As an alternative, the HCO yields, at any given conversion, can be determined from the LCO plots in Figure 8, using the relationship:

HCO = Conversion - LCO

Figure 9
Effect of Conversion on Heavy Cycle Oil Yields

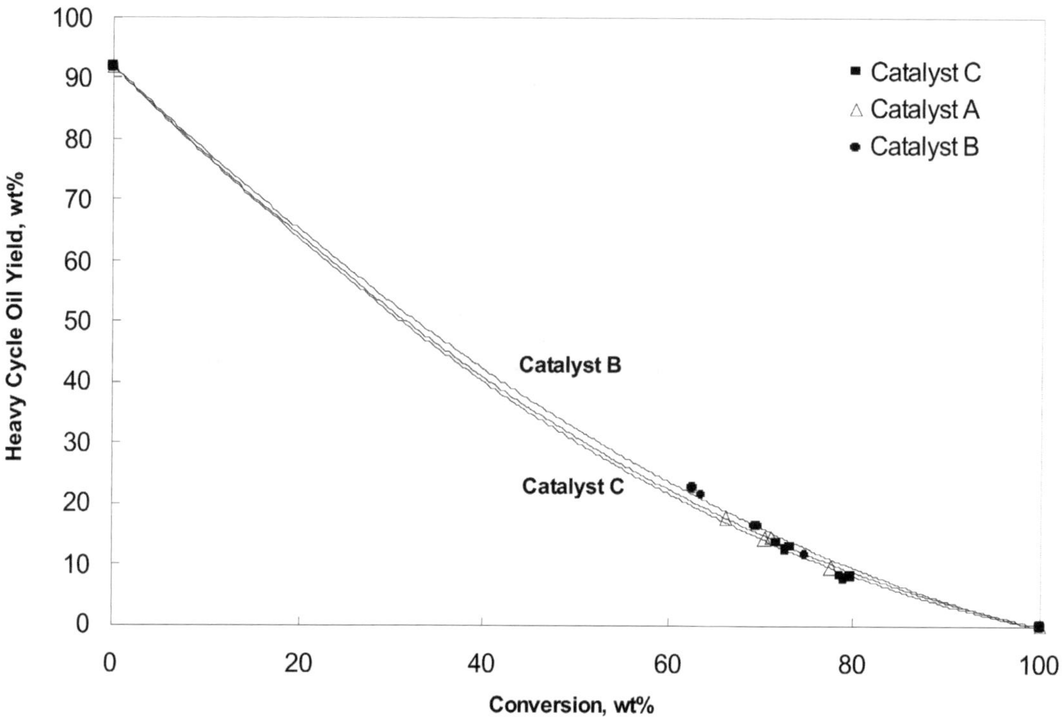

How to Compare Catalyst Test Results

Having completed the testing of a series of catalyst, the question then becomes: How best to compare the results from different catalysts to arrive at a decision as to which catalyst best meets the requirements of the intended refinery.

One school says that results should be compared at constant conversion. Another group says – no, no -- commercial FCCU operations are heat balanced and thus results should be compared at constant coke make.

The constant coke argument has some intuitive appeal, but would only be true if the type of coke measured in the laboratory test was the same type of coke obtained in a commercial FCCU . To settle this argument, it is necessary to consider some basic concepts related to coke formation.

The coke on catalyst prior to regeneration is a combination of two components.

$$C_{On\ Cat} = C_{Hard\ Coke} + C_{Soft\ Coke}$$

1) Hard coke is the catalytically produced coke. This highly carbonaceous material, containing only 3-4 wt% hydrogen, is firmly attached to the acid sites present on the zeolite and on the active matrix components of the catalyst, and is a result of the cracking reactions that took place on the catalyst. The hard coke is essentially unaffected by stripping.

2) Soft coke, itself, consists of two components:

a) Hydrocarbons lightly adsorbed on the surface of the larger entrance pores (the macro-pores) of the catalyst and on the external surface of the catalyst particle. Adsorbed soft coke is sometimes referred to as “extractable” coke, since it can be removed from the catalyst in a laboratory test by extracting it with a solvent, such as carbon tetrachloride.

b) The hydrocarbons that are in the vapor space in the catalyst pores and between the catalyst particles.

The combined hydrogen content of these two soft coke components is in the range of 11 – 12 wt%.

The quantity of this soft coke material is strongly influenced by stripping conditions, the better the stripping, the less soft coke.

After the catalyst has been exposed to the hydrocarbon feed, it is stripped with steam or nitrogen in both the commercial FCCU and in the laboratory cracking tests to reduce the amount of carbonaceous material on the catalyst, prior to regeneration. Under the very severe stripping conditions which exist in the laboratory test (typically a stripping time of 10 – 15 minutes), the material remaining on the catalyst after stripping is essentially all hard coke.

Commercial stripping, however, is much less intensive, with stripping times typically in the range of 30 – 45 seconds. As a result, in addition to the hard coke going to the regenerator, a significant amount of soft coke is also present in the coke going into a commercial regenerator. The amount of soft coke going to a commercial regenerator will depend upon the stripper conditions, but typically is in the range of 25% of the total coke. The hydrogen content of the coke in a commercial, well stripped FCCU, is typically around 6 wt%, reflecting the presence of both hard and soft coke.

Since the type of coke measured in the laboratory tests is different from the type of coke occurring commercially, both testing concepts (constant conversion or constant coke) are wrong. Since the coke produced in the laboratory test is only hard coke, the comparison must take this into consideration. This is illustrated in Figure 10.

Figure 10
Technique for Determining Proper Conversion for Catalyst Comparison

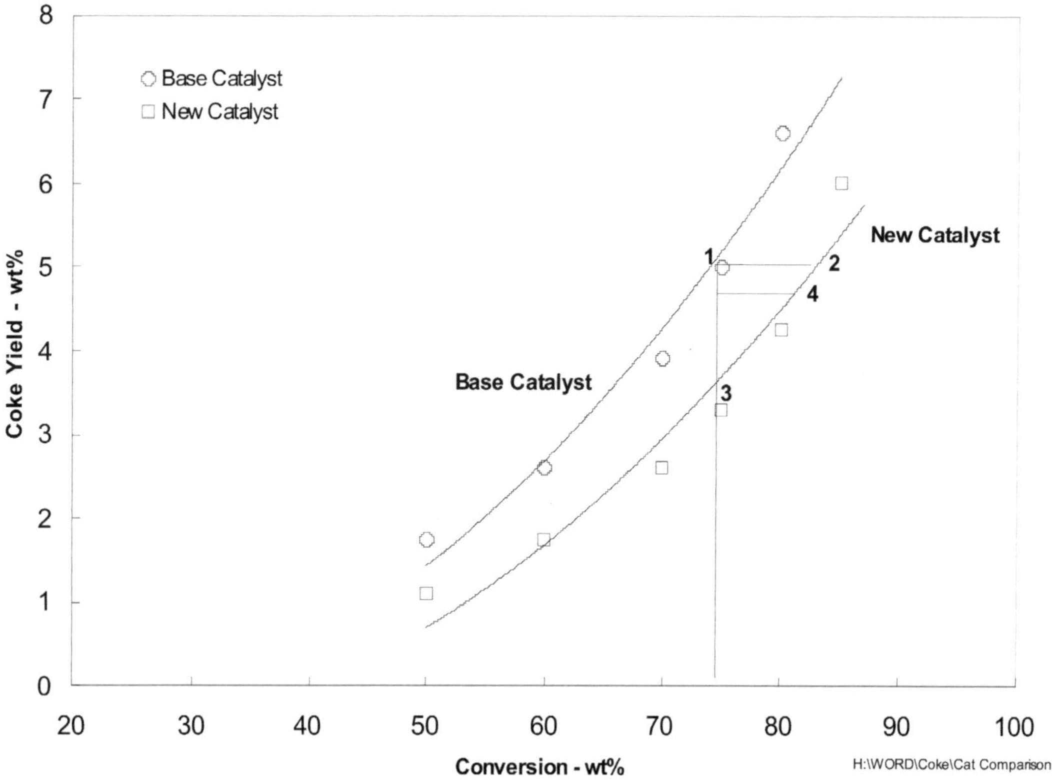

For the demonstration case shown above, using a hypothetical data set, where the base case and a proposed new catalyst have equal activity but different coke selectivity characteristics, i.e., the same conversion at a given C/O, but different product yields, the base case has a hard coke yield of ~ 5.0 wt% (Point 1) at 74% conversion. The new case has a lower hard coke yield of~ 3.6% (Point 3) at 74% conversion.

For constant stripping conditions, yielding a constant amount of soft coke, the lower hard coke yield would result in a lower total coke yield at constant conversion. In commercial operations, to stay in heat balance (constant coke yield), the catalyst circulation rate would have to increase, increasing conversion to provide the necessary increase in hard coke and soft coke. Thus constant conversion is not correct. For a proper comparison, the new catalyst would have to run at a somewhat higher C/O and somewhat higher conversion than constant conversion would dictate.

The constant hard coke comparison of the two catalysts is equally faulty. If one catalyst produces less hard coke than the other at constant conversion, constant hard coke would have to be achieved by increasing the conversion, via an increase in C/O, i.e. going from Point 1 at 74% conversion to Point 2 at 83% conversion.

However, since increasing C/O also increases soft coke in a commercial operation, the total coke yield in a commercial operation at 83% conversion would exceed the base case coke yield. Thus, to achieve a constant coke yield in a heat balance operation, the hard coke in the more selective case would have to be less than the base case hard coke, not equal. The proper laboratory comparison would require that the new catalyst case would have to run at a somewhat lower C/O and somewhat lower conversion than a total hard coke comparison would require.

Thus the proper comparison of laboratory data for the two catalysts lies somewhere in-between the conversion required for constant conversion and the conversion required for constant coke, starting from the base case conversion and coke.

Since the hard coke is typically ~ 75% of the total coke, at a constant total coke yield (as dictated by the heat balance), the actual change in hard coke would only be ~75% of the coke change required to bring the ACE or MAT result to a constant coke level. Thus the proper conversion for the catalyst comparison is at the conversion that gives that 75% ACE or MAT coke change.

This point is also illustrated in the example in Figure 10. The new catalyst at 74% conversion has a hard coke yield of 3.6%, which is 1.4 wt% lower than the 5.0% hard coke yield of the old base catalyst. The actual coke yield of the new catalyst, for comparison purposes should then be 3.6 + 0.75(1.4) = 4.6. At 4.6 wt% coke yield, the new catalyst conversion is ~ 81 wt% (point 4 in Figure 10), which is the conversion to use to compare product yields of the new catalyst with yields from the base catalyst at 74 wt% conversion.

Conclusion

A simple procedure is described, that provides reliable correlations of small scale laboratory FCC test results, even though there may only be a very limited amount of data and the data may be badly scattered .

The method uses the yield of the individual products at zero conversion as an anchor point through which the regressed data for each product yield must pass. In these regressions, the anchor point at zero conversion is not necessarily 0 wt%, and techniques are disclosed as to how to position such anchor points. For the correlation of LCO and HCO yields, an additional anchor point of 0 wt% at 100% conversion is also applied.

The coke yield, and all the hydrocarbon yields of the C_4 and lighter products are correlated against second conversion. For many of these products, second order conversion provides a linear correlation with product yield.

Gasoline yield is obtained by difference, using the relationship:

$$\text{Gasoline} = \text{Conversion} - (\text{Coke} + \Sigma\ C_4\text{-})$$

Since the laboratory tests and commercial unit operations have fundamentally different coke yields, a new method is proposed to allow for the interpretation of laboratory test results at conditions relevant to potential commercial FCC unit operations. This procedure facilitates the accurate comparison of various catalysts and their potential commercial cracking performance.

References

(1) ASTM Standard D 5154-91 "Standard Test Method for Determining the Activity and Selectivity of Fluid Catalytic Cracking (FCC) Catalysts by Microactivity Test"
(2) F.G.Ciapetta and D.S.Henderson, O. & G.J., 65, (42), 1967 p.88
(3) W.R.Gustafson, ACS Div. Petrol. Chem. 14 (3), 1969 B 56-57
(4) D.M.Stockwell, W.S.Wieland, and F.L.Himpsel, "Catalyst Evaluation using Fixed Fluid Bed Reactors" Paper presented at Spring 1998 AiChE National Meeting, New Orleans, LA, March 3-12
(5) J.F.Crocoll and R.D.Jaquay, Petro./Chem.Eng.,C-24, Nov. 1960
(6) D.M. Nace, I&EC, Prod Res. & Dev., 8, 24, 1969
(7) V.W.Weekman and D.M.Nace, AIChE J, 16, 3, May 1970
(8) C.Vergel-Hernandez, Appl. Catalysis A, 220, 2001, pg 265

H:\WORD\ Coke\ Catalyst Testing Paper-Short Version - Resized

63b Laboratory Evaluation of Gasoline Sulfur Reduction Additives

Julie A Francis

Removing sulfur from gasoline is recognized as one of the more urgent concerns facing refiners today. Environmental legislation regarding the sulfur content in the gasoline pool is becoming more stringent as is demonstrated by the fact that the US EPA requires that sulfur levels in refinery gasoline must average no more than 30 ppm by 2006 and other countries are already pushing for a 10 ppm limit. RESOLVE technology provides flexibility to the refiner and softens the economic impact of producing lower sulfur gasoline as it reduces the sulfur level in gasoline without affecting yields or entailing major capital investment.

The evaluation of gasoline sulfur reduction additives is complicated by the gasoline sulfur removal activity of other factors that may be inherently present in the system, either due to the base ecat or in the feedstock. It is known that some metals, especially vanadium, play a large role in the reduction of gasoline sulfur and this may dominate over an additive effect. Other factors such as feedstock properties, base case technology and reaction conditions can also significantly affect the sulfur content in gasoline. As a consequence, traditional deactivation methods where active metals are applied to a catalyst are not compatible with elucidating the gasoline sulfur reduction ability of an additive that would be evident in a refinery. The contents of this paper will focus on investigating the effects of different parameters such as metal content, base case technology, additive type and deactivation, feedstock properties, reaction conditions and test protocol on gasoline sulfur reduction and FCC Gasoline sulfur reduction additive performance. A unique gasoline sulfur reducing additive test protocol will be suggested that correlates well with commercial additive performance.

INTEGRATING ALBEMARLE RESOLVE DESULFURIZATION TECHNOLOGY WITH NOVEL PETRO-CANADA PROCESS CONCEPTS IN COMMERCIAL FCCU OPERATIONS

Kelly Benham
Petro-Canada
Canada

Chris Kuehler
Albemarle Catalysts Company LLC
USA

ABSTRACT

The gasoline and distillate sulfur regulations promulgated throughout the world to reduce tailpipe emissions are now strongly impacting refinery operations and investments. FCC gasoline is recognized as the principal contributor of sulfur to the gasoline pool and has become the focus for meeting the new specifications. The difficulty in removing sterically hindered sulfur species in the FCCU cycle oil drives up the hydrogen and investment costs for treating the distillates.

Although installation of pre- and post-treatment facilities is planned by many refiners, other non-capital approaches such as undercutting are being evaluated to meet interim and future sulfur levels. Even when expensive treatment facilities are installed, operating costs can be lowered and the flexibility of the facility increased with improvements in the ability to remove sulfur in the FCCU.

In this paper, we detail Petro-Canada's experience in integrating Albemarle's RESOLVE sulfur reduction technology with a combination of innovative process ideas. These concepts include heavy naphtha recycle, coprocessing of hydrogen donor feeds, and recycle of LCO to a specially designed stripper reactor. Special attention is paid to the interaction of deep FCC feed hydrodesulfurization with the FCC performance. The results demonstrate that very low FCC gasoline sulfur levels can be achieved without significant capital investment through novel approaches to recycle, creative integration of CF`- FCCU designs and operations, and application of state-of-the-art sulfur reduction additive technology.

An added benefit of the RESOLVE 950 sulfur reduction technology is the substantial elimination of sulfur oxides in the FCC flue gas. This has been observed in Petro-Canada operations and numerous other RESOLVE 950 applications around the world.

This paper is an elaboration and extension of the work presented at the Akzo Nobel Scope 2004 Symposium[1]

INTRODUCTION

The shift to the production of ultra clean fuels and chemicals from refineries combined with a focus on minimizing greenhouse gases warrants revisiting the integration of the FCCU within a conventional refinery setting. Evolving technologies can allow refiners to restructure their processing and phase into the heteroatom free products demanded by the market place. Additive technology such as RESOLVE can be utilized both in the transition process allowing for additional time for directions in the market to become clearer and structurally as part of an integrated solution. Integration of improved desulfurization catalyst technology directly into the design of the FCCU offers the potential to simplify the refinery processing scheme and provide some interesting advantages in many applications.

In Canada, between July 2002 and January 1 2005, refiners were required to meet an interim average gasoline sulfur specification of 150 wppm. Starting January 1, 2005 the specification was lowered to 30 wppm. The phasing in of the gasoline sulfur specifications will be followed by a distillate maximum sulfur specification of 15 wppm June 1, 2006. This has resulted in a critical examination of the effectiveness of different approaches and future product demands.

From a refiner's perspective, there are a significant number of unknowns in moving forward with capital expenditures. What appears to be an obvious solution today may not be tomorrow. Evolving technologies both to produce and utilize the products may significantly change the demand and product slate from the refineries. The push forward into a hydrogen economy may not happen as quickly as many have anticipated. In North America a shift from gasoline to distillate may not happen. Practical solutions for making the transition to more environmentally friendly products may be achievable with the industry's base infrastructure.

Feedstock quality will also influence the refining infrastructure. For example, evolving technologies allow for low hydrogen and high sulfur content tar sands bitumen to be viable feedstocks. Tar sands provide a long term security of supply. Infrastructure built into the primary upgrading will influence how a refiner adapts by reconfiguring refining complexes to process feedstocks derived from tar sands.

General criteria for the evolution are:

1. Effective utilization of hydrogen and the subsequent balancing of carbon in the products in any configuration
2. Energy efficiency
3. Minimization of CO2 generation
4. Minimization of gaseous and particulate emissions

FCCU CONFIGURATION

The FCCU is a carbon rejection and hydrogen transfer device. The process tailors the carbon distribution based on the hydrocarbon structures in the feedstock and the drive towards equilibrium in the cracking process. Historically the FCCU has been viewed as a relatively inexpensive gasoline and light olefin generator that now has significant application as a residual oil upgrader.

With improvement of bulk desulfurizing efficiency within the FCCU process, the FCCU could fill a role as a pseudo hydrocracker operation. The process would generate high olefinicity LPG, a low carbon number high hydrogen content stream for fuel cells, a low hydrogen content alkylbenzene stream for chemicals, and a tailored narrow boiling cycle oil that is significantly easier to integrate into existing refinery hardware. Optionally, the cycle oil could be eliminated completely negating the need for additional hydrogen and associated energy and CO2 generation. The FCCU would retain its carbon rejection flexibility.

Tailoring the FCCU product distribution to eliminate the 330 to 430 °F boiling range improves the quality of the gasoline, eliminates or reduces subsequent processing costs, and drops the driveability index of FCCU gasoline from about 1300 to 1000. Lower values of the driveability index mean improved cold-start and warm-up performance.

Further tailoring of the FCCU product distribution to remove the 600 to 700 °F cut eliminates the sterically hindered LCO components that are very difficult to hydrotreat. Ideally these low hydrogen content components could be utilized as coke and eliminate the hydrogen, energy, and capital required to upgrade this stream into the high hydrogen content fuels.

The 700°F+ FCCU slurry has a niche in heavy oil upgrading and coke related products. The high boiling nature of the FCCU slurry allows it to act as a liquid catalyst in some heavy oil upgrading processes[2]

PETRO-CANADA OPERATIONS

Petro-Canada operates three refining complexes in Canada. Each of these three refineries have significantly different configurations and operating objectives. Technology development and infusion of phased capital over a number of years provided sequential steps on the four FCCUs in moving towards a bulk processing configuration described in this paper.

The Petro-Canada Edmonton refinery is located in western Canada. This refinery is landlocked and has a development plan based on replacing the depleted conventional crude with locally produced tar sands bitumen and synthetic crudes derived from the tar sands. One of the crude trains in this refinery has operated since 1983 on 100% synthetic crude produced by Syncrude Canada in nearby Fort McMurray. Petro-Canada is a part owner in the Syncrude McMurray upgrader. Due to the very low hydrogen and high sulfur and nitrogen content of the bitumen to be processed going forward, the base technology selected to achieve the 2005 low sulfur gasoline target was a 1900 psi cat feed hydrotreater (CFHTU).

As part of the CFHTU integration work, the Edmonton FCCU catalyst was migrated to a mix of 90% Horizon 57 and 10 % Resolve 750 from 100% Horizon 57. Horizon 57 catalyst is based on Albemarle's Topaz technology. RESOLVE 750 is a component of the RESOLVE desulfurization technology detailed further on in this paper. This change provided a 26% reduction in gasoline sulfur for the 150 wppm phase in period and an equilibrated catalyst sample consistent with the rest of the Petro-Canada operations to serve as the basis for pilot plant development work. The results of this pilot work with Albemarle on a wide range of CFHTU feedstocks are discussed in this paper and illustrate an extension of the process.

In eastern Canada , Petro-Canada operates a refining complex in Montreal. This facility is the largest asphalt producer in eastern Canada. A large proportion of the crude slate is asphaltic sourced offshore. The technology implemented to achieve the 2005 low sulfur gasoline target was IFP Prime G. FCCU gasoline sulfur reduction hardware was incorporated directly into the FCCU in 1998 as part of the methodology used to phase in expansion of that unit[3].

In central Canada, Petro-Canada operates a relatively simple refinery in Oakville which supplies gasoil to a Petro-Canada lubes and white oil producing complex. The Oakville refinery has two crude units and two small FCCUs. Similar to many small North American refineries, an excessively large capital expenditure was projected to upgrade the refinery to produce low sulfur fuels using conventional technologies.

The approach taken for the Oakville Refinery was to meet the shorter term 150 wppm sulfur phase down with further development and implementation of FCCU based technologies. Alternate supply options would be utilized to meet the local market demand when the more stringent gasoline specifications came into effect January 2005. In 2002, Petro-Canada modified the Oakville #1 FCCU with a project very similar to the 1998 Montreal work. The design also incorporated the ability to send a heavy naphtha (HN) recycle stream to a 700 psig distillate desulfurization unit (DDS) and return the vapor and gasoline from the DDS stabilizer tower back to the FCCU.

Parallel developments on the four Petro-Canada FCCUs allowed for a platform to further develop the FCCU operation discussed above. The two Oakville FCCUs have been used to benchmark the RESOLVE 750, 850 and 950 systems in hardware systems that allowed a wide range of interactive conversion and product distillation combinations and also process a wide range of feedstocks. The Edmonton and Montreal FCCUs compliment this database with additional hardware and feedstock variations. All four units have operated with a common base catalyst system and are equipped with the Petro-Canada RTD system.

BASE FCCU SULFUR DISTRIBUTION

The immediate issue requiring refinery process modifications is meeting the market place demand for low sulfur gasoline. As has been well documented in the literature, the FCCU contributes over 90% of the sulfur in an FCCU based refinery gasoline pool.

Reaction Model

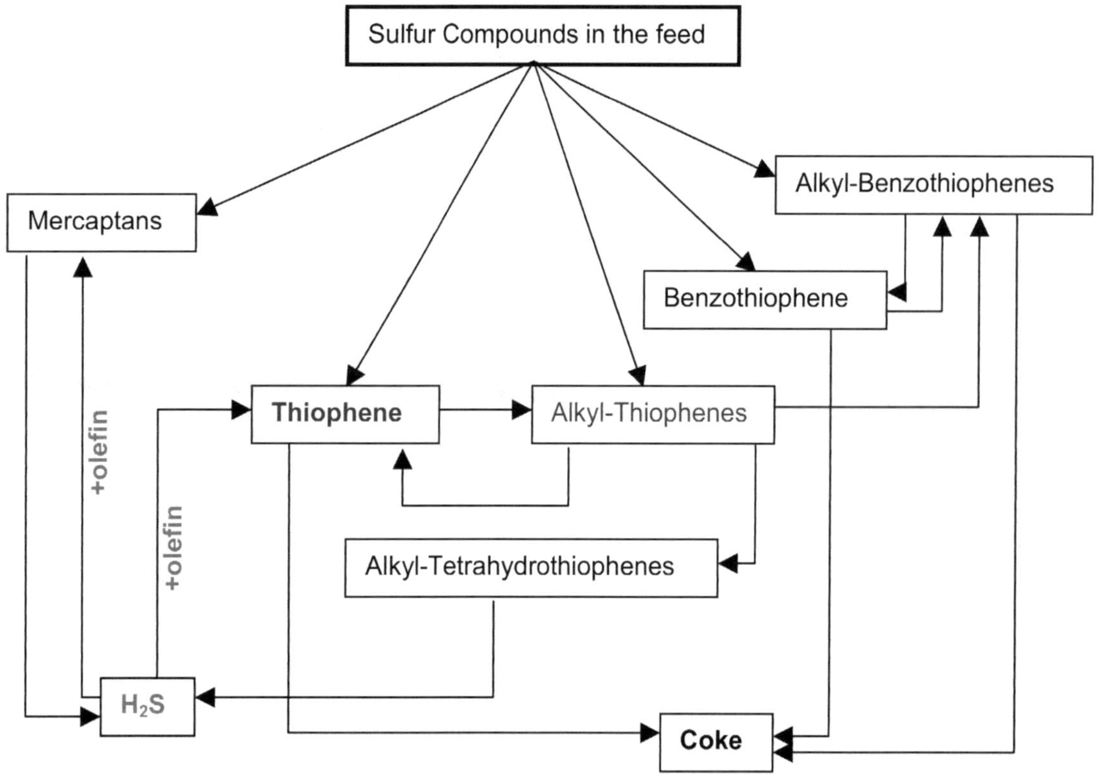

Figure 1: Gasoline Sulfur Species Formation Pathways

Figure 1 shows the reaction pathways postulated for the creation of sulfur species in the gasoline boiling range[4]. By analogy, sulfur species will be generated in the other FCCU products through recombination of H2S with olefins or molecular rearrangement during cracking. Petro-Canada has done pilot plant studies using model compounds to develop a model for relative coking rates and sulfur distributions. This work confirms the potential for addressing cycle oil sulfur and quality issues within the FCCU process.

The two key objectives for adjusting the sulfur reaction pathways to enable the FCCU to be a more efficient bulk desulfurizer and hydrogen management tool are:

1. minimize the potential for olefin and H2S recombination
2. maximize the potential for thiophenes and thiophenes in aromatic complexes to be converted to coke

Feedstock Aromatic Sulfur Species

Figure 2 illustrates the sulfur profile obtained for a number of FCCU gasolines sampled from the three Petro-Canada refineries. The FCCU gasolines were cut in 45 degree F cuts in a

TBP column and characterized. The volumetrics, qualities and compositions reported in this paper represent the average for the individual 45 degree F cut.

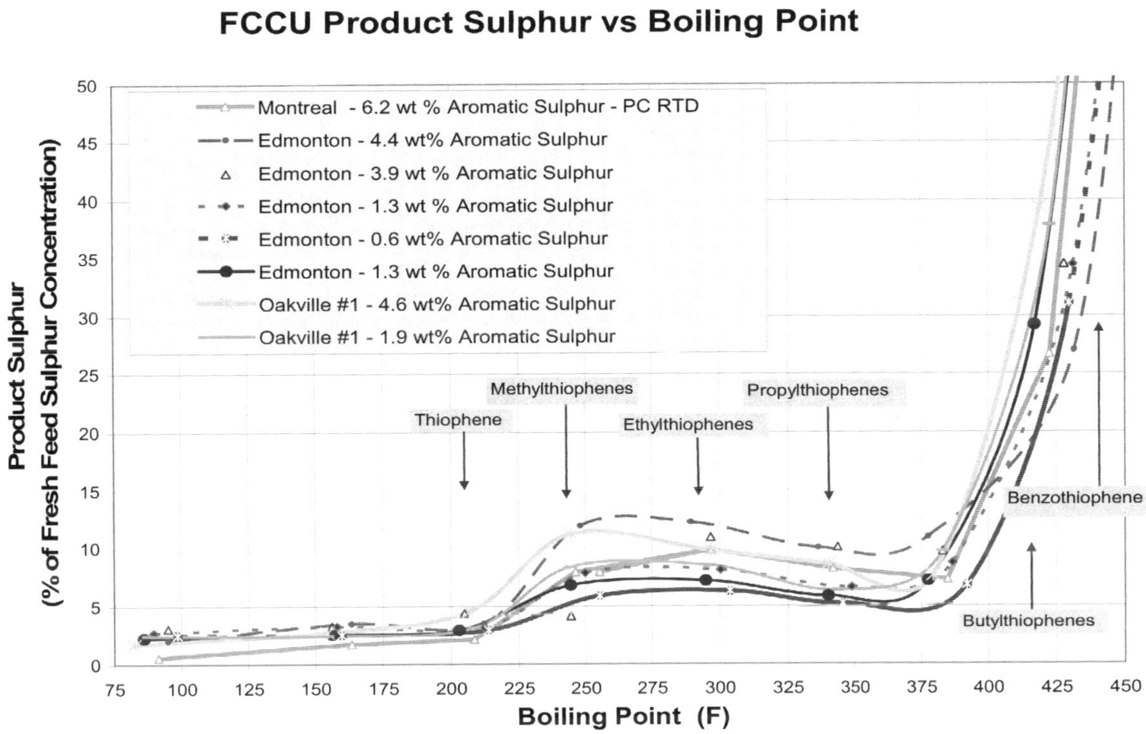

Figure 2: FCC Product Sulfur Distribution

The data in Figure 2 represents three different Petro-Canada FCCUs operating with variations in catalysts, hardware and operating conditions in 1999 and 2000. The bulk aromatic sulfur species content in the FCCU feedstock has been used to differentiate the feedstock qualities to these operations and was determined by mass spec analysis.

Figure 2 illustrates the three distinct sulfur distribution regions common to all the FCCU gasoline benchmarks. As has been well documented in the literature, the major concentration of sulfur is found in the back end of the gasoline boiling range and is contributed by the benzothiophenes. A sulfur peak is observed in the gasoline at about 257°F. The mid gasoline boiling range peak and the associated plateau between about 266°F and 347°F is due primarily to the alkylated thiophenes in the gasoline. The height of the sulfur peak at about 257°F for a given FCCU, catalyst system and hardware configuration is a function of the aromatic sulfur species in the feedstock.

Figure 3 shows the relationship between the aromatic sulfur species in the FCCU feedstock and the plateau heights for six sets of data from the Edmonton FCCU operation. There is a significant increase in gasoline sulfur as the amount of aromatic sulfur species of the feedstock is increased. The sulfur content of the FCCU gasoline boiling between 257 and 347°F increases about 1.4 % of feed sulfur concentration for every 1 wt% increase in the feedstock aromatic sulfur. The baseline operation represents a system with conventional hardware, a high zeolite conventional gasoil catalyst and feedstock blends comprised of virgin gasoils, delayed coker gasoils and slurry, and hydrocracker bottoms.

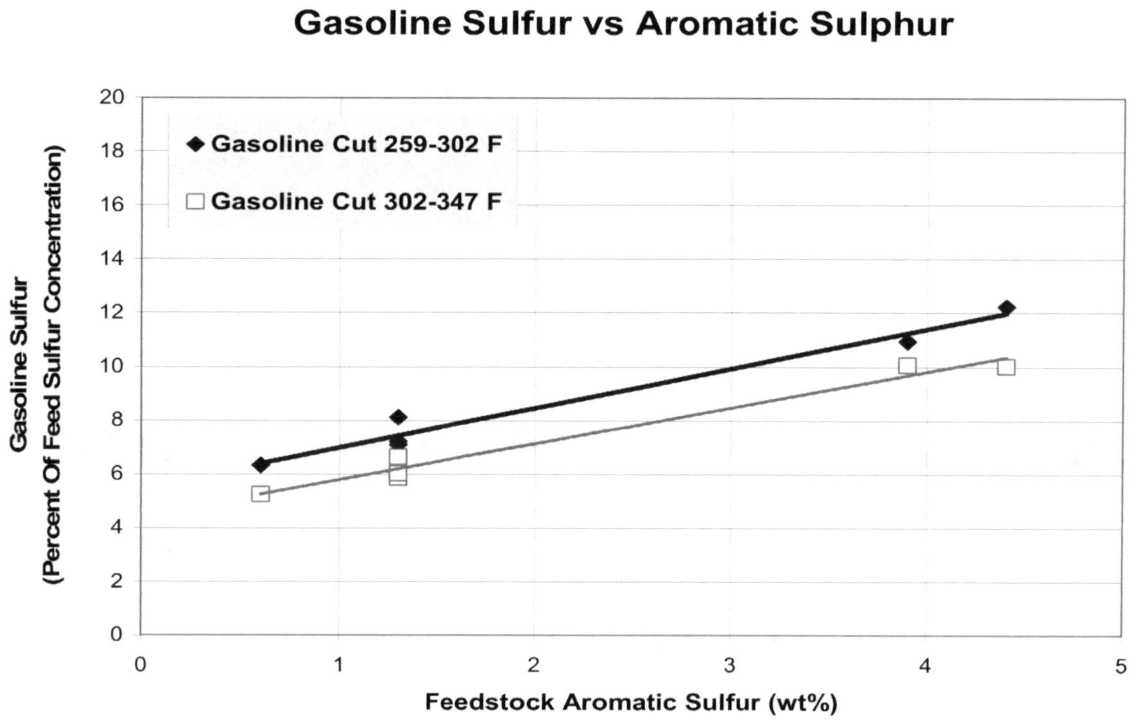

Figure 3: Gasoline Sulfur vs Feedstock Aromatic Sulfur

Benchmarking the four Petro-Canada FCCUs based on the aromatic sulfur criteria resulted in standardizing the catalyst systems to high alumina Albemarle Topaz type systems by mid 2002. Relative to the Edmonton baseline shown in Figure 3, the Oakville #1 FCCU data for a similar low metals operation had sulfur gasoline benchmarks about 5 % lower for a 1.5 wt% aromatic sulfur feedstock and 20% lower for a 4.6 wt% aromatic sulfur content feedstock. With the switch to the higher alumina Topaz system, the Edmonton FCCU gasoline sulfur dropped about 17% on a typical 3.5 wt% aromatic sulfur feedstock. These data are consistent with literature and illustrate the interaction of base catalyst design and feedstock aromatic sulfur content to gasoline sulfur[5].

As shown in Figure 2, the Montreal FCCU gasoline with metals on ecat and PC RTD shows substantially lower gasoline sulfur throughout the gasoline boiling range. A large part of this sulfur reduction could be due to the much higher vanadium level on the Montreal catalyst. As shown in Figure 1, another key factor could be the reduction for the opportunity of olefins to recombine and form mercaptans and thiophenes. This is in addition to the observation that the RTD generates less heavy boiling gasoline components as previously reported.[6]

Tracing Reaction Pathways with H2S Injection

Adding incremental sulfur directly to the riser in the form of H2S results in incremental sulfur being incorporated throughout the gasoline boiling range. Co-processing of sulfur containing species or H2S directly results in sulfur being incorporated into the FCCU products as shown in Figure 1.

A test was run on the Petro-Canada Oakville #2 FCCU where about 30% more H2S was introduced into the riser relative to the amount of H2S generated by processing the feedstock. An overall increase in the mercaptans and thiophene sulfur species of about 30% was observed with a 12% increase in mercaptans, an 83 % increase in thiophene and a 20 % increase in alkylated thiophenes. These results suggest that the thiophene peak observed in Figure 2 can be largely influenced by controlling both the amount of H2S and olefin in contact with the catalyst. This is consistent with elevated gasoline sulfur observed at elevated riser temperatures.

CARBON NUMBER DISTRIBUTION

Hydrocarbon feedstocks undergo thousands of reactions within the FCCU. The 430°F- product yielded is similar, independent of the feedstock. Petro-Canada has undertaken cracking studies on model compounds and observed that even various straight chain paraffin pure components generate the typical carbon number and specie distribution observed when processing conventional FCCU feedstocks. Similar results have been reported in the literature. There are variations in the product distribution based on hydrogen content and specific structures in the model compound cracked, but the cracking process approaches an equilibrium and the differences in the cracked products becomes less as the cracking process time is extended. Employing the concept that the FCCU process will move towards an equilibrium several studies were undertaken on the four commercial units.

Figure 4 illustrates the typical aromatic distribution found in the FCCU gasoline. The aromatic distribution is made up of an alkylbenzene distribution as shown in Figure 5 and less desirable higher boiling aromatics illustrated in Figure 6. The asphaltic feed is characterized by elevated alkylbenzenes but about the same profile for the undesirable components shown in Figure 6.

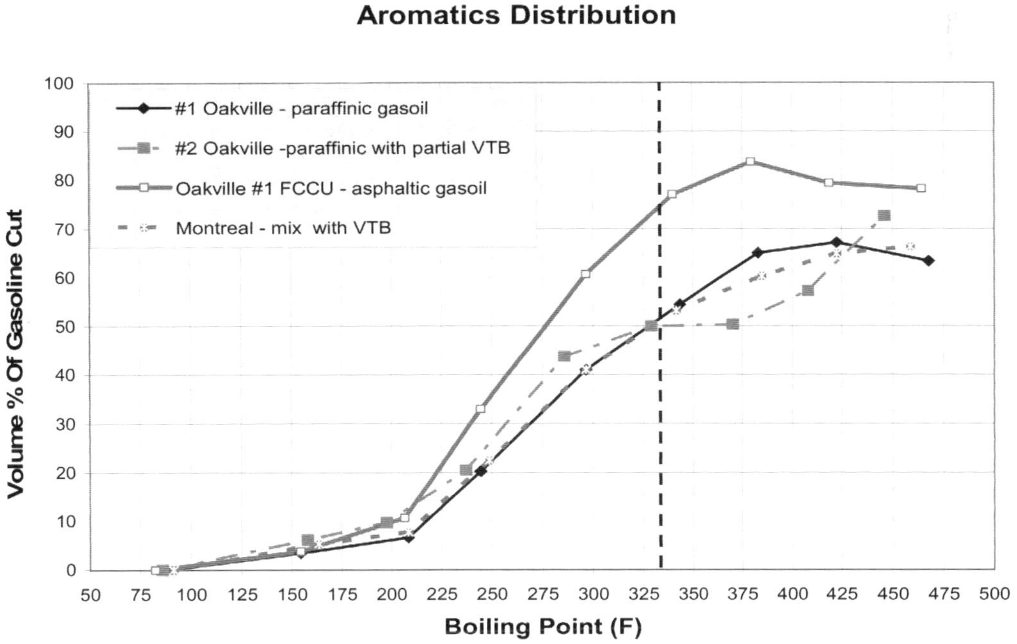

Figure 4: Gasoline Aromatics Distribution

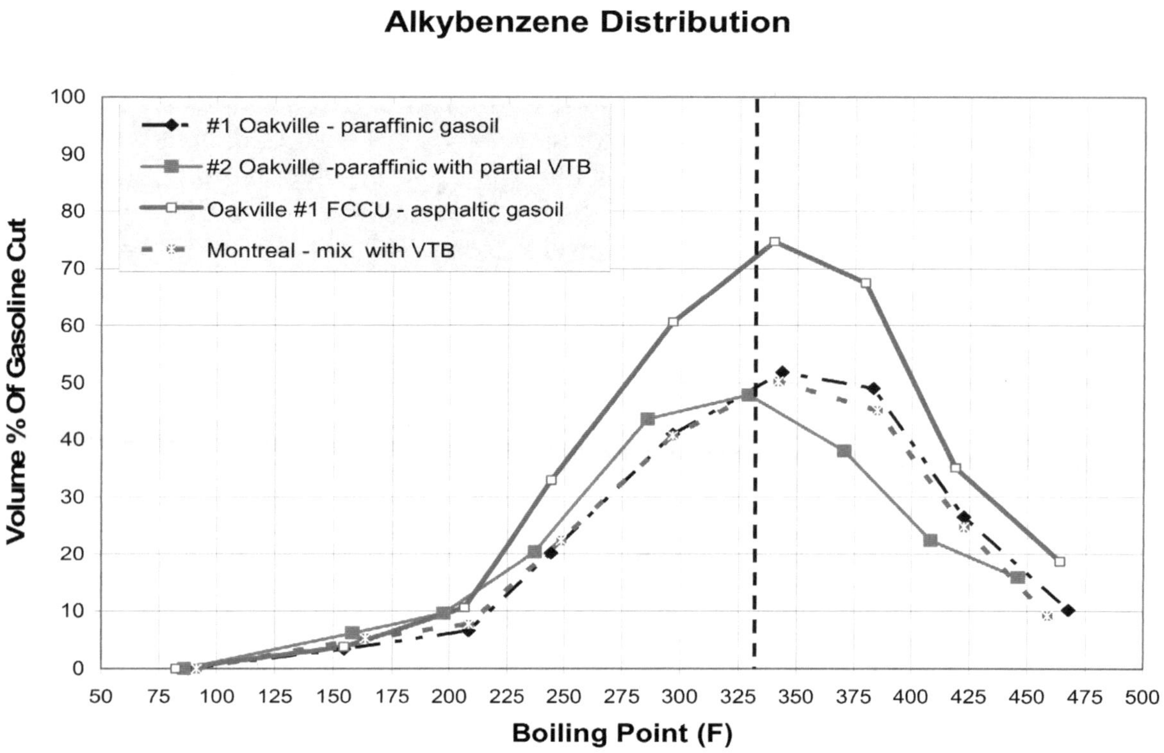

Figure 5: Gasoline Alkylbenzene Distribution

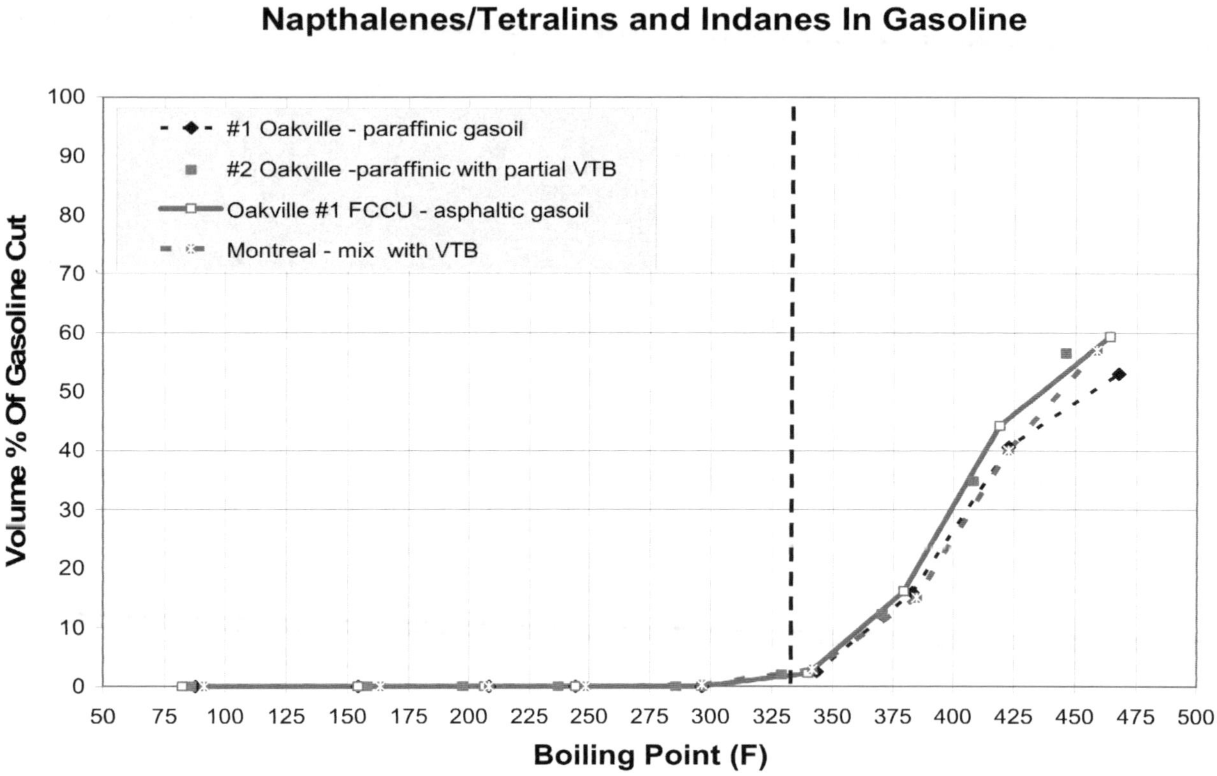

Figure 6: Napthalenes, Tetralins and Indanes in Gasoline

Modified Gasoline Aromatics Distribution

Recycling of the back end of the gasoline back into the riser results in a combination of cracking and adjusting the FCCU product equilibrium such that the 330°F+ recycled material is almost completely eliminated from the FCCU reactor products. This process removes the C10+ alkylbenzenes, tetralins, naphthalenes, indenes, indanes from the gasoline pool and reduces the driveability index of the FCCU gasoline from about 1300 to 1000.

As shown in Figure 7, the reintroduction of the light FCCU product recycle stream results in the generation of lower boiling gasoline. Very little C4- product is generated. The Oakville #2 FCCU was operating in a low severity gasoline + distillate product mix mode for the distribution shown in Figure 7. For feedstocks that are processed at higher severity the reintroduction of the Heavy Naphtha (HN) back into the cracking system results in the generation of primarily lower boiling alkylbenzenes. The Road Index blending values of these lower boiling alkylbenzenes formed are in the range of 96 to 110.[7]

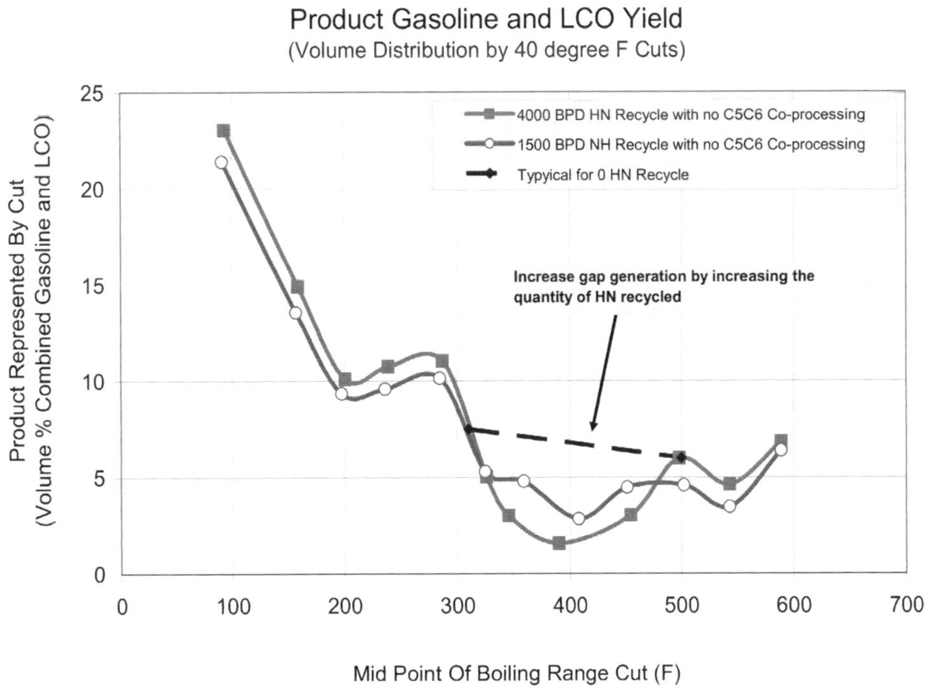

Figure 7: Effect of HN Recycle on Product Distribution

As illustrated in Figure 2, the highest sulfur concentration in the FCCU gasoline is in the 388°F+ boiling range. The reprocessing of the back end of the gasoline through the FCCU typically results in the elimination of more than half of the sulfur from the net gasoline product without the addition of any other sulfur removal mechanism such as a gasoline sulfur reduction additive. The percentage of sulfur removed by this process is increased for feedstocks with low aromatic sulfur concentrations since the sulfur content in the back end of these gasolines represents a greater percentage of the total sulfur in the gasoline.

Integrated HN reactor product recycle back into the cracking system has been tested on all four FCCUs. The cut point for the HN product recycle material has ranged from about

302°F to 482°F. The recycle has been added ahead of, with, and after the main feedstock injection point. The recycle has been blended with various other streams before reintroduction into the cracking system. A net recycle product has been withdrawn from the system.

HN recycle has been withdrawn with variations in the number of fractionation trays between the product recycle draw and the net gasoline product and cycle oil product. The number of fractionation stages between the various draw points influences the width of the cut recycled and the ability to fractionate out the heavier boiling sulfur species.

The commercial testing indicates that the efficiency of tailoring out the material boiling in the 330 to 500°F range can be significantly greater than that shown in Figure 7. Catalyst activity and composition have an impact on the process efficiency. Recycling the light alkylbenzenes back into the reaction system without co-processing high hydrogen content materials has a minimal impact on the net alkylbenzene yield. Figure 8 shows the redistribution of the reactor net yields is to both lighter gasoline and 482 to 572°F cycle oil.

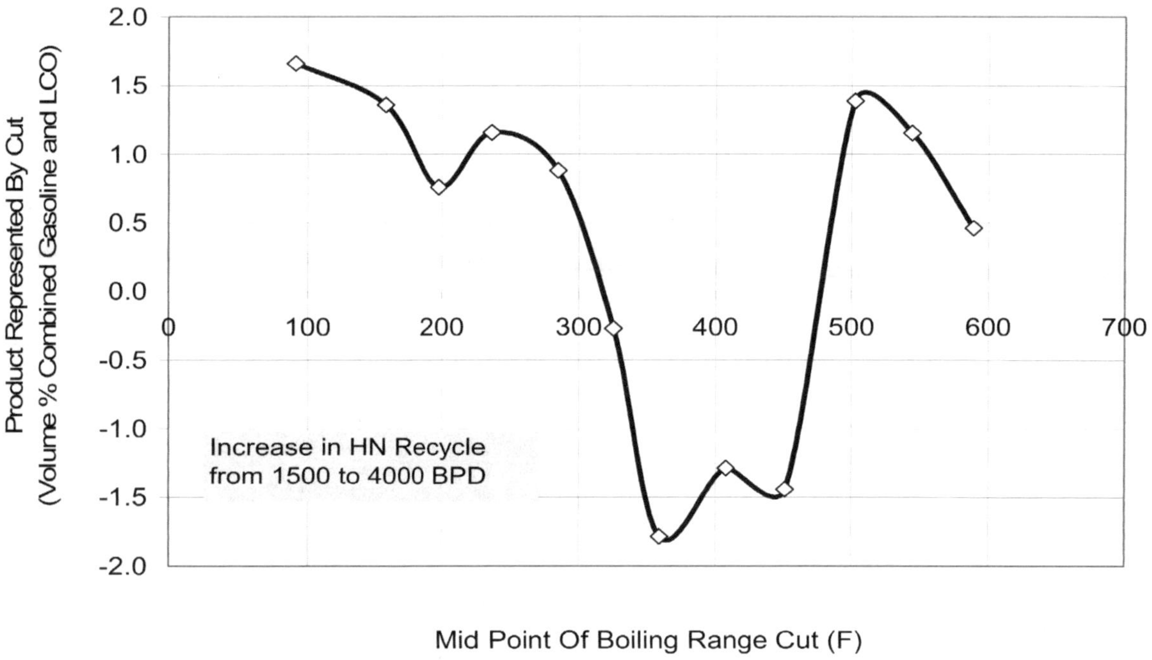

Figure 8: Product Profile Shift with HN Recycle

Typical gasoline densities are reduced by about 0.03 SG and typical FCCU gasoline selectivities obtained are similar to non recycled operations and range from low 80s for paraffinic feedstocks to high 70s for asphaltic feedstocks.

Carbon Distribution Shifts with Saturate C5C6 Co-processing

The impact of co-processing virgin crude C5C6s in the commercial FCCU illustrates another unconventional processing scheme for taking advantage of the FCCU product equilibrium. This process is particularly effective when used in conjunction with the product recycle process discussed above. The percentage of C3C4s generated from this kind of feedstock is similar to the base FCCU feedstock. Only about 40% of the C5s and 31 % of the C6s remain in the 104 - 207˚F boiling range of the original feedstock. The yield profile shift obtained when co-processing the C5C6s relative to that generated by the base feedstock alone is shown in Figure 9. There is a shift to higher carbon number structures in the gasoline with some additional LCO generated in the 446˚F boiling range. This process is utilized as a mechanism to reduce net RVP and increase the octane in the refinery gasoline pool.

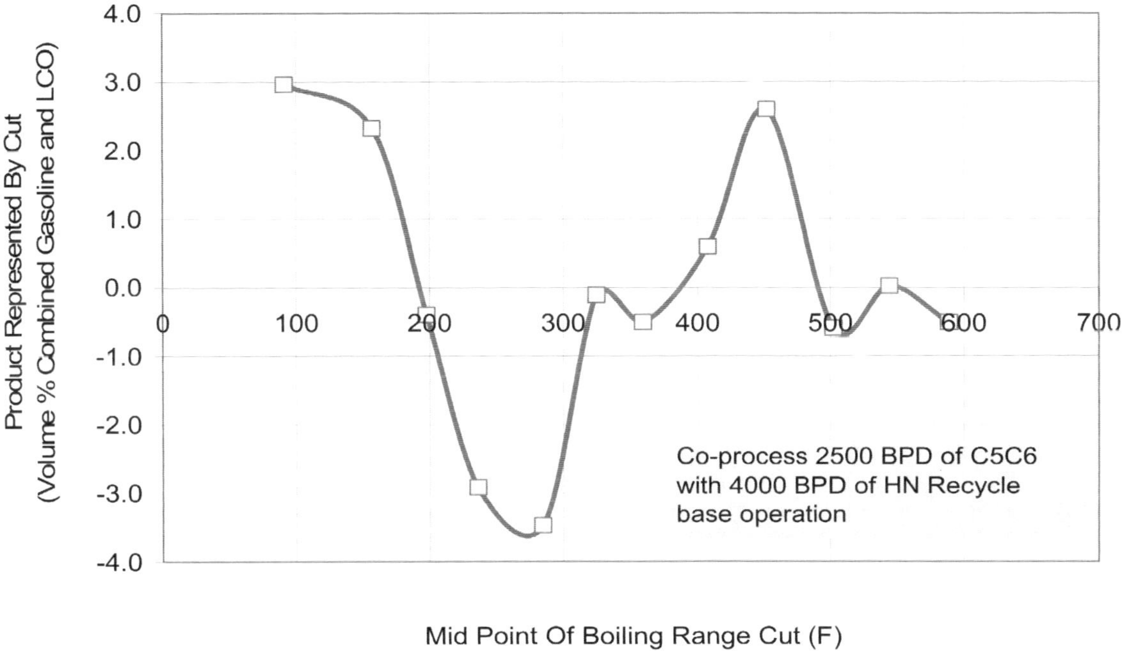

Figure 9: Product Profile Shift with C5C6 Co-processing

Heteroatom Changes with Saturate C5C6 Co-processing

Figure 10 illustrates the reduction of sulfur in the 338˚F- boiling range of the gasoline. The introduction of saturate C5C6s into the FCCU feed results in not simply a suppression of the sulfur concentration of the FCCU gasoline boiling below 207˚F, but also about a 50 % reduction in the sulfur concentration of all the gasoline boiling above 207˚F. A similar trend is observed in the cycle oil boiling range of the FCCU products with about a 20% reduction in the sulfur concentration observed.

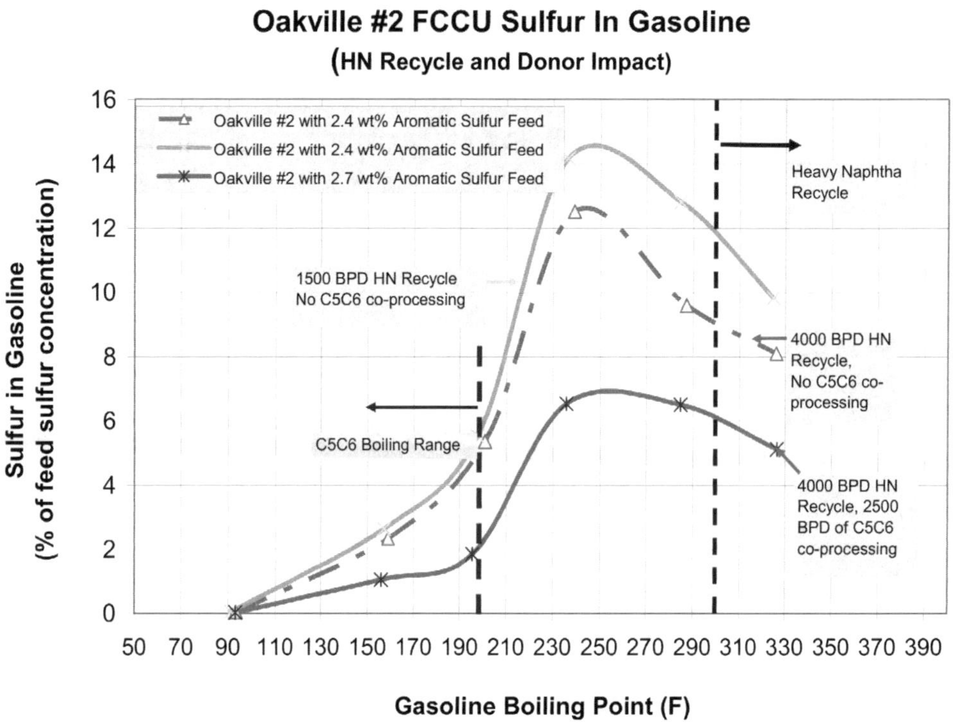

Figure 10: Effect of Co-processing C5C6 on Sulfur Distribution

The gasoline sulfur concentration at the 248°F peak region is also about half the concentration of the base operation with no C5C6 co-processing. The IBP of the HN recycle stream only extended down to about 302°F product. About a 50% sulfur reduction is observed in the 207 to 302°F gasoline product. This suggests the elevated hydrogen in the C5C6 stream was actively involved in determining the entire FCCU product sulfur distribution.

Figure 11 shows the corresponding changes in the olefinicity in the 45°F gasoline cuts. Light gasoline olefinicity is substantially less than the base operation. However, the olefinicity is slightly higher in the 248 to 302°F boiling range where the large reduction in the sulfur peak is observed. This combined with the reduction in net reactor product shown in Figure 9 supports the idea the olefin and H2S recombination pathway shown in Figure 1 could be significantly impacting the sulfur distribution. This phenomenon is consistent with general observations that catalysts that generate more olefins and ZSM-5 promoted catalyst systems result in higher sulfur concentration in the gasoline[8].

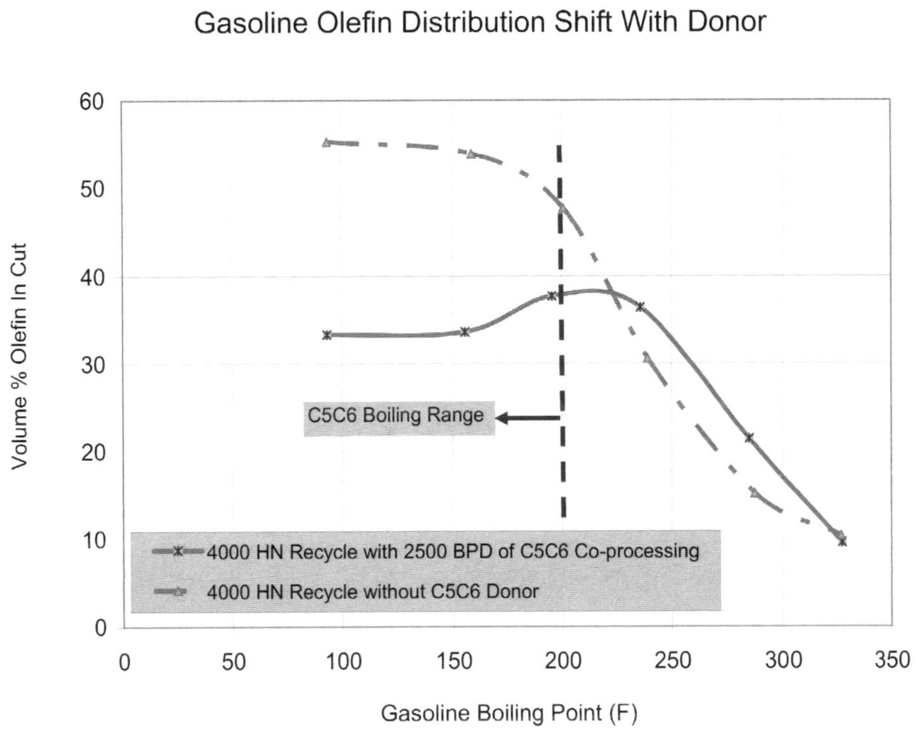

Figure 11: Effect of C5C6 Co-processing on Gasoline Olefinicity

Figure 12 illustrates that the hydrogen donor effect noted with the sulfur distribution also appears to influence the nitrogen concentration in FCCU gasoline. The recycle of the HN gasoline cut back into the FCCU cracking process and the co-processing of the C5C6 virgin crude result in incremental reductions of the nitrogen in the gasoline.

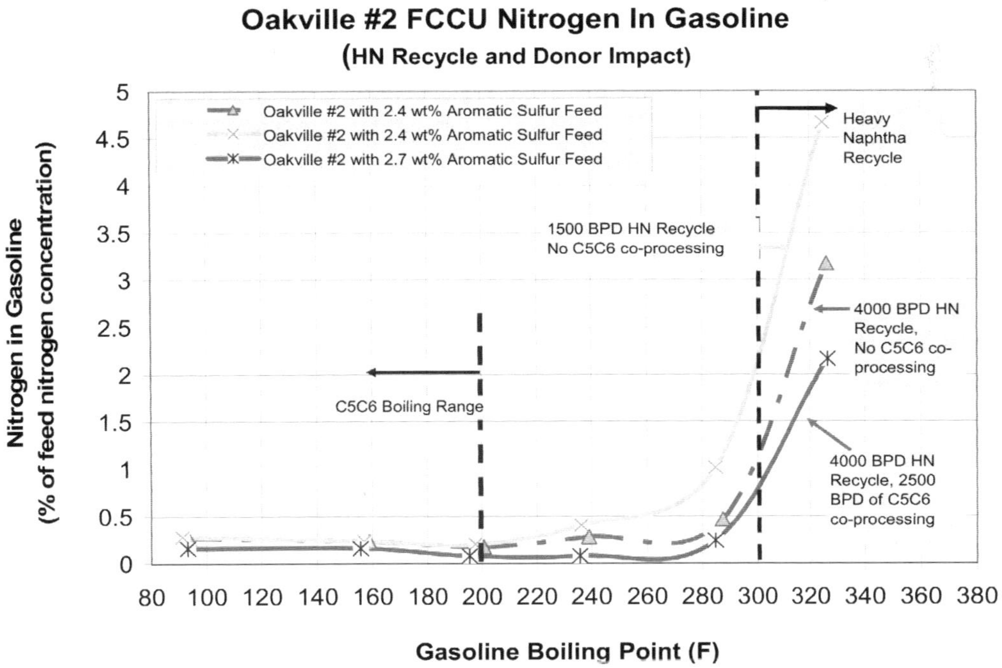

Figure 12: The Effect of C5C6 Co-processing on gasoline nitrogen distribution

The above commercial examples demonstrate:

1) The sulfur concentration in the FCCU gasoline derived from virgin crude feedstocks can be benchmarked utilizing the aromatic sulfur content of the feedstock
2) The base sulfur distribution can be stabilized with a more consistent lower sulfur concentration profile by removing supplemental sulfur sources such as H2S and by utilizing a base catalyst technology similar to the Topaz system.
3) The introduction of feed sources traditionally believed to be non-reactive in the FCCU such as saturate C5C6s will influence both the net carbon and heteroatom distributions obtained from the FCCU. The net product of the C5C6s co-processed is a carbon distribution approaching the base FCCU product carbon distribution.
4) The quantity of high boiling component of the FCCU gasoline can be effectively cut out of the net FCCU product by recycling thereby removing more than 50% of the sulfur in the gasoline and focusing the effort on removing incremental sulfur from the gasoline on the mercaptans and lower boiling thiophenes. The tailoring of the FCCU gasoline reduces the net aromatics content of the gasoline, increases the quantity of C9 and lower alkylbenzenes such that the remaining heavy naphtha has elevated octane, an N+2A above 100, and has reduced concentration of sulfur and nitrogen. This now improves the feed quality for heart-cut reforming.

The modifications described above can be used as a platform for driving the desulfurization within the FCCU. If the bulk desulfurization level achieved in the FCCU is significant enough, then a number of simple low capital cost refinery integration configurations can be employed to meet a staged product sulfur elimination. In the limit, the sulfur levels could be low enough to implement radically different low cost post treat systems to reach a zero sulfur product.

RESOLVE CATALYTIC SULFUR REDUCTION TECHNOLOGY

Albemarle's RESOLVE FCC catalytic sulfur technology has been well documented by Humphries, *et. al.*[9]. This technology leads the industry with more than fifty commercial applications, and has full range gasoline sulfur reductions of up to 50% without undercutting. Significantly higher sulfur reductions can be achieved depending on the extent to which a refiner is able to reduce gasoline end point. Reductions in cycle oil sulfur of up to 30% have also been achieved. Both commercially and in the lab, it has been shown that the reduction is not only achieved with easily convertible molecules like mercaptans, but also in the more constrained species such as (substituted) thiophenes and benzothiophenes.

This technology has evolved into a family of products covering a range of sulfur reduction needs. The applicability of the products is best illustrated in Figure 13.

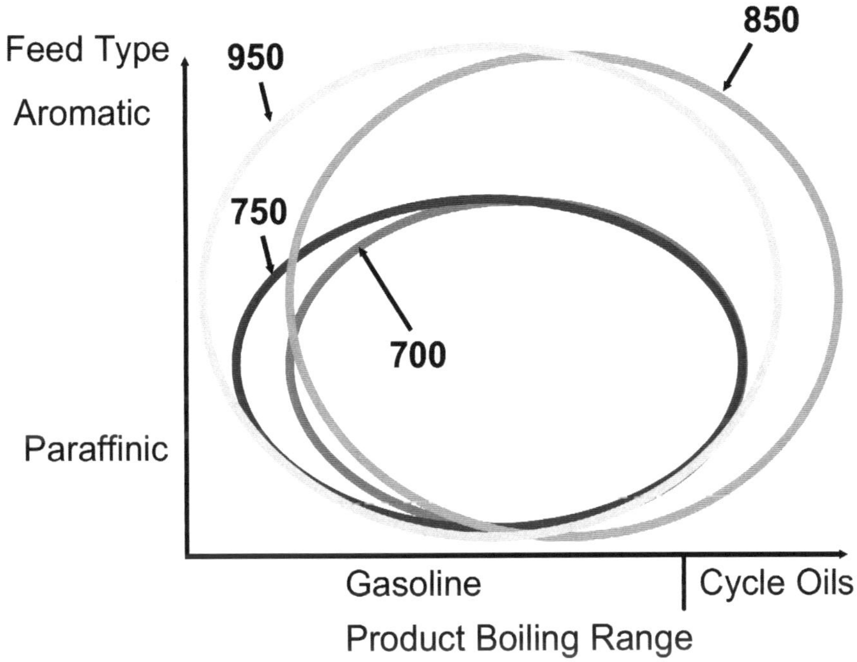

Figure 13: Applicability of RESOLVE technology

It is quite apparent from this diagram that there is a lot of flexibility for the RESOLVE system. Blending of the additives expands the application opportunities.

Finally, it should be noted that the RESOLVE family of additives has no negative activity or selectivity effects. The reasons for this differ among the different additives, but at a minimum, there will be no increase in fresh catalyst addition. In many cases, fresh additions may be reduced in correspondence with additive usage.

APPLICATION OF RESOLVE 950 AT PETRO-CANADA

Since 2001, Petro-Canada has benchmarked RESOLVE 750 on three of it's FCCUs, RESOLVE 850 on one unit and RESOLVE 950 on 3 FCCUs. The first commercial testing of the Albemarle RESOLVE 950 system has been underway in the #1 FCCU at the Petro-Canada Oakville refinery since late 2003. The concentration of RESOLVE 950 in the Oakville #1 FCCU has been stepped up over 2004 and maintained at 35 wt% in the fresh catalyst mixture since the end of June 2004. Figure 14 shows the tracking of the RESOLVE 950 in the Oakville #1 FCCU ecat being stepped up from 10% to 35% since the beginning of 2004.

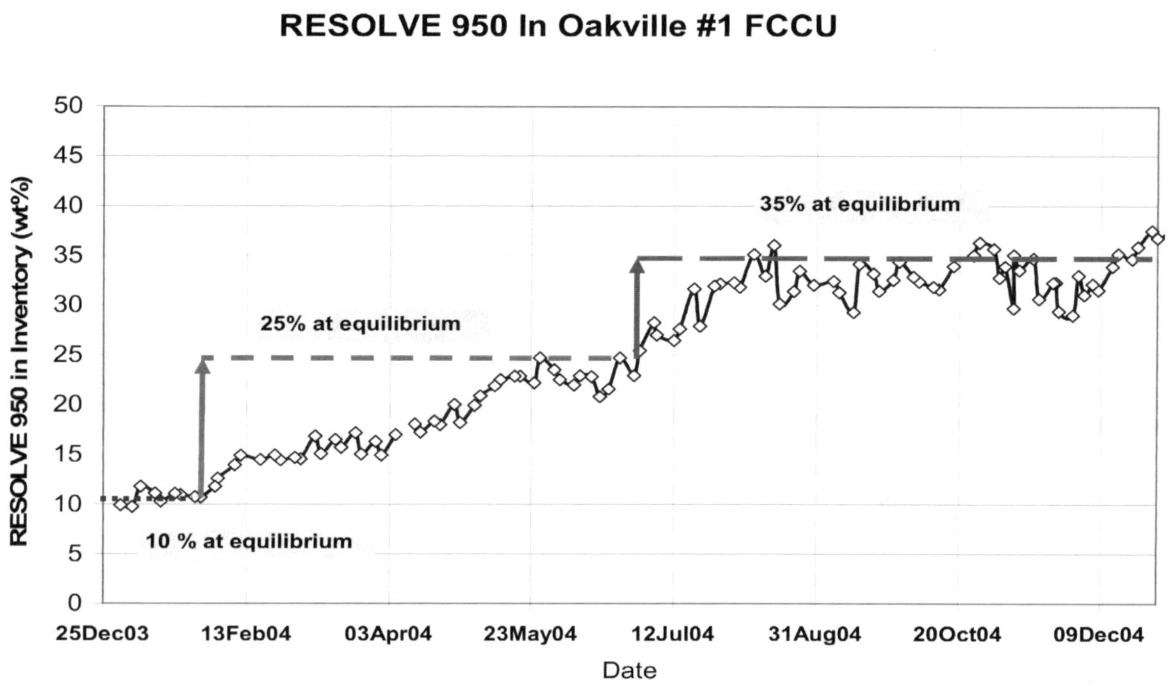

Figure 14: RESOLVE 950 Usage at Oakville #1 FCCU

Figure 15 shows a 30 day rolling average of the total catalyst addition rates to the Oakville #1 FCCU including the RESOLVE 950.

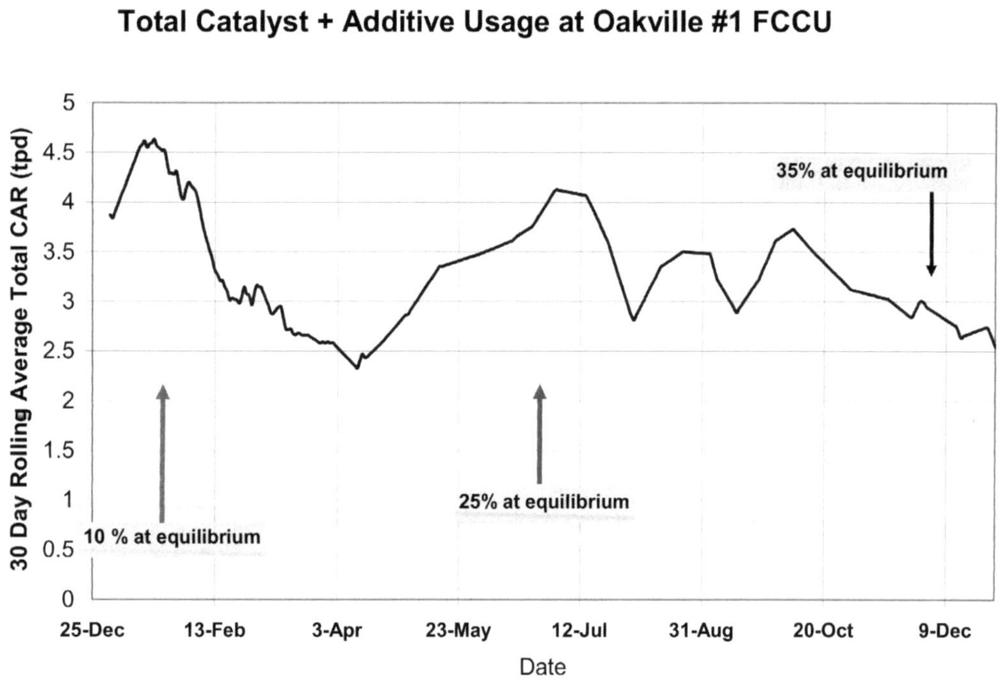

Figure 15: Oakville #1 FCCU 30 Day Rolling Average Catalyst + Additive Usage Rate

During this time period, two step changes were made in the host catalyst system. The rare earth on zeolite and zeolite content were increased in the beginning of the year with an octane barrel host system and then decreased with a lower rare earth octane catalyst system at the end of April. Zeolite content of this catalyst was essentially double that of the catalyst used at the beginning of the year. The rare earth on catalyst marker is shown in Figure 16.

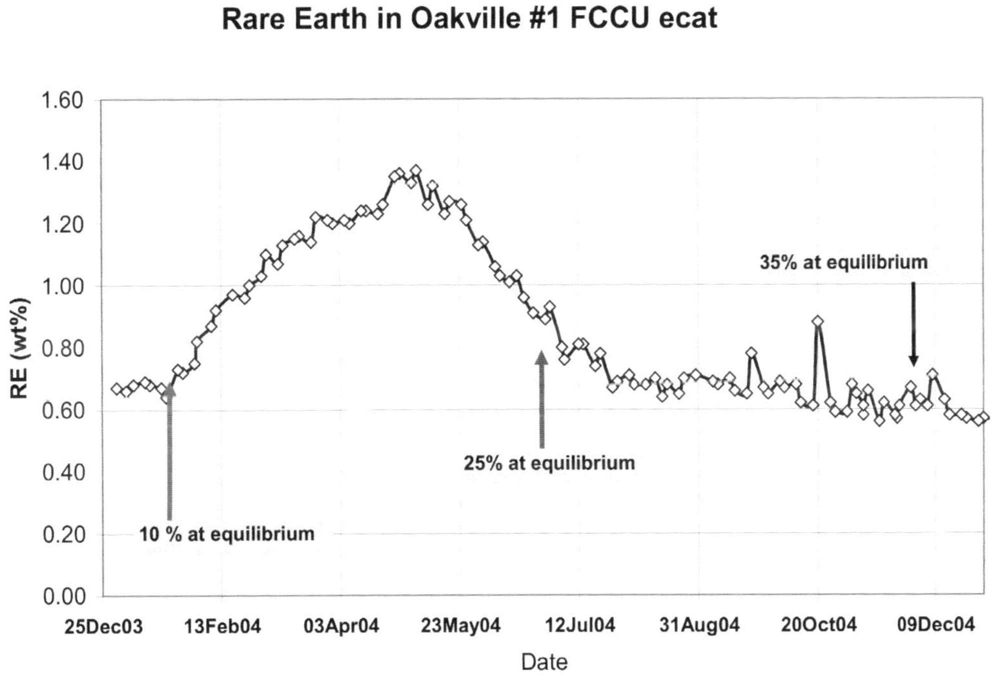

Figure 16: Rare Earth in Oakville #1 FCCU Equilibrium Catalyst

As developed earlier in this paper, hydrogen transfer has a large impact on the desulfurization characteristics within the FCCU. Low rare earth FCCU catalyst systems exhibit higher sulfur in gasoline. This has been identified by others[9] and observed in Petro-Canada Oakville and Montreal FCCU operations. The Oakville #1 FCCU began using an octane-barrel host catalyst in the beginning of 2004. In late April, the Rare Earth on zeolite of this system was lowered substantially, consistent with an octane catalyst. Experience has shown that moving from an octane-barrel to an octane catalyst increases sulfur in gasoline in the range of 0.5 to 1.0% relative to FCCU feed sulfur concentration.

Figure 17 shows that through 2004 the ecat MAT ranged from about 71 to 75. The operation at the end of 2004 with both 35% RESOLVE 950 and an octane catalyst had a stable MAT of about 72. The MAT activity correlates well with the RE and surface area data presented in Figures 16 and 18.

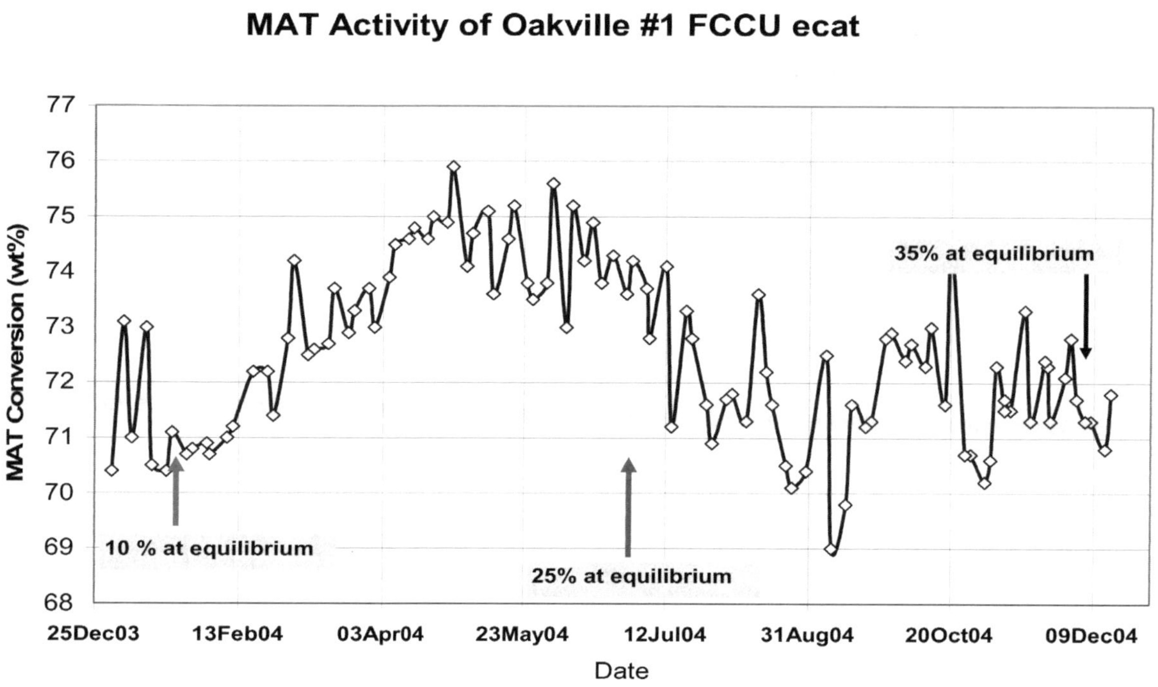

Figure 17 MAT Activity of Oakville #1 Equilibrium Catalyst (wt%)

The Oakville #1 FCCU ecat surface area exhibits a similar pattern relative to the MAT data. The surface area ranged from 125 to 150 m^2/g, with about a 132 m^2/g SA at the end of the 2004. This is shown in Figure 18.

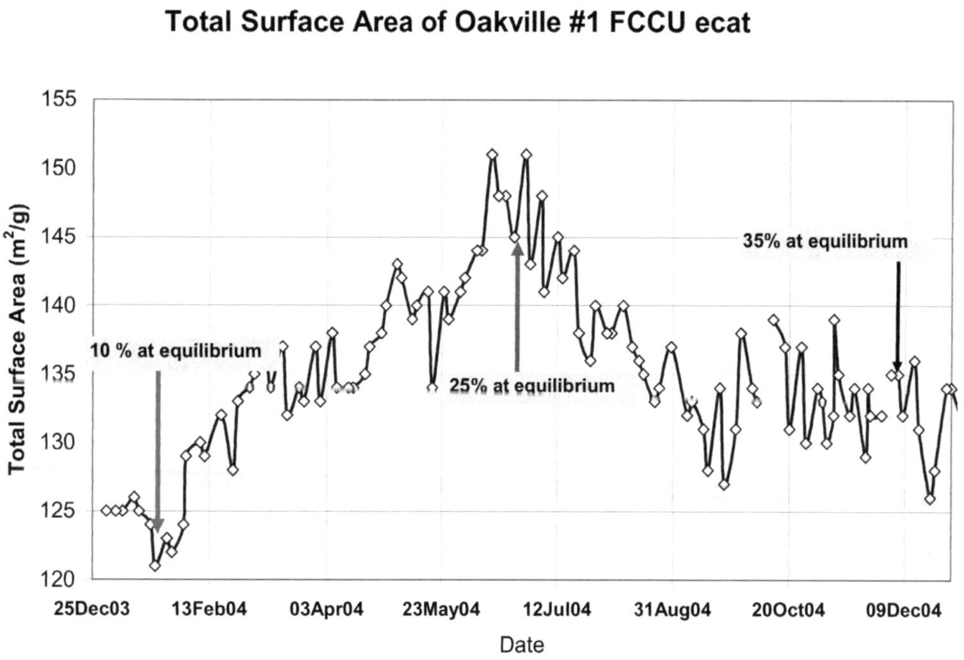

Figure 18: Total Surface Area (m^2/g) of Oakville #1 FCCU Equilibrium Catalyst

The Oakville #1 FCCU processes blended heavy fuel oil (HFO) as part of the feed slate. Figure 19 shows that the level of vanadium on ecat increased over 2004 to about 2500 wppm largely in correspondence with reduced catalyst addition rate. It is recognized that higher levels of vanadium can result in reduced gasoline sulfur, but the effect in this case would be minor compared to the reductions observed for RESOLVE 950.

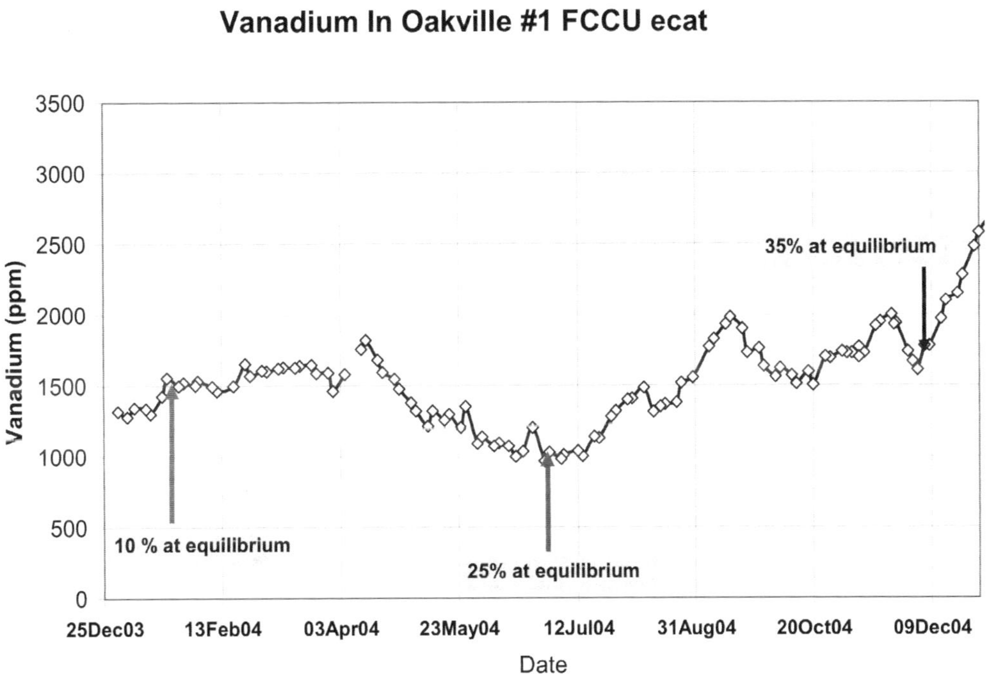

Figure 19: Vanadium on Oakville #1 FCCU Ecat (ppm)

INTEGRATION WITH PETRO-CANADA PROCESSING PLATFORM

Figure 20 illustrates the desulfurization levels achieved with the RESOLVE 950 in combination with an octane barrel catalyst system and the HN recycle operation. The relationship does not reflect the use of a hydrogen donor. The net desulfurization achieved for a given feedstock would be the base desulfurization achieved with the HN recycle and then the percentage desulfurization shown in Figure 20 applied to the remaining sulfur.

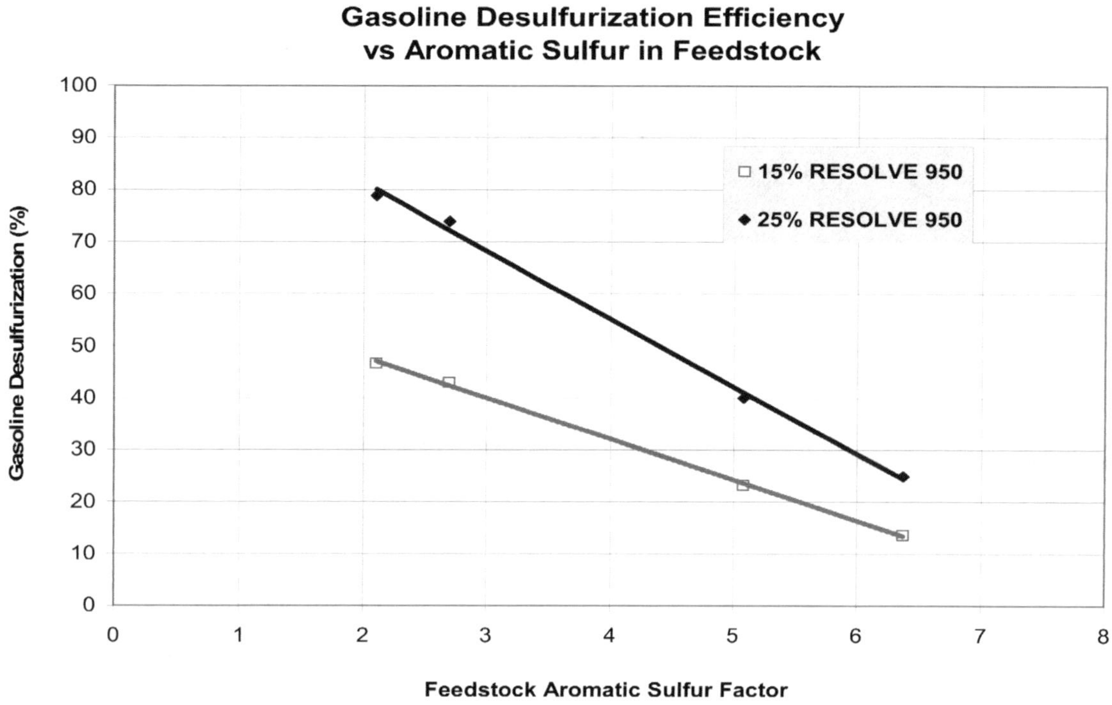

Figure 20: Gasoline Desulfurization Efficiency of RESOLVE 950

The desulfurization level achieved over and above the sulfur reduction obtained with the platform described above is very dependent upon the aromatic sulfur content of the FCCU feedstock. Extremely high levels of desulfurization are achievable with virgin feedstocks containing low levels of aromatic sulfur. Desulfurization levels for a typical sweet gasoil with an aromatic sulfur content in the feed of about 2 wt% will be about 82% with about 25 % RESOLVE 950 in inventory.

In contrast, virgin feedstocks that contain higher levels of aromatic sulfur and typically higher base sulfur levels will exhibit substantially lower desulfurization efficiencies. Typical gasoils processed from asphalt operations at the Oakville Refinery and operating with the same 25 % RESOLVE 950 would only exhibit about a 40% reduction in gasoline sulfur due to the RESOLVE 950. The net desulfurization efficiency of the additive and the recycle platform would be about 71%.

Figure 20 was constructed from benchmark data from the Oakville #1 FCCU, Oakville #2 FCCU and Albemarle pilot work. The pilot work was done on a gasoil representative of paving grade asphalt crude operation which is the worst of the gasoils processed at Oakville. The gasoil had a specific gravity of 0.927 and a sulfur concentration of 19890 wppm. The aromatic sulfur concentration tie-in for this feedstock was 6.37 wt%.

Laboratory testing, shown in Figure 21 indicates that all three gasoline desulfurization additives, RESOLVE 750 , 850 and 950 , are active in reducing the full range gasoline sulfur. At 70 wt% fresh feed conversion, gasoline sulfur dropped from about 3000 wppm to 2460 wppm with the inclusion of 20 wt% RESOLVE 950. The 18 percent reduction on full range gasoline sulfur was consistent with commercial observations.

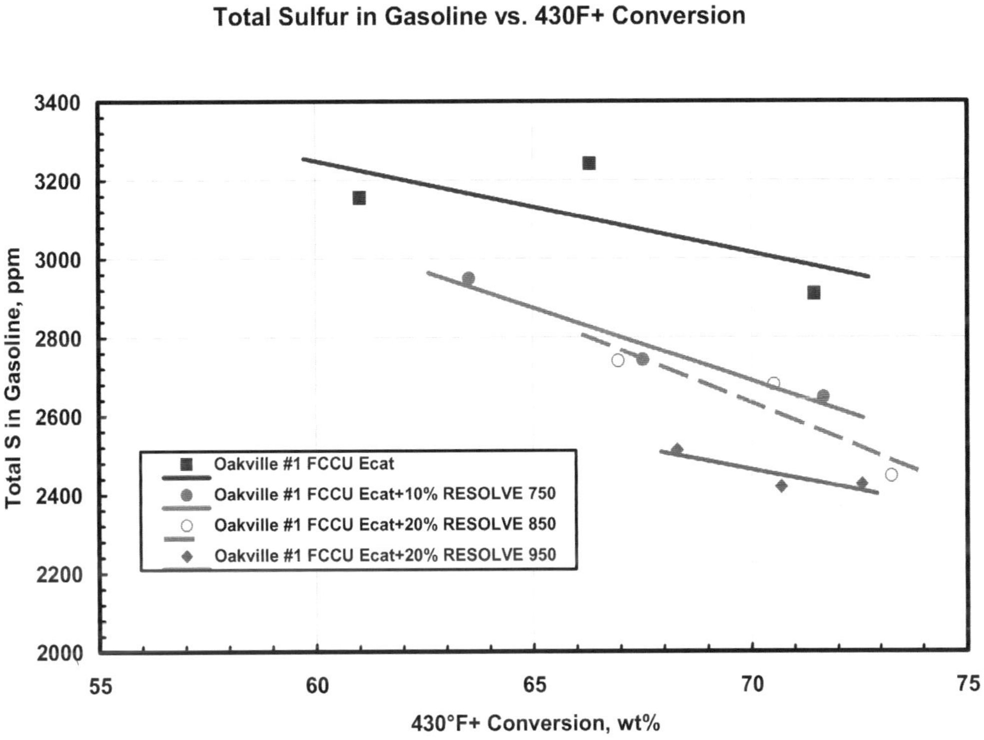

Figure 21: Laboratory Testing of RESOLVE Additives

Figure 22 shows the data from the Oakville #1 FCCU processing asphaltic gasoil. The unit data covers blended feedstocks with aromatic sulfur concentrations ranging from 4.55 to 5.61 wt%. The average base sulfur reduction associated with the HN recycle platform for these feedstocks was 52 wt%. About 34% desulfurization was achieved with 24% RESOLVE 950 on the remaining gasoline sulfur. The net desulfurization achieved in the commercial operation was 70% as indicated by the line showing the combined impact on Figure 22. The effect of incremental RESOLVE 950 in the unit inventory is linear for the range examined in the unit.

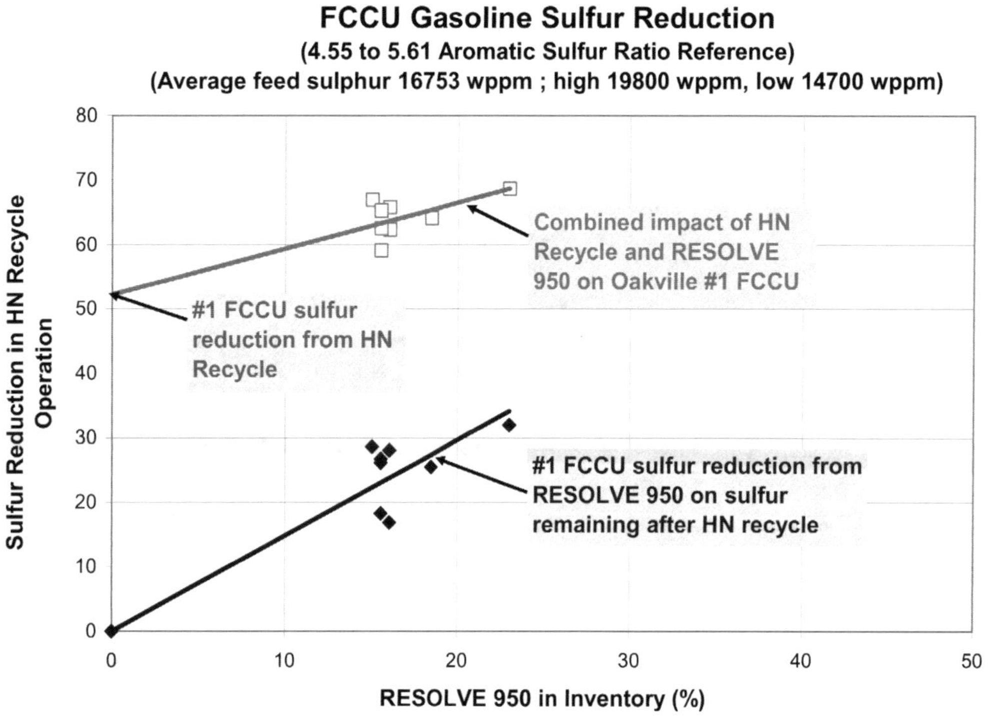

Figure 22: The Effect of RESOLVE 950 on an Asphaltic Feed

Figure 23 shows the data Oakville #1 FCCU processing primarily sweet gasoil. The Figure shows the base desulfurization associated with the HN recycle operation for the 2.1 wt% aromatic sulfur content average feedstock was about 60 wt%. An additional 60 % desulfurization was achieved with 20 % RESOLVE 950 on the remaining gasoline sulfur. The net desulfurization achieved in the commercial operation was 85% as shown by the combined impact line on Figure 23. As in the data set for the asphaltic gasoil, the effect of incremental RESOLVE 950 in the unit inventory is linear for the range examined.

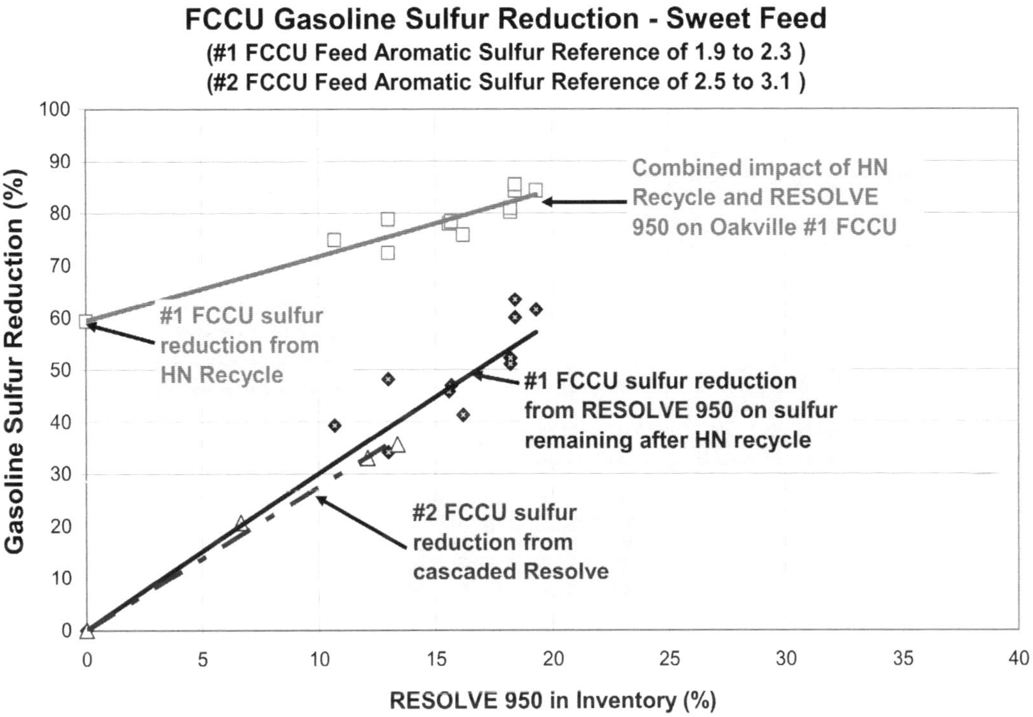

Figure 23: Effect of RESOLVE 950 on a Sweet Gasoil

Figure 23 also shows the data for the Oakville #2 FCCU processing a feedstock mix including VTB. The majority of the RESOLVE 950 in the Oakville #2 FCCU inventory has cascaded from the Oakville #1 FCCU unit. The desulfurization effect in this unit is also relatively linear with the desulfurization efficiency being slightly lower for the 2.7 wt% aromatic sulfur reference feedstock.

Linear extrapolation of the data sets shown in Figures 22 and 23 along with the Albemarle pilot data discussed above were used to construct the desulfurization vs aromatic sulfur content relationships shown in Figure 20. Figure 20 indicates that the incremental gasoline desulfurization over and above that associated with the HN recycle and an octane barrel host catalyst can easily be in excess of 90% for lower aromatic sulfur content feedstocks.

A more detailed look at the Oakville #1 FCCU operation with the system changed out to 35% RESOLVE 950 in the ecat is shown in Figure 24. In this plot, the impact of adjusting the cutpoint for the HN recycle back to the riser is shown. In this operation a small flow roughly equivalent to the net HN production is drawn from the FCCU main fractionator and routed to the Distillate Desulfurization (DDS) unit. The volume of this flow is typically in the range of 10% of the total FCCU gasoline. The 400- F hydrotreated vapor is returned from the DDS stabilization column directly to the FCCU vapor recovery system. The HN recycle significantly reduces the volume of the back end of the gasoline as illustrated in Figures 7 and 8 by adjusting the net carbon distribution from the FCCU. It however does not completely eliminate the total volume of heavy gasoline due to fractionation inefficiencies. Pulling a small volume of heavy gasoline side draw provides a lot of flexibility to numbers shown below. Incremental processing and capital associated with operating with no gasoline in the LCO cut are minimized.

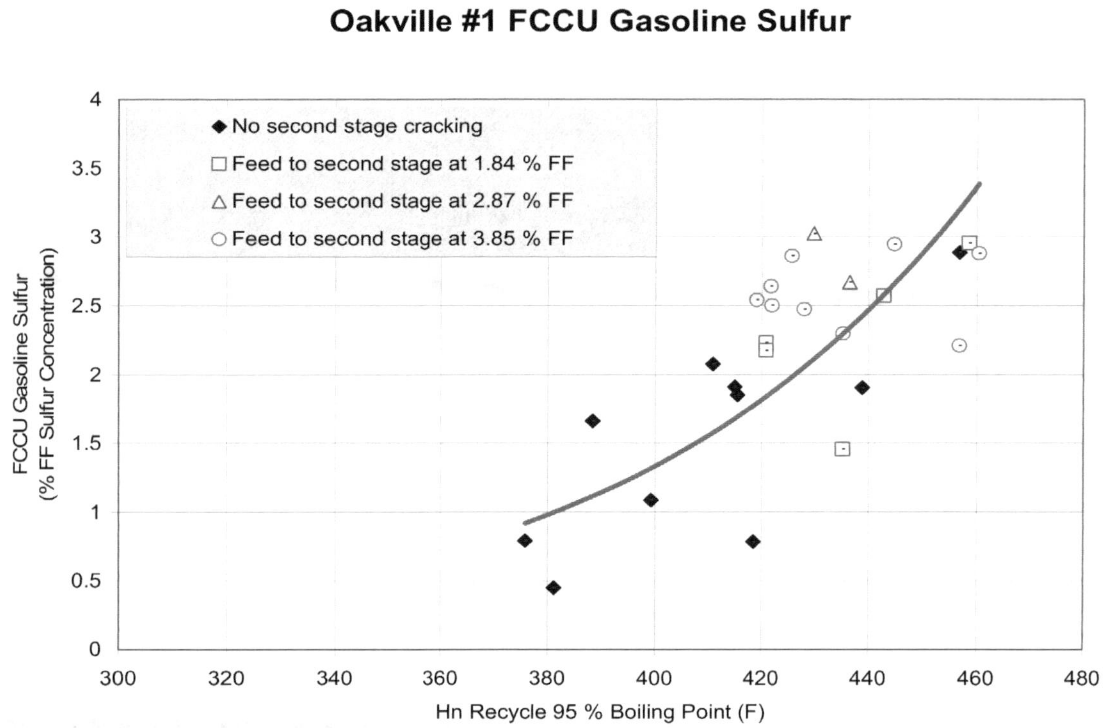

Figure 24: HN recycle Cutpoint Effect On FCCU Gasoline Sulfur

Figure 24 indicates that most of the remaining sulfur is related to benzothiophene in the back end of the gasoline. Even with the low hydrogen transfer associated with the low rare earth octane catalyst and no co-processed hydrogen donor, net gasoline sulfur levels of 1% on feed sulfur are observed. As shown in Figure 10, an increase in hydrogen transfer substantially reduces the sulfur content of the lower boiling gasoline. The utilization of a higher hydrogen transfer catalyst system and /or a hydrogen donor system would further lower the FCCU gasoline sulfur concentrations observed in Figure 24.

The Oakville #1 FCCU was operated between 78 and 80 % 420- F volumetric conversion for the data shown in Figure 24 and the gasoline selectivity averaged 83.5 %. The higher end conversions resulted from converting LCO in the second stage delta coke control operation.

Figure 25 shows the same data as Figure 24, but is expressed in terms of wppm sulfur in gasoline. Sulfur concentrations as low as 30 wppm are achieved with feedstock sulfur concentrations ranging from 0.37 to 0.60 wt%.

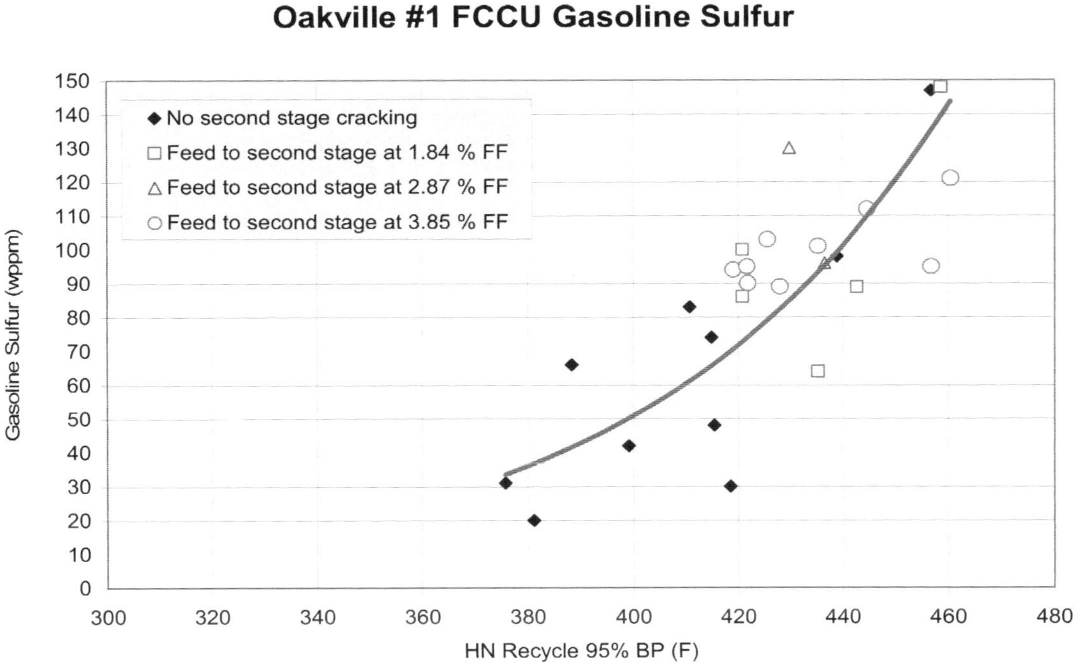

Figure 25: HN recycle Cutpoint Effect On FCCU Gasoline Sulfur

The typical LCO and slurry oil sulfur concentration reductions observed with RESOLVE additives in the Oakville FCCUs have been about 20 and 30 % respectively. This is consistent with the pilot plant predictions. Recycling of the LCO such as in the delta coke management system that will be discussed later results in further sulfur reduction in the LCO.

The integration of RESOLVE is also useful in dealing with refinery and pipeline slop. Slop is routinely charged to the riser in the Oakville #1 FCCU. The naphtha component in the slop typically contains about 1500 wppm sulfur. The sulfur removal with RESOLVE 950 in the system is nominally about 70% but depends upon the sulfur species in the slop.

Commercially, improved bottoms conversion and higher activity were observed using the RESOLVE 950 additive. Analyses of the equilibrium catalyst data along with catalyst addition rates and additive concentration taken from Figures 14 through 19 for the Oakville #1 FCCU during 2004 are given in Table 1. Fresh FCC catalyst addition rates are obtained by multiplying the total addition rate by the additive concentration.

Table 1: Comparison of Equilbrium Catalyst Data at 10 and 35% RESOLVE 950

	12/2003	12/2004
% RESOLVE 950	10	35
Fresh FCC Catalyst Addition Rate (tpd)	3.6	1.7
Fresh FCC Catalyst Zeolite Content (relative)	100	200
Ecat Surface Area (m^2/g)	124	132
Ecat RE (wt%)	0.67	0.59
Ecat V (ppm)	1300	2600
Ecat MAT Activity (wt%)	71.5	71.5

Equilibrium catalyst MAT activity was the same for these periods. While zeolite in the host catalyst was doubled, catalyst additions were essentially halved. In general, increasing zeolite content is not as effective for increasing activity as is increasing catalyst addition rate. At the same time vanadium level has doubled. RE on zeolite for the fresh FCC catalyst used during the 35% RESOLVE 950 period is actually lower than during the 10% period although the equilibrium analyses do not quantitatively reflect this because of transitional effects.

These data would indicate that at a minimum, the RESOLVE 950 caused no dilution effect. There is a strong argument for a significant activity enhancement that compensates for the negative influences described previously. One likely cause of this is preferential trapping of vanadium by RESOLVE 950. Density separation of equilibrium catalysts containing RESOLVE 950 have confirmed this.

Physical retention had been demonstrated to be about the same as the base catalyst by close correspondence of additive marker calculations with addition data and analysis of catalyst fines from the fluegas circuit.

Equilibrium catalyst containing RESOLVE 950 was cascaded to the Petro-Canada Montreal FCCU in November 2004. The sulfur in FCCU gasoline results from the Montreal FCCU are very consistent with the Oakville results. A maximum concentration equivalent to 9% RESOLVE 950 was reached in the Montreal FCCU ecat. A light FCCU gasoline sulfur reduction in the range of 35% was observed. Although the Montreal FCCU has similar process features as outlined for Oakville operations, the results reflect operating with no HN recycle.

The experience with RESOLVE 950 in the Petro-Canada system can be put in context with other usage in refineries around the world as summarized in Figure 26.

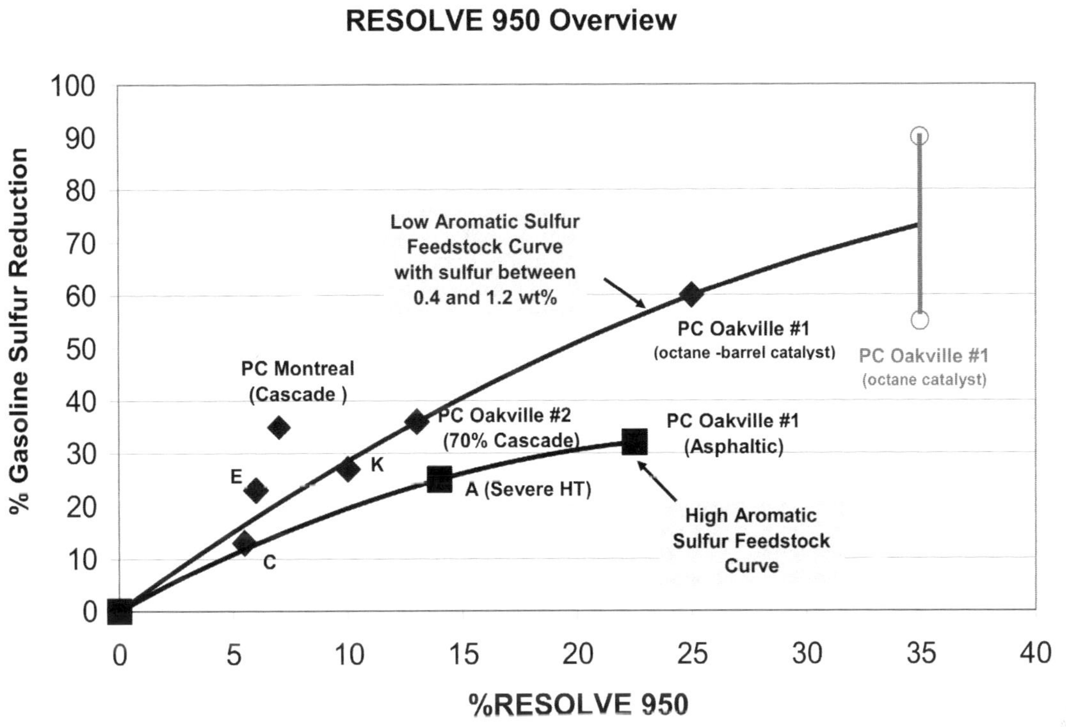

Figure 26: Overview of Worldwide RESOLVE 950 Applications

The general performance picture is quite consistent although there is variation in performance due to cut point and hydrogen transfer environment as reflected by the bar in Figure 26 for 35% RESOLVE operation at Oakville #1 FCCU. The elevated gasoline sulfur reduction shown in the Petro-Canada Montreal FCCU operation is associated with a LCCG cut with a 280 F 95% BP and using an octane barrel catalyst. The steep slopes exhibited by the Oakville #1 FCCU in Figures 24 and 25 and the Montreal data reflect the ability to address the sulfur species boiling below the benzothiophenes in the FCCU gasoline.

The reductions shown in Fig. 26 are due solely to RESOLVE 950 and further improvement can result from using the process options described earlier. It is very interesting to note that the cascaded RESOLVE 950 appears to retain the effectiveness of fresh additive. No negative yield or activity effects have been observed in any of these applications.

In November 2004, #2 Oakville FCCU was shut down permanently as part of the conversion of the Oakville site to a multifunctional terminal.

REGENERATOR EMISSIONS REDUCTION WITH RESOLVE 950

In addition to reduction of sulfur in the liquid products by RESOLVE 950, substantial reductions in regenerator sulfur oxide emissions have been observed in every application where measurements were made. Referring to Figure 26, Refinery K observed a drop in SOx from 400 vppm to 25 vppm. In Refinery A, SOx emissions dropped from 60 vppm to less that 10 vppm on a severely hydrotreated FCCU feedstock. It is interesting to note that Refinery A operates their regenerator in a partial combustion mode which is especially challenging for SOx reduction.

In the Oakville #1 unit, stack testing conducted recently to evaluate the effectiveness of RESOLVE 950 at the 35% level for SOx, NOx and condensables reduction showed consistent total SOx concentrations between 1 and 2 vppm. Relative to historical data tied to decant sulfur concentrations and standard industry correlations[10] the SOx reduction with RESOLVE 950 was 99.6%. No other SOx reduction technology was utilized. During this testing, HFO comprised about 15% of the total fresh feed to the FCCU. These results are summarized in Table 2

Table 2: Summary of RESOLVE 950 SOx Reduction Experience

	RESOLVE Loading (wt%)	SOx without RESOLVE (vppm)	SOx with RESOLVE (vppm)	SOx Reduction (%)
Refinery K	9	400	25	94
Refinery A (partial comb.)	14	60	9	85
Oakville # 1	35	420	1 – 2	99.6

This Oakville #1 FCCU example illustrates an effective way of converting VTB and HCO into energy without SOx emissions. In this case, the HN recycle also serves as the heat balance control and energy transfer mechanism.

Further testing of the particulate fines in the fluegas confirm no emissions issues.

The Petro-Canada Montreal FCCU was benchmarked with 7 % RESOLVE 950 in inventory for fluegas condensables where the fluegas was sampled at 248 and 350 F using variations on the EPA 5 methodology. There was no increase in particulates measured at the 248 F reference conditions relative to the 350 F conditions. This result indicates the effective elimination of the fluegas condensables generated from SO3 in the fluegas. It also indicates that the SOx capture efficiency is retained even when the RESOLVE 950 is cascaded from one unit to another and no fresh is added.

BENCHMARKING FEEDS USING AROMATIC SULFUR SPECIES

The desulfurization efficiency that has been demonstrated for the FCCU correlates very well with the aromatic sulfur content of the feedstock. Virgin crudes have a characteristic pattern for FCCU hydrocarbon precursors that significantly influence the feedstock conversion and product distribution achieved with the FCCU. Virgin crudes also have a characteristic pattern for sulfur and aromatic sulfur content. Low boiling components of the crude are very low in aromatic sulfur content while high boiling crude components are high in aromatic sulfur content. More aromatic asphaltic crudes have more aromatic sulfur than paraffinic crudes at a given boiling point in the crude. Hydroprocessed feedstocks exhibit aromatic sulfur levels that correlate with the gasoline precursors in the feedstock and an adjustment for the specific aspects of the hydroprocessing step.

Conventional Crude

Figure 27 shows the typical relationship for sulfur in virgin crude relative to boiling point for a paraffinic and an asphaltic crude benchmarks. The sulfur level of the asphaltic crude increases at a much faster rate than the sulfur in the sweet paraffinic crude.

Figure 28 shows that the benchmark crudes exhibit a similar pattern for the aromatic sulfur content relative to boiling point.

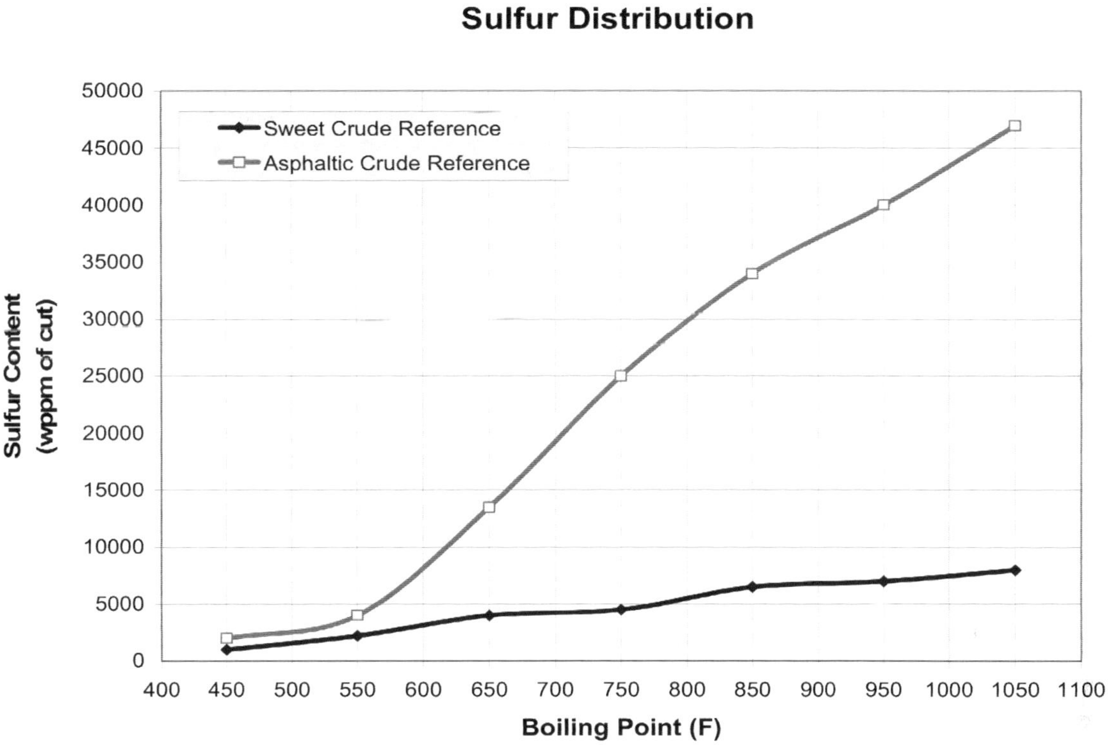

Figure 27: Sulfur Distribution in Virgin Crude

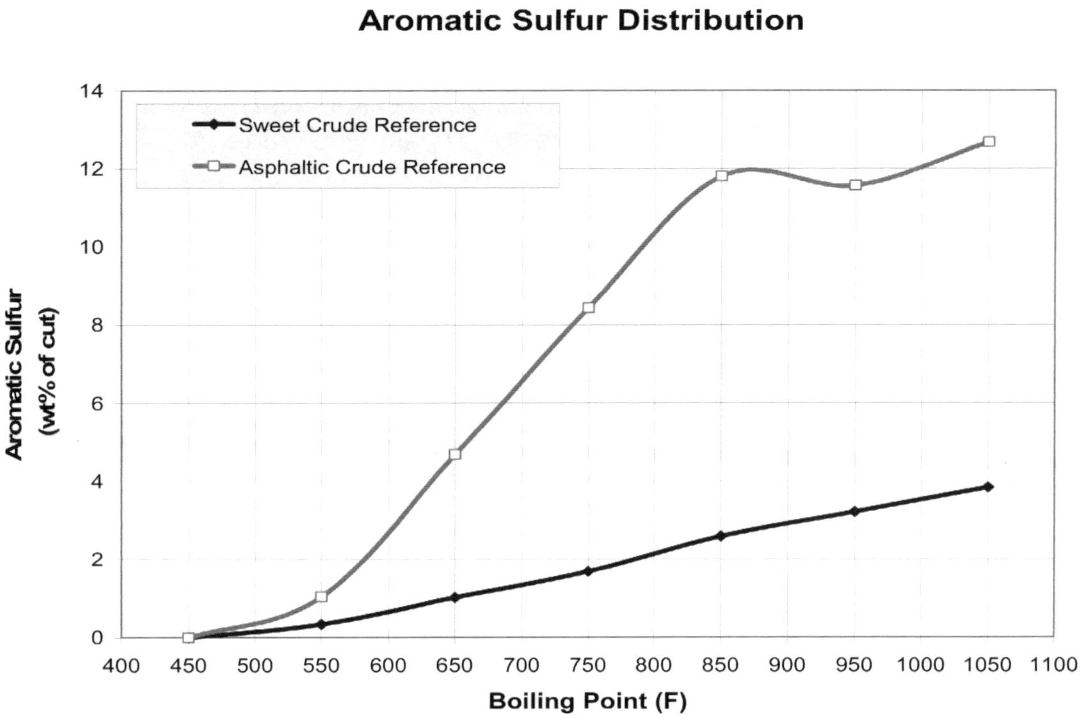

Figure 28: Aromatic Sulfur Distributions in Virgin Crudes

Table 3 shows the range of typical feed qualities processed by the two FCCUs in the Oakville refinery. The feedstock precursors are defined by mass spec molecular types. The gasoline precursors are calculated as the sum of the paraffins , cycloparaffins and monoaromatics in the feedstock. The two FCCUs tend to run at 430°F- conversion levels several percent higher than the gasoline precursor level in the feedstock with the unit 430°F- conversion increasing slightly as the average carbon number of the feed is decreased.

Table 3: Conventional FCCU Feedstock Properties at Oakville Refinery

	Sweet Gasoil	Sweet VTB	Asphaltic Gasoil PG 64-22	Asphaltic Gasoil Flux
Precursors (wt%)				
Gasoline	81.2	42.5	70.7	75.0
LCO	10.2	12.8	16.3	15.0
Coke and Slurry	8.6	44.7	13.0	10.0
Average Carbon Number	25.8	47.1	23.0	21.4
Vol % boiling below 650°C	25.7	0	31.6	40.7
Aromatic Sulfur (wt%)	1.7	6.5	6.1	4.8
Sulfur (wppm)	4425	15800	22900	18300

The asphaltic gasoils contain a large component of 650°F- crude and have aromatic sulfurs in the range of that contained in the benchmark sweet crude VTB. The sweet gasoil has relatively low aromatic sulfur content at about 1.7 wt%. At the 150 wppm average gasoline pool specification, a large amount of asphaltic gasoil and sweet VTB can be processed.

Table 4 illustrates the result of blending 50/50 distillate and 650°F- gasoil from the benchmark sweet crude. Given an aromatic sulfur content of 1.2 wt%, and a net feed sulfur of 3592 wppm, a 50 wppm FCCU gasoline could be generated by dropping the gasoline sulfur to 1.4 % of the feed sulfur. Based on the above desulfurization efficiencies, this could be accomplished with a 67% desulfurization efficiency from the RESOLVE 950 using the above configuration. This would require about 20 wt% RESOLVE 950 in the ecat when an octane barrel catalyst is used. Incremental amounts of RESOLVE 950 allow for processing feedstock mixes with higher sulfur and aromatic sulfur content.

Table 4: Potential FCCU Feedstock Blend

	Sweet Gasoil	Sweet Distillate	Blend to 50% 650°F-
Precursors (wt%)			
Gasoline	81.2	93.6	85.3
LCO	10.2	6.0	8.8
Coke and Slurry	8.6	0.4	5.9
Average Carbon Number	25.8	15.8	22.5
Vol % boiling below 650 F	25.7	100	50.0
Aromatic Sulfur Ratio	1.7	0.3	1.2
Sulfur (wppm)	4425	1886	3592
Density	.89	0.83	0.87

Figure 29 shows the Oakville #1 gasoline desulfurization performance expressed as absolute sulfur in the full range gasoline. The low end of the data set for operation with low aromatic sulfur feeds reflects about 20 wt% RESOLVE 950 in the ecat and an octane barrel host catalyst.

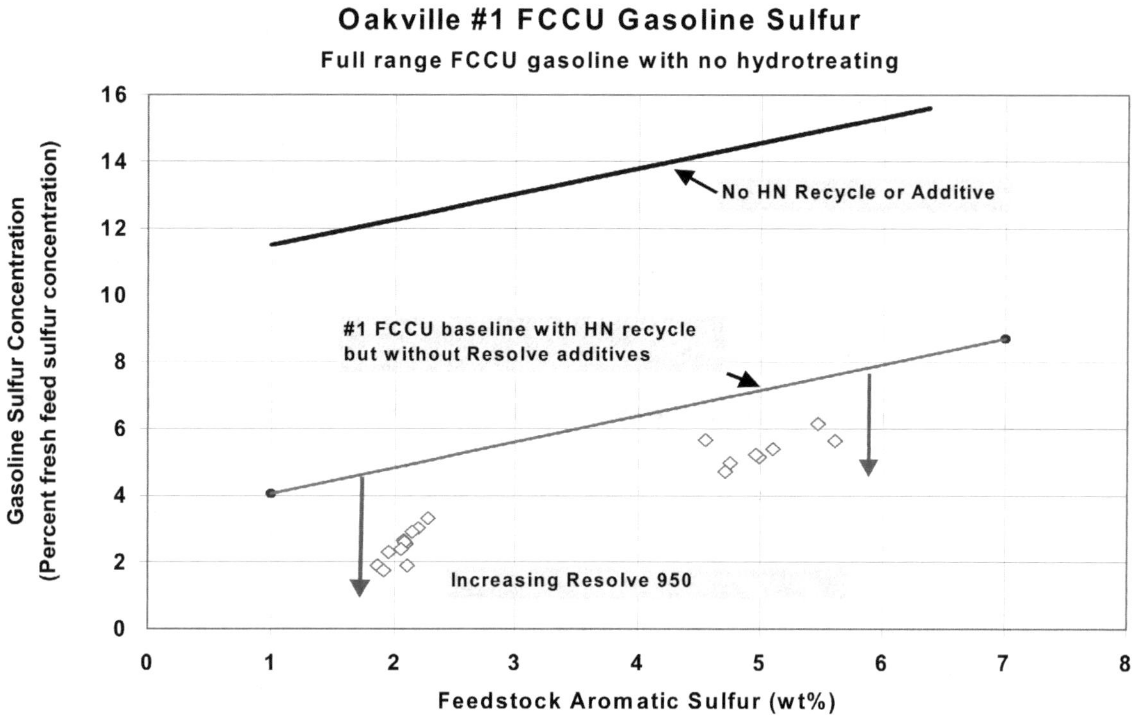

Figure 29: Gasoline Sulfur Reduction at Oakville #1 FCCU

Synthetic Crude

Table 5 compares synthetic crude components derived from tar sand and available from Syncrude in Fort McMurray to distillates from conventional crudes. The hydrotreated synthetic crude is low in both sulfur and aromatic sulfur. Similar to the blend of 50/50 sweet conventional crude distillate and gasoil discussed previously, yield similar to light sweet gasoil operation could be achieved. About 13 % RESOLVE 950 in ecat would be required to generate a 50 wppm sulfur content FCCU gasoline from this feedstock with the integrated system.

Table 5: Non Conventional FCCU Feedstock Properties

	Syncrude 392˚F+	Syncrude Gasoil 675˚F+	Syncrude distillate 392-675˚F	Sweet distillate 400-650˚F	Asphaltic distillate 400-650˚F
Precursors (wt%)					
Gasoline	80.2	70.4	90.9	93.6	89.3
LCO	12.1	17.6	6.1	6.0	9.0
Coke and Slurry	7. 7	12.0	3.0	0.4	1.7
Aromatic Sulfur (wt%)	1.4	2.1	0.6	0.3	1.1
Sulfur (wppm)	1700	2700	500	1886	5127
Carbon Number	21.1	28.9	15.1	15.8	15.3
Density	0.914	0.932	0.895	0.834	0.869

Hydrotreated Feedstocks

Petro-Canada has constructed a 1900 psig FCCU feed hydrotreater (CFHTU) at its Edmonton refinery as part of a program to shift the refinery conventional crude train from conventional light crude to tar sands based feedstocks. The addition of the CFHTU was driven by the depletion of sweet crude in this land locked area as well as the huge strategic potential in tapping the local tar sands. A second crude train is dedicated to processing bottomless synthetic crude from Fort McMurray and the neat gasoil is processed in a 2700 psig hydrocracker.

The Edmonton CFHTU is a 1900 psi unit designed to severely hydrotreat feedstocks derived from the Canadian tar sands. This severity is well in excess of conventional systems with total operating pressures less than 1400 psig used to process poor quality conventional feedstocks. The full range gasoil feedstock derived from the tar sands is expected to have a specific gravity of about 0.96 and sulfur and nitrogen contents in the 22000 wppm and 3100 wppm range respectively.

Petro-Canada's in-house pilot studies had indicated severe heat balance issues associated with processing product from this type of CFHTU operating at high desulfurization efficiencies. Albemarle and Petro-Canada conducted a study in which various severely hydrotreated feeds were cracked using an ecat containing 10% RESOLVE 750 generated in the Edmonton FCCU. This study provided the sulfur distribution in the products and relative coke yields for the present sweet gasoil, the hydrotreated conventional feedstock, the proposed tars sands based feedstock, and the bottoms of the Edmonton refinery 2700 psig syncrude gasoil cracker. The catalyst system is very similar to the low rare earth Topaz systems utilized in the other three Petro-Canada FCCUs with a base of 10% RESOLVE 750 to position the catalyst to take advantage of the lessons learned on the non hydrotreated feedstocks discussed earlier.

Table 6: Base and Hydrotreated Feed and Product Property Inspections

	Sweet Gas Oil	HT #1	HT #2	HT#3	Tar Sands HT	Syncrude HC Bottoms
% HT	0	90.7	92.9	97.3	99.1	99.5 +
Density	.896	0.891	0.891	0.891	0.909	0.895
ACE Conversion (wt%)	78.3	86.6	86.3	87.9	82.4	86.1
Precursors (wt%)						
Gasoline	79.5	85.9	85.7	87.0	85.2	90.4
LCO	9.7	7.12	7.12	6.61	7.27	5.02
Coke and Slurry	10.8	6.96	7.13	6.41	7.52	4.59
wt % boiling below 650°F	24.6	0	0	0	0	0
Aromatic Sulfur (wt%)	2.78	0.69	0.65	0.73	1.80	1.01
Benzothiophenes	1.08	0.09	0.03	0.06	0.54	0.18
Dibenzothiophenes	1.38	0.51	0.62	0.62	1.18	0.80
Tribenzothiophenes	0.32	0.09	0.00	0.06	0.08	0.02
Sulfur (wppm)	7182	743	561	215	200	65
CFHTU LHSV	NA	1.5	0.8	0.8	.50	1.0
Gasoline Sulfur (wppm)	698	26	26	19	8	16
Light Cut (183°F-)	44.7	6.0	3.5	3.0	1.5	1.5
Mid Cut (183-350° F)	369.1	13.7	16.9	13.3	5.6	11.4
Thiophene	54	6.5	6.5	6.0	2.5	6.0
Thiophene (% of Mid Cut)	15	48	39	45	44	53
Heavy Cut (350-430°F)	284.6	6	6	3	1.5	1.5
Benzothiophene	208.6	6	6	3	1.5	1.5
Benzothiophene (% Heavy Cut)	73.3	100	100	100	100	100
Gasoline Sulfur (% feed sulfur)	9.7	3.5	4.8	8.9	4.0	24.0
LCO S (430-650°F) (wppm)	21977	2607	2386	1008	425	232
LCO S (% feed sulfur)	306	351	425	469	213	357
HCO S (650°F+ Cut) (wppm)	13997	2338	2500	1041	100	100
HCO S (% feed sulfur)	195	315	446	484	50	154

Table 6 shows the relevant feedstock and property data. The base feedstock is a representative virgin crude gasoil mix containing about 24.6 volume percent material boiling below 650°F. Hydrotreated gasoils #1 through 3 represent three levels of hydrotreating of the base gasoil using variations in LHSV and operating temperature. All hydrotreated gasoils are cut at 650°F. The desulfurization of the 650°F plus conventional gasoil ranged from 90.7 to 97.3 percent and the hydrotreated feed sulfur ranged from 743 to 215 wppm.

Full range gasoline sulfur ranged from 26 to 19 wppm with most of the sulfur in the 183 to 350°F mid cut. The percentage of the feed sulfur routed to the gasoline increased with increased feedstock desulfurization in the CFHTU pilot plant.

The net desulfurization efficiency for the two tar sands sourced gasoils is over 99 %. The gasoline sulfur for the 2700 psi hydrocracker bottoms is 24% of feed sulfur . The cycle oil sulfur concentration is higher relative to the base sweet gasoil feedstock in all the cases except the low LHSV 1900 psi tar sands operation. The elevated thiophenes and the reduced benzothiophenes and mercaptans in all the hydrotreated cases suggest the sulfur formed is undergoing recombination reactions with the olefins and generating the majority of the thiophenes and alkylthiophenes. The cracking studies for all the feedstocks indicate that the thiophene concentration in the gasoline increases with conversion.

The data suggests the minimum sulfur level that can be achieved by increasing the feed desulfurization level will be limited until the cycle oil sulfur levels are reduced sufficiently. Alternatively, to achieve very low gasoline sulfur levels, the FCCU would have to be set-up to inhibit the olefin and H2S recombination reaction.

The observation on the hydrotreated feedstocks shown in Table 6 is similar to the H2S injection and hydrogen donor examples for conventional crude processing discussed earlier. Table 6 also shows that even with severe hydroprocessing of the FCCU feedstocks, the sulfur level in the LCO contains hard to treat sulfur species and is too high in sulfur content to blend to a 15 wppm sulfur content product.

The chemical hydrogen uptake is expected to be 70% higher to achieve 98% desulfurization vs 90% for the conventional gasoil feedstock. As can be seen in the data using the mechanism postulated, the complementary approach of improving this bulk desulfurization in the FCCU may be very beneficial.

Typical operation of the Edmonton FCCU since the start-up of the CFHTU confirmed the pilot work. The average feedstock blend sulfur has been about 571 wppm similar to the #2 HT case in Table 6. The FCCU operates with a 430- conversion close to 90% with gasoline and LCO sulfur concentrations averaging 25 and 2231 wppm respectively. The average as cut gasoline final boiling point was 402F with an IBP on the LCO of 416F. The above results were achieved with a blended feedstock comprised of CFHTU product, syncrude hydrocracker bottoms and straight run syncrude gasoil. The FCCU catalyst has a higher rare earth on zeolite content than the pilot plant catalyst and does not contain any RESOLVE additive due to the very low gasoline sulfur contents achieved. Most of the 6 wppm benzothiophene shown in Table 5 would be expected to be in the LCO cut at these typical operating conditions.

The sulfur generated from the operation as SOx continues at about 10% relative to the feedstock sulfur content. This represents a significant reduction in the absolute quantity of SOx relative to the base operation prior to the CFHTU. However, SOx levels are still substantially higher than that achieved at Oakville without feedstock pretreatment and utilizing the additive technology.

As shown by the severely hydrotreated operation represented by A on Figure 27, the additive technology has an impact on gasoline sulfur content for these hydrotreated feedstocks. However, the gasoline sulfur levels are already very low and there is only a small amount of sulfur left in the product. The application of the sulfur control system may be more appropriately focused on optimizing the degree of feed pretreatment and further reducing SOx emissions from these systems.

Tar Sands

Figure 30 shows the relationship between the gasoline precursors and aromatic sulfur concentration in tar sands derived materials. The amount of aromatic sulfur component in the gasoil component and light VTB is a little higher than conventional asphaltic crude. Processing bitumen VTB in the slurry phase hydrocracker increases the gasoline precursors and produces a gasoil with similar aromatic sulfur relative to gasoline precursors that are observed in the bitumen. Subsequent hydroprocessing at various severities and pressures further reduces the aromatic sulfur.

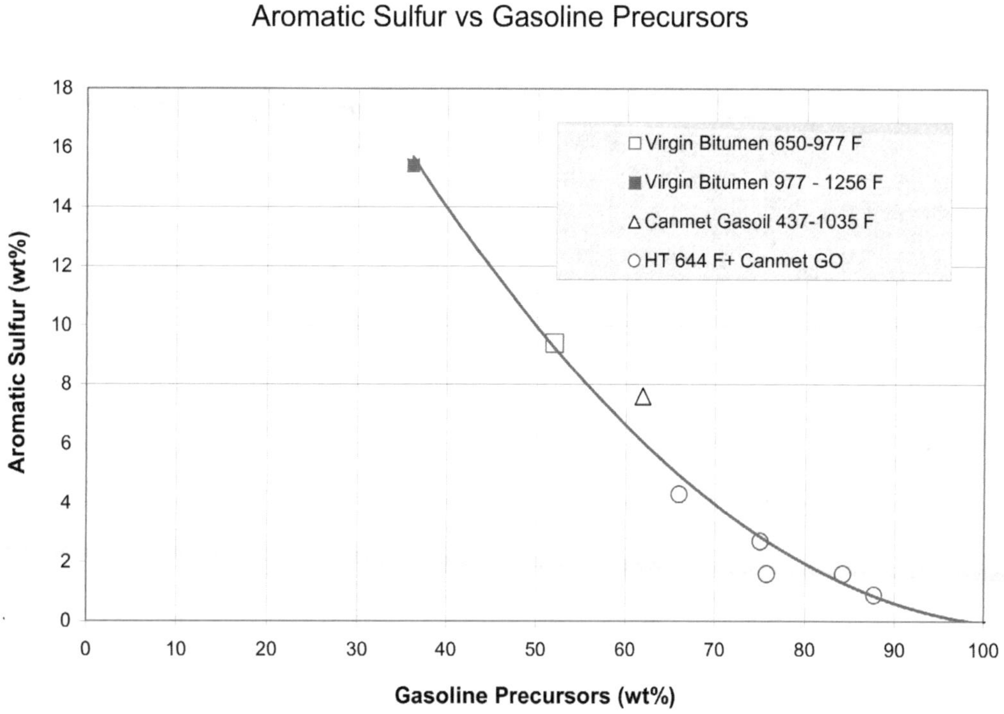

Figure 30: Aromatic Sulfur vs. Gasoline Precursors

In slurry phase upgrading of tar sands, in excess of 30% of the bitumen VTB can be converted into high hydrogen content naphtha with sulfur contents in excess of 5000 wppm[11]. Recombining the gasoil and naphtha from the slurry phase hydrocracker produces a feedstock with hydrocarbon precursors similar to the sweet conventional gasoil shown in Table 3. This blended feedstock becomes a suitable feed for the FCCU in the context of the carbon distribution equilibrium and hydrogen management concepts presented earlier. This bulk hydroprocessing protects and simplifies the downstream treatment and processing.

FCCU COKE YIELD

The feed quality impact on coke yield in an FCCU can be expressed by the following equation:

Coke on catalyst = A * time ^ N

where A = feedstock coking index
N = contact time factor

Figure 31 shows the correlation between the coke and slurry precursors in the feed and the relative coking index achieved with an MAT reaction system. Adjusting the feedstock to the FCCU to generate a very low aromatic sulfur feedstock results in a substantial reduction in the feedstock coking index.

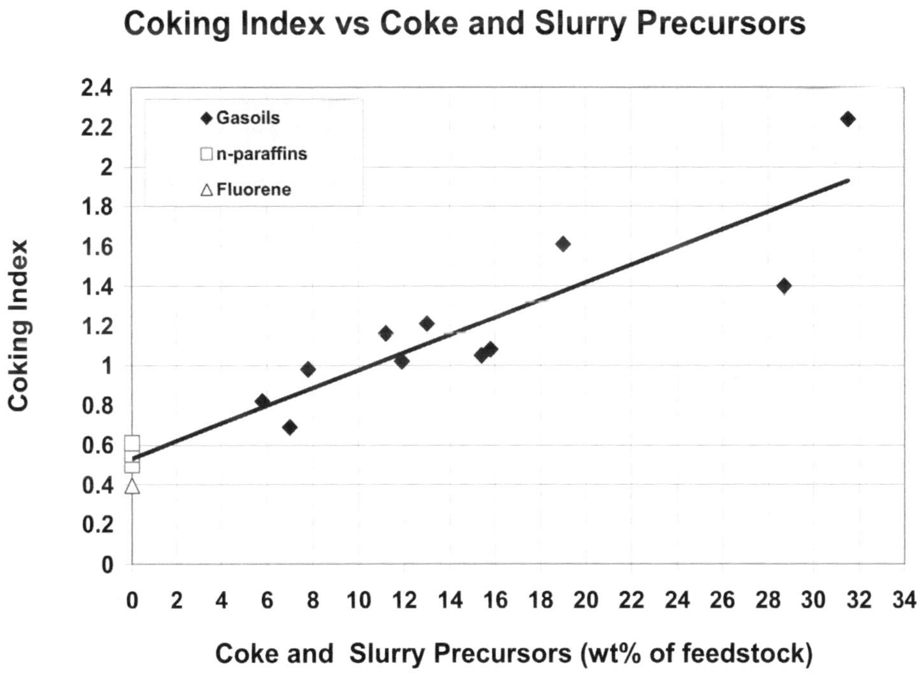

Figure 31: Coking Index vs. Coke and Slurry Precursors

Figure 32 indicates that as the hydrotreating severity is increased, the quantity of coke and slurry precursors is reduced for all operations examined. At desulfurization levels above 98%, there is a rapid drop off in the coke and slurry precursors for all feedstocks. This rapid drop off can result in both steady state heat balance issues as well as instability issues.

The low coking index of the low aromatic sulfur content feedstocks derived from virgin crudes or through high pressure hydrotreating of very poor feedstocks presents a significant problem for the FCCU heat balance. The coking index for these feedstocks could be a fraction of what is required to support the unit heat balance. Driving to very low sulfur concentrations in the CFHTU gasoil to facilitate the production of low sulfur distillate can create issues.

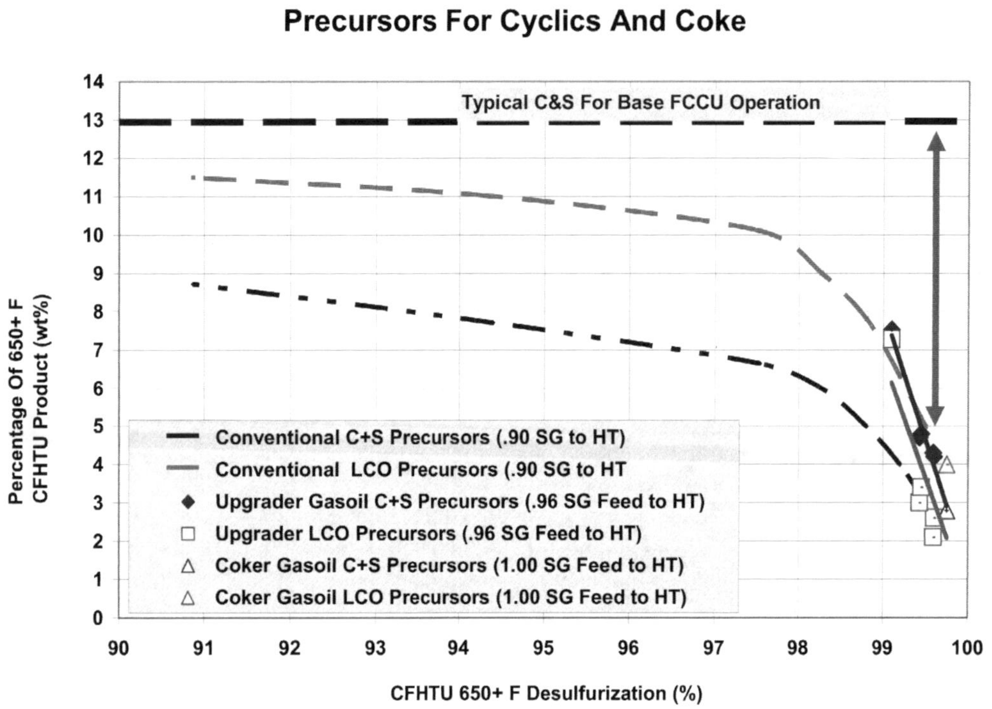

Figure 32: Precursors for Cyclics and Coke

Figure 33 indicates that the migration to low aromatic sulfur feedstocks increases the gasoline precursors in the FCCU feedstock. At a given severity, the FCCU has to operate at higher conversion levels. Independent of feedstock source, the FCCU will produce very high conversion levels at high desulphurization rates. This could have a significant impact on downstream processing capability.

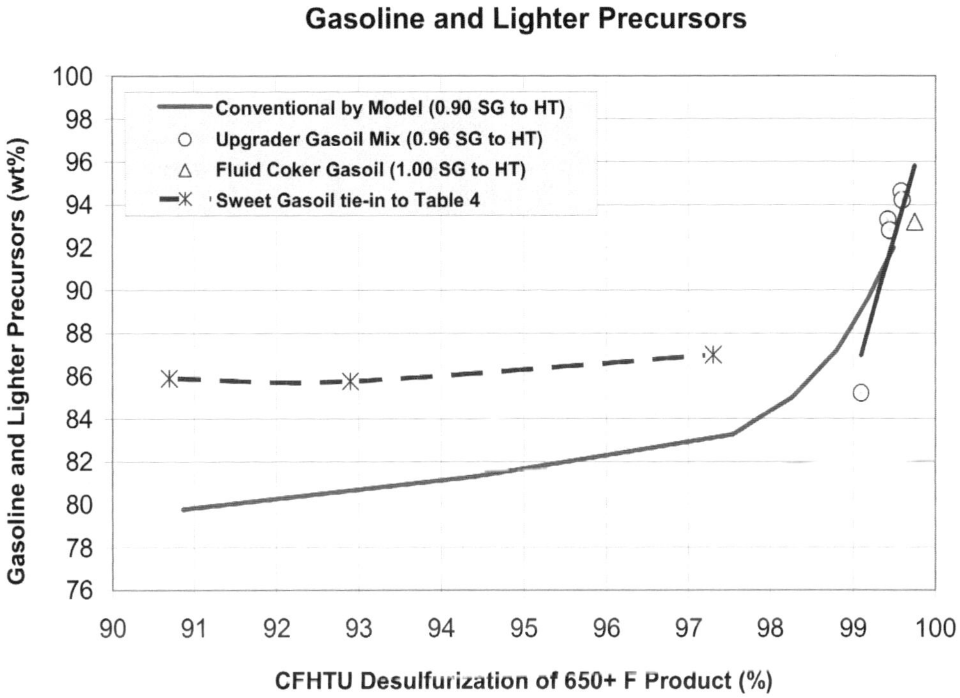

Figure 33: Gasoline and Lighter Precursors as a Function of Desulfurization

FCCU DESIGN CHANGES

The Petro-Canada FCCUs employ a proprietary Riser Termination Device (RTD) developed by Petro-Canada[12] and licensed by Shaw Stone and Webster which results in efficient disengaging of catalyst and product vapors. Non-selective post-riser reactions are minimized resulting in low gas make and delta coke. The RTD system has an integrated degassing system to minimize the amount of hydrocarbon reaching the stripper. The unit coke balances typically have been maintained by blending the base feedstocks with coker slurry, HFO or VTB to adjust coke and slurry precursor levels of between 10 and 16 wt%. The units typically operate with delta cokes of about 0.6 wt % resulting in cat/oils in the 8 to 10 range.

In order to process severely hydrotreated feedstocks or very low aromatic sulfur feedstocks adjustments have to be made to the FCCU processing scheme. Coke and slurry precursors levels are insufficient to generate a comfortable heat balance with some of these low sulfur feedstocks.

The approach taken by Petro-Canada was to build on the advantages of the efficient riser disengager system to rapidly separate riser products. The condensed aromatics produced by the FCCU cracking process are recovered from the fractionation system and injected into the stripper to generate coke to adjust the unit heat balance. This second stage cracking system is added below the RTD in the top of the conventional stripper. The introduction of the LCO into the long contact, high cat to oil, dense bed cracking system is targeted to convert the majority of the low hydrogen LCO stream into coke. The high cat/oil ratio (in the range of 100) combined with very low levels of coke on the catalyst entering the dense bed contacting zone should enhance the reduction of sulfur by the RESOLVE additive for the non-coked vapors generated from the LCO and routed to product recovery. This process was commissioned in May 2004 in the Edmonton refinery to prepare for commissioning the CFHTU in June 2004.

The delta coke control process was also commissioned in the Oakville #1 FCCU in November 2004. Figures 24 and 25 show that there is negligible effect on gasoline sulfur when charging low hydrogen content and higher sulfur feedstocks to a point after the RTD. By removing the dependency of the FCCU on coke generated from feedstock contact in the riser, the application of FCCU process is broadened to encompass a wide range of feedstocks.

The changes provide many processing advantages[13]:

1) Independent heat balance control for a fuel deficient system.
2) Lower delta coke in the riser providing more selective catalytic processing at higher catalyst activity
3) Rapid separation of the olefin and H2S at the end of the riser which reduces sulfur recombination reactions
4) Utilization of the low hydrogen content product for fuel and providing sufficient time for the polyaromatic coke to be formed from the light cycle oils.
5) Partitioning of the olefin exiting the riser from the sulfur contained in the fuel charged to the stripper to minimize sulfur recombination reactions
6) Ability to process higher sulfur content feedstocks and process higher aromatic sulfur feedstocks
7) Co-processing of low carbon number feedstocks for improved net carbon distribution, heteroatom removal and hydrogen management
8) Bulk processing of a wider range of feedstocks in the FCCU and the associated elimination of the complexity and inefficiency of additional processing steps
9) Segregation of feedstock based on aromatic sulfur content
10) Direct disposal of low quality, high aromatic sulfur feedstocks such as coker slurry in the second stage system

The integration of recycle streams from the main fractionator allows for:

1) Tailored carbon distribution product and flexibility in hydrogen production within the refinery.
2) Isolation of low hydrogen content aromatics produced in first pass cracking so that they can be exposed to severe cracking at long residence time and very high cat/oil.
3) Sulfur and nitrogen removal as polar compounds are preferentially converted to coke
4) The previous items enhance the sulfur reduction efficiency of the RESOLVE technology
5) Energy efficiency

CONCLUSIONS

The technology approaches discussed in this paper offer the potential to integrate the variations into the FCCU process that will allow refineries to more efficiently produce ultraclean fuels and chemicals. Distinctions with a hydrocracking approach become increasingly blurred.

Utilizing a combination of carbon rejection, carbon distribution tailoring, hydrogen transfer and significantly improved heteroatom removal, there is the potential to simplify the processing scheme, improve the refinery energy efficiency and optimize the hydrogen balance.

Hydrogen management continues to be a major issue. Elimination of the FCCU heavy gasoline and cycle oils reduces the need for subsequent processing, hydrogen and energy utilization. Elimination of the FCCU 330°F+ naphtha also offers benefits in terms of providing an improved 100+ N+2A naphtha reformer feedstock and allows for flexibility to increase the crude unit naphtha cut point to generate more hydrogen.

Efficient desulfurization of low aromatic sulfur containing feedstocks within the FCCU reduces or eliminates the need for gasoline post treatment with conventional processes and positions the product for simple low energy utilization final clean-up approaches. Octane losses associated with post treat options are also eliminated. Alternatively pretreatment severities can be reduced with the associated lower hydrogen and capital requirements.

The entire RESOLVE line has proven effective in Petro-Canada's commercial experience when used in conjunction with the processing platform described in this paper. RESOLVE 950 achieved high levels of sulfur reduction for low aromatic sulfur content feedstocks in addition to providing cracking activity and yield benefits. The development program allowed Petro-Canada to cost effectively manage the 150 wppm gasoline phase in period at the Oakville refinery and better position Petro-Canada for sulfur in fuels management in the future.

RESOLVE 950 also proved to be particularly effective for the reduction of SOx and condensables in the regenerator flue gas, in the case of Oakville #1 FCCU virtually eliminating sulfur emissions.

The ability to readily crack distillates into low sulfur gasolines and subsequently separate out the low hydrogen content aromatics provides a route to phase into the hydrogen fuel cell market. As the fuel cell market evolves the process provides improvement in the quality of the gasolines generated during this interim period. The FCCU based North American refining system would be able to utilize this approach to minimize the costs and changes required to meet the future market needs. This approach will also facilitate the integration of increased volume of feedstocks derived from tar sands into the refining system.

---

[1] Benham, K., Bridle, J., Kuehler, C., and Zelonis, R., "Integrating RESOLVE Desulfurization Technology with Novel Process Concepts in Commercial FCCU Operations", presented at the Akzo Nobel Scope 2004 Symposium, June 2004

[2] U.S Patent, 5,755,955 "Hydrocracking of heavy hydrocarbon oils with conversion facilitated by control of polar aromatics", Benham, K. et al, May 26 1998

[3] Benham, K., Letzsch, W. ,"Riser Termination Developments at Petro-Canada", S&W 4th FCC Forum, May 2000

[4] Corma, A., Martinez, C., Ketley, G., Blair, G., Applied Catalysis A: General 208, 135-152 (2001)

[5] Humphries, A., "Catalytic Strategies For Controlling Sulfur In Gasoline", Akzo Nobel Catalyst Seminar, Toronto, 1994

[6] Benham, K. ,"Riser Termination Developments at Petro-Canada", S&W 5th FCC Forum, May 2002

[7] Keyworth, D., "Controlling Octane through FCC Catalyst Selection", Ketjen Catalyst Seminar, Calgary, July 15, 1988

[8] Lappas, A., Valla, J., Vasalos, I., Kuehler, C., Francis, J., O'Connor, P., Gudde, N., Applied Catalysis A: General 262, 31-41 (2004)

[9] Humphries, A., Kuehler, C., 'Meeting Clean Fuels Objectives with the FCC', AM-03-57, NPRA Annual Meeting, San Antonio, TX, 2003

[10] Huling, G., McKinney, J. and Readal, T., Oil & Gas Journal, 73-79 May 19 1975

[11] Benham, K., Pruden, B. ,"Canmet Residium Hydrocracking: Advances Through Control Of Polar Aromatics,", 1996 NPRA Annual Meeting, March 1996

[12] U.S Patent, 6,692,552 "Riser termination device", Benham, K., Feb 17, 2004

[13] U.S Provisonal Patent, "Integrated Desulfurization And FCCU Process", Benham, K., June, 2004

Paper 63d – 8th International Conference on Refinery Processing
American Institute of Chemical Engineers, April 12, 2005

Intra-particle Mass Transfer and Contact Time Effects in Fluidized Catalytic Cracking

David M. Stockwell
Engelhard Corporation Research and Development, 101 Wood Avenue, Iselin, NJ 08830

Certain yield changes in FCC due to changes in catalyst contact time, reactor configuration, and/or catalyst properties are consistent with the hypothesis that mass transfer in the pores external to the zeolite is additionally limiting the catalysis. In this paper we attempt to validate this hypothesis by evaluating catalyst effectiveness factors and their parametric sensitivity to catalyst contact time. The rate constants required to obtain a given conversion vary as the reciprocal of catalyst contact-time, so that the potential importance of mass transfer increases for processes with shorter contact times. If molar expansion or the complexities of catalyst age distribution and deactivation by coking are considered, conventional gas oil properties and catalyst pore volumes lead to the prediction that vapor-phase gas-oil cracking is either diffusion-limited at the riser entrance, or very nearly so, for contact times and conversions typical of many of today's operations. This is consistent with commercial process data. Calculations for fixed bed testing (MAT) predict that diffusion limitations will not occur at the most common catalyst contact times however. Experimental data agree directionally but appear to be more sensitive to contact time and less sensitive to pore volume than implied by the models.

KEYWORDS: FCC, catalytic cracking, contact time, effectiveness factor, Thiele modulus.

Introduction

In the recent evolution of the FCC process, vapor and catalyst contact time as well as catalyst porosity have received much attention in their respective domains of the art and literature. Short (vapor) contact time (SCT) designs providing rapid catalyst-vapor disengagement have reduced post-riser vapor residence time at high temperatures and eliminated most of the post-riser cracking, reducing dry gas and increasing gasoline yields.[1] More conversion must then take place within the riser. There are also now process designs[2,3,4,5,6] with catalyst-oil contact times less than one second. Higher riser outlet temperatures and equilibrium catalyst activities are employed by these processes and imply a higher rate of cracking is taking place.

Some have consistently asserted that catalyst porosity is important for maximizing yields of primary cracking products.[7,8] Mitchell et al.[9] showed convincing pilot plant results for an aromatic resid feed, but lower boiling gas oil feeds are most common. Engelhard has presented data showing that increasing macroporosity can increase selectivity to primary cracking products in FFB testing with gas oil,[10,11] and commercial FCC results confirm this result. Lopez-Isunza et al.[12] found that gasoline yields at 773 K were better for catalysts screened to smaller particle sizes.

The literature appears to be lacking however an assessment of FCC effectiveness factors for the non-zeolitic pore volume. Effectiveness factors for model compound cracking in the zeolite micropores have been determined.[13] We employ the simplest and most readily understood reaction engineering models in order to illustrate the role of catalyst contact time, correlate reactor performance and estimate the conditions necessary for the onset of diffusion limitations. If present, such limitations will reduce the yields of the primary cracking products.

Theoretical Development

Kinetics. Weekman and Nace[14] showed that an effective second order kinetic model of lumped feed and product groups could fit zeolitic cracking catalyst data well; molar expansion was neglected. For the sake of generality, we retain the expansion coefficient ε_A, which accounts for the molar expansion of gas during reaction.[15] The specific rate of reaction in terms of these parameters and the conversion X_A is then expressed as

$$r'_A = kC_{A0}^2 \frac{(1-X_A)^2}{(1+\varepsilon_A X_A)^2} \tag{1}$$

Batch reactor. Since the catalyst and vapor travel at nearly the same velocity in a riser, we take the frame of reference of the moving catalyst and approximate the riser as a batch reactor having a non-deactivating catalyst and the same batch time as the riser catalyst contact time. Equation (2) describes the reactor performance, where W represents the weight of the catalyst, N_{A0} the number of moles of feed in the batch, and τ_{Cat} the reaction or catalyst contact time.

$$\frac{Wk_2' C_{A0}^2 \tau_{Cat}}{N_{AO}} = F_{Batch}(X_A, \varepsilon_A) \tag{2}$$

$$F_{Batch}(X_A, \varepsilon_A) = \frac{(1+\varepsilon_A)X_A}{1-X_A} + \varepsilon_A \ln(1-X_A) \tag{3}$$

The weight ratio of catalyst to oil is a parameter commonly used in FCC descriptions, and this is proportional to the molar cat/oil ratio W/N_{A0} in Equation (2). The function $F(X_A, \varepsilon_A)$ becomes equal to the conventional second order kinetic expression used in FCC when expansion is neglected ($\varepsilon_A = 0$).

Plug flow reactor. The differential material balance for second order kinetics with expansion in a steady-state plug flow reactor is well established.[15] Catalyst deactivation violates the assumption of steady state, and the implications of this will be discussed later. For the moment we ignore deactivation so that the results are to be regarded as describing the time-averaged behavior. The solution then has the same form as Equation (2), except that initial moles of A in the batch and the reaction time are replaced by the molar feed rate to the reactor. If we make the substitution

$$F_{AO} = \frac{N_{AO}}{\tau_{Cat}} = \frac{total\,moles\;feed}{oil\;\;delivery\;\;time} \tag{4}$$

into the plug flow reactor performance equation, the result is identical to the form of Equation (2), except that we now have

$$\begin{aligned} F_{Plug}(X_A,\varepsilon_A) &= 2\varepsilon_A(1+\varepsilon_A)\ln(1-X_A) \\ &+\varepsilon_A^2 X_A + (1+\varepsilon_A)^2 \frac{X_A}{1-X_A} \end{aligned} \tag{5}$$

Note that the group $W\tau_{Cat}/N_{A0}$ is the reciprocal of the weight hourly space velocity after appropriate conversion of units. When expansion is neglected (ε_A=0), this expression also reduces to the conventional second order activity expression.

Rate constants from cracking processes. In order to calculate and compare the rate constants associated with real cracking processes using these ideal reactors, we apply the ideal gas law and rearrange Equation (2) to obtain Equation (6).

$$k^{'} = \left(\frac{1}{\tau_{Cat}}\right)\left(\frac{N_{AO}}{W}\right)\left(\frac{RT}{y_{AO}\pi}\right)^2 F_i(X_A,\varepsilon_A) \tag{6}$$

From this Equation we see the common features of these reactors are that the rate constants required to obtain a given conversion can be represented as inversely proportional to the catalyst contact time, this contact time being either the batch (riser) residence time or the oil delivery time in the laboratory bench-scale reactors. The cumulative catalyst to oil ratio W/N_{A0} also appears as a parameter in the expression, even for the case of the flow reactors.

Effectiveness factor. The assumption of a steady-state does not strictly apply under the transient conditions of the batch reactor, or to flow reactors where catalyst deactivation occurs. The steady-state approximation will apply however if little conversion or deactivation occurs on the time scale of the initial penetration of gas oil feed into the catalyst. Since the time constant for catalyst deactivation is numerically similar to the catalyst contact time in each of the reactors, it will be sufficient if the diffusive penetration time of the oil into the catalyst is much less than the catalyst contact time. O'Connor and Houtert[7] considered the initial transient diffusion of resid feed into a catalyst, showing that the characteristic dimensionless parameter of the solution was the Fourier number, $N_F = D_e\, t / r^2$. For typical values of effective diffusivity of gas oil in steam and catalyst radius, we estimate that fully vaporized gas oil penetrates the catalyst completely in 2.5 milliseconds, so that the pseudo-steady-state hypothesis and conventional effectiveness factors may be applied in most cases.

The Thiele modulus for second order kinetics in a spherical catalyst is

$$\phi_2 = \frac{r}{3}\sqrt{\frac{k'\rho_{cat}C_s}{D_e}} \tag{7}$$

Since the numerical dependence of the effectiveness factors (η) based on appropriate ϕ_n are nearly the same for first and second order kinetics,[16] we have employed the equation for first order kinetics in the results below. Once significant limitations set in, Equation (6) is no longer a valid representation of the intrinsic rate, and Equation (7) is not reflective of the true Thiele modulus. Because we have not made corrections for this, those estimates of η_1 which we report that are significantly departed from unity will be too high.

The second order Thiele modulus contains the exterior concentration, and so the modulus is highest and the effectiveness factor lowest at the entrance of the reactor, where $C_S=C_{A0}$. Therefore, we report the effectiveness factors estimated for the entrance of the riser or plug flow reactor.

Parameter estimation. The properties of the hypothetical catalysts used in this study are typical of commercial FCC catalysts. The average microsphere diameter was taken as 75 μm and the base case mercury pore volume as 0.25 L/kg. The Chapman-Enskog theory and the random pore model[16] were used to estimate the effective diffusivity of binary mixtures of component A (gas oil) in steam or nitrogen with the molar expansion due to cracking ignored. For the gas oil we assumed a molecular weight of 400, a density of 0.91 kg/L, a molar volume at boiling of 0.56 L/mol, a critical temperature of 866 K and a process temperature of 811 K. We used an expansion factor of 3 for cracking and "bottoms conversion" instead of the conventional conversion to gasoline and lighter. Most commercial bottoms yields are in the 3-10 vol% range, or 90-97 vol% conversion on an LCO and lighter plus coke basis.

Results

Figure 1 shows the effectiveness factor results we obtained for batch reactors operating at C/O=5 for batch contact times between 0.3 s and 40 s. Four cases were created by alternating between expansion factors of 0 or 3 and conversions of 90 or 95%. Overall the results indicate, perhaps surprisingly, that the prototypical FCC process running with a 2 s contact time riser and less than 10% bottoms yield is either close to being or is already substantially limited by intra-particle mass transfer.

Each of the parameters ε_A, τ_{cat}, and X_A is significant to the degree of diffusion limitation. Second order plots of MAT data using the plug flow reactor model using either the gasoline or bottoms basis for conversion, and either $\varepsilon_A=3$ or $\varepsilon_A=0$ showed R^2 values of 94-99%. Since the corresponding kinetic model fits very well, it is now at least plausible that $X_A>90\%$ with $\varepsilon_A=3$ is a reasonable model for catalytic cracking, so that the conventional 2 s riser is diffusion limited according to Figure 1. $\varepsilon_A=0$ is equally plausible however, which would shift the critical contact time for the onset of diffusion limitations to smaller values.

The performance equations above only strictly apply in the absence of endotherms and deactivation for catalysts of uniform activity. None of these assumptions are in fact true in a commercial FCC unit. We will not attempt a rigorous mathematical description of these additional complexities here, instead choosing to consider each of these effects individually and to estimate their individual and combined impacts with the simplest models possible. With regard to the impact of endotherms of cracking, the thermal enhancement of the riser entrance rate constant, k^0, versus the average rate constant, k', is no more than 20%. The non-uniformity of properties in equilibrium catalysts was considered earlier.[17] Using those methods and typical fresh and "equilibrium" gas oil catalyst properties, we estimate that the front end of

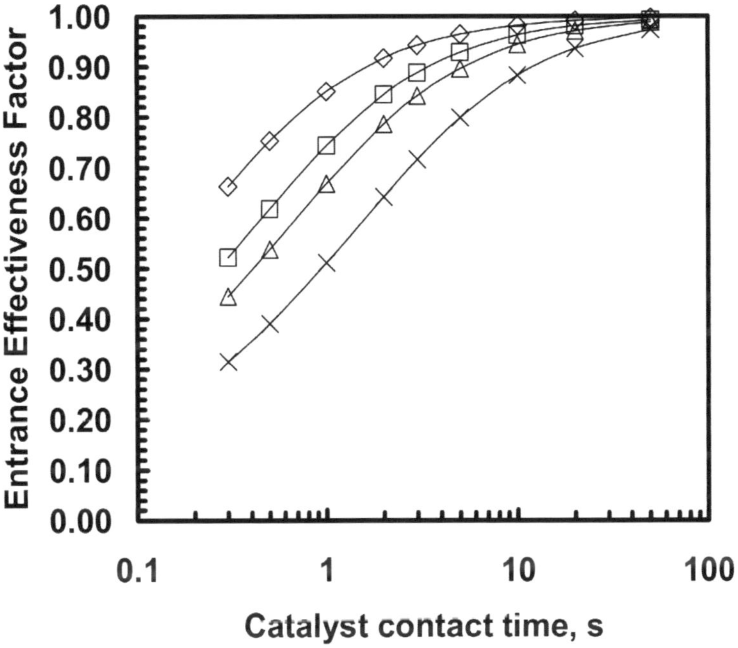

Figure 0. Effectiveness factors for cracking in batch reactors at 2 atm but different conversions (X_A) and expansion factors (ε_A). Symbols for (X_A, ε_A) values: ◊ for (90%, 0); □ for (95%, 0); Δ for (90%, 3); × for (95%, 3).

the catalyst age distribution, represented by the 2-3 wt% range, has about 1.96 times more activity than the averaged activity of the whole age distribution. With regard to deactivation by coking, a worst case scenario can be taken as a catalyst losing 90% of its activity by the end of the catalyst contact time. The initial cracking rate constant is then ca. 2.8 times higher than the time-averaged rate constant when an exponential decay model is used.

The accumulating impact of these non-idealities is estimated in Figure 2, using the base case catalyst and 95% bottoms conversion. When we model the front end of the age distribution also affected by an endotherm, the initial activity of this newest age fraction of catalyst is estimated to be 2.4 times higher than the average activity. About 29% of the catalyst effectiveness is predicted to be lost at 2 s contact time. If we additionally consider severe deactivation by coking, the rate constant for that portion of the catalyst at the riser entrance is about 6.7 times higher than the average rate. This leads to at least a 48% loss in catalyst effectiveness, the actual loss being higher because of the inaccuracy of Equation (6). The severity of deactivation by coking we elected is very high, and so the actual performance should be between the non-deactivated and severely deactivate curves. The example illustrates that the non-idealities involved in the real cracking processes make it more likely

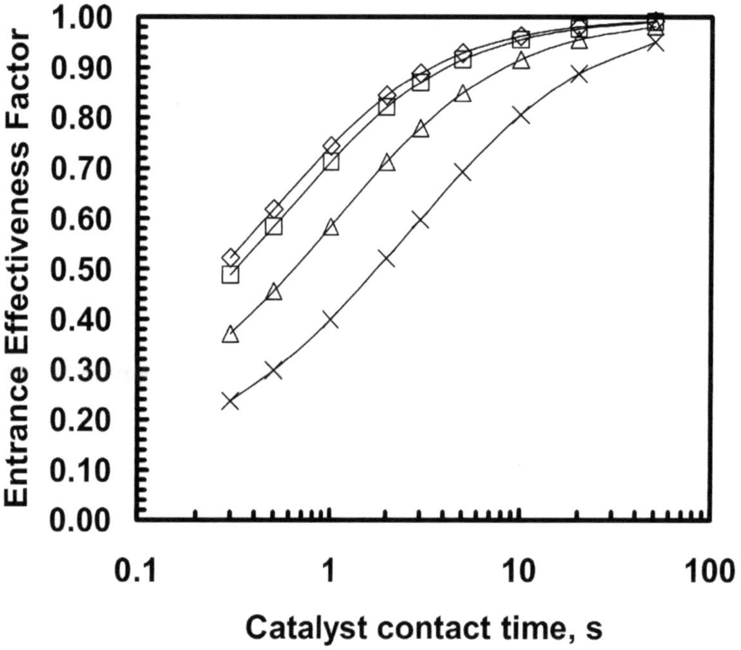

Figure 2. Effectiveness factors for gas oil cracking in steam in a riser reactor at 2 atm pressure, 0.25 L/kg pore volume, X_A=95%, ε_A=0, and C/O=5, while considering three non-idealities. The ideal base case (◊, k°) is additionally affected by a 30 K endotherm (□, 1.2 x k°), or an endotherm on the front of the age distribution (Δ, 2.4 x k°), or endotherm on the front end of the age distribution with a 90% loss of activity by coking (×, 6.7 x k°).

that diffusion limitations are present at the riser entrance, if only momentarily. These cases also reinforce the general conclusion above that the 2 s riser is either close to being or is already substantially limited by intra-particle mass transfer for at least some of the catalyst.

Comparisons were made of the η values for ideal batch and plug flow, each constrained to X_A=95% at ε_A=3 and C/O=5, as a function of catalyst contact time. Severe deactivation is much more likely in the latter type of test. The results predicted that MATs run above roughly 30 s contact time should not be affected by mass transfer limitations, either for ε_A=3 without deactivation, or ε_A=0 and deactivation. Unpublished 30 s contact time MAT data obtained by us for catalysts of increasing pore volume show no convincing selectivity effects, but such changes are known to give large improvements commercially. Although the model predictions are not consistently and completely in accord with experimental data, the general relationship between contact time and catalyst effectiveness described here does provide an explanation for the inability of the MAT to detect important selectivity effects found in the riser. Performance ranking reversals between MAT and pilot unit testing have been reported earlier,[18] although the explanation differed, and those results are in accord with the present model as well.

Engelhard has recognized the issues of short contact time cracking and has been supplying catalyst solutions for this for some time.[19,20] The DMS technology represents the first catalysts truly designed for short contact time. Initial commercial results[11] showed that Engelhard catalysts with optimized porosity and structure significantly improve LPG/gasoline, LCO/bottoms and coke/activity ratios, and improve conversion via heat balance. These results have since been validated in over 100 FCC units. These commercial results are consistent with intra-particle mass transfer limitations and inconsistent with conventional MAT results.

Discussion

While it does not affect our main conclusion, there is uncertainty in the parameter values to adopt in the kinetic models. If $\varepsilon_A=3$, as would be the case if cracking were truly a second order reaction, then severe limitations would often be predicted, especially when coking and age distribution are considered. Such predictions do not appear to be consistent with experiment, unless they led to initially very rapid coking and deactivation, a swift reduction in turnover frequency, and relief of the diffusion limitation during the later portions of the catalyst contact time. We have not yet been able to exclude this scenario.

Engelhard's experimental FFB data are consistent with severe exponential decay, but we do not have data for very short times on stream upon which to claim more aggressive decay than the simple exponential. The concentration of aromatics, Conradson carbon, and nitrogen in the feed will influence the deactivation rate, and may therefore indirectly play a role in determining the degree of initial diffusion limitations, under the constraints of constant C/O and X_A, by demanding a higher initial catalyst activity in order to achieve the targeted time-averaged conversion. Such severe decay by coking may not occur in the riser however.

Separately, the batch reactor model does not provide for the catalyst slip of a riser, leading to concerns about the suitability of that model. Even in the presence of catalyst slip however, the catalyst to feed-plus-product weight ratio is constant going up the riser. In the absence of deactivation then, the calculated average rate constant is not affected. In the presence of deactivation, any increase in catalyst residence time due to catalyst slip will increase deactivation by coking, resulting in a sharper activity profile. The initial rate constant for cracking must therefore be higher with catalyst slip, at given conversion and C/O, than without catalyst slip. The effectiveness factor at the riser entrance will then be the same or lower with catalyst slip than without catalyst slip, so that the conclusions are not generally affected. The vapor residence time in the riser must be employed in the Figures under these scenarios however, not the catalyst contact time.

Other FCC unit parameters may also have an impact. Non-ideal feed distribution at the riser mix point may make effectiveness problems worse for a portion of the catalyst. While the yield shifts anticipated for these non-idealities could be plausibly attributed solely to back-mixing, Engelhard's commercial experience has shown that a catalyst with porosity and structure optimized for SCT also provides benefits for FCC units without that modern hardware. If effectiveness factors were unity in these non-SCT units, the catalyst change should not have made any difference. Other pilot unit results obtained at different contact times also substantiate our conclusions, having shown large differences in yields between catalysts at 1 s contact time that had shown little or no difference in performance at longer contact times.[21]

While the η model estimates activity loss due to transport limits, selectivity changes are more consistently observed in our fluid bed testing when catalyst pore volume is varied than are activity changes. We have made no attempt to model the impact of diffusion limitations on selectivity, but some experimental selectivity changes appear to be more sensitive to contact time than the η model for rate seem to suggest. This may again be a result of initially steeper than exponential initial coking and deactivation. Further work in this area would be helpful.

Conclusions

Effectiveness factors have been estimated for the entrance of the FCC riser by applying the pseudo-steady state approximation to an idealized model. Two plausible hypotheses proposed based on the details of the kinetic model and deactivation of the catalyst lead to the conclusion that FCC is either diffusion limited or close to the edge of diffusion limitations at the prototypical 2 s contact time used in the industry. Selectivity trends in the refinery and the laboratory are directionally consistent with our conclusion. The model also explains in general the inability of the MAT to detect important selectivity changes that will occur in the riser, although quantitative simulation of the steep sensitivity of laboratory selectivity data to contact time has not been attempted, and may not be well represented by the models.

Nomenclature

ε_A	fractional volume change on complete conversion of A, dimensionless
π	total absolute pressure in the reactor, atm
ρ_{Cat}	density of an individual microsphere, kg/m^3
τ_{Cat}	catalyst contact time, or equivalently, oil delivery time for a plug flow reactor, or catalyst contact time for a riser reactor, or batch time for batch reactor, s
C_{A0}	initial concentration of reactant A, mol/L
C_S	concentration of reactant A at the exterior surface of the catalyst, mol/L
D_e	effective diffusivity of component A (gas oil) in the catalyst, m^2/s
F_{A0}	molar feed rate of reactant A to a plug flow reactor, mol/s
F_{Batch}	batch reactor performance equation integral, dimensionless
F_{Plug}	plug flow reactor performance equation integral, dimensionless
k'	average second order rate constant based on mass of catalyst, L^2 gas/mole A/kg catalyst/s
k^o	initial second order rate constant, L^2 gas/mole A/kg catalyst/s
N_{A0}	amount of reactant A in a batch reactor or fed to a MAT reactor, mol
N_F	Fourier number characteristic of penetration time, $D_e\, t / r^2$, dimensionless

R	ideal gas constant, L-atm/mol/K
r	radius of the catalyst microsphere, 35 x 10^{-6} m
r_A'	rate of reaction based on mass of catalyst, mol/kg catalyst/s
t	elapsed time during diffusion into the catalyst, s
T	absolute temperature, Kelvin
W	weight of catalyst in a batch or plug flow reactor, kg
X_A	fractional conversion of component A (gas oil), dimensionless
y_{A0}	mole fraction reactant A in the feed mixture, dimensionless

References

1. Upson, L. L., Hemler, C. L., and Lomas, D. A., Unit Design And Operational Control: Impact On Product Yields And Product Quality. Studies in Surface Science and Catalysis. 1993, 76, 257.
2. Bartholic, D. B., U. S. Patent 4,985,136, 1991.
3. Letzsch, W. S., and Gerald, E. U. S. Patent 5,662,868, 1997.
4. Johnson, A. R., Gartside, R. J., Ross, J. L., and Duncan, D. A., U. S. Patent 5,976,355, 1999.
5. Lomas, D. A., U. S. Patent 6,010,618, 2000.
6. Hemler, C. L., Sexton, J. A., Schnaith, M. W., Sexson, P. A, and Bartholic, D., Short-Contact-Time FCC, State of the Technology. AIChE 1998 Spring Meeting, New Orleans.
7. O'Connor, P. O. and van Houtert, F. The role of diffusion in bottoms conversion. Akzo Catalyst Symposium. 1986.
8. Bruno, K., Fletcher, R., Hakuli, A., and Imhof, P. NPRA, 2003 paper AM-03-58.
9. Mitchell, M. M., Hoffman, J. F., and Moore, H. F. Residual Feed Cracking Catalysts. Studies in Surface Science and Catalysis. 1993, 76, 293.
10. Stockwell, D. M. The Role of Porosity in the Cracking Selectivity of FCC Catalysts. Third Intl. Conf. on Ref. Proc. AIChE. 2000, 131.
11. McLean, J. B. and Stockwell, D. M. NaphthaMax™ - Breakthrough FCC Catalyst Technology for Short Contact Time Applications. NPRA, 2001, AM-01-58.

12. Lopez-Isunza, F., Moreno-Montiel, N. Quintana-Solorzano, R., Moreno-Mayorga, J. C., and Hernandez-Beltran, F. Modelling Diffusion, Cracking Reactions and Deactivation in FCC Catalysts. Studies in Surface Science and Catalysis. 2001, 133, 293.

13. Atias, J. A. and de Lasa, H. Adsorption and catalytic reaction in FCC catalysts using a novel fluidized CREC riser simulator. Chemical Engineering Science. 2004, 59, 5663.

14. Weekman, V. W. and Nace, D. M. Kinetics of Catalytic Cracking Selectivity in Fixed, Moving and Fluid Bed Reactors. AIChE Journal. 1970, 16(3), 397.

15. Levenspiel, O. Chemical Reaction Engineering, 2nd edition. Wiley: New York, NY, 1972.

16. Smith, J. M. Chemical Engineering Kinetics, 3rd Edition. McGraw-Hill: New York, NY, 1981.

17. Stockwell, D. M., and Wieland, W. S. Simulation of FCC Equilibrium Catalyst Age Distribution. Prepr. ACS Petr. Div. 2000, 45(2), 310.

18. Young, G. W. Realistic Assessment of FCC Catalyst Performance in the Laboratory. Studies in Surface Science and Catalysis. **1993**, 76, 257.

19. McLean, J. B. and Bovo, E. B. Improved FCC Yields Through A Combination Of Catalyst And Process Technology Advances. *NPRA*, **1994** paper AM-94-44.

20. McLean, J. B. and Bovo, E. B., FCC Catalyst Trends: Responding to the Challenges of the 1990's. *NPRA*, **1995** paper AM-95-64.

21. Stockwell, D. M., Wieland, W. S., and Himpsl, F. L., Catalyst Evaluation using Fixed Fluidized Bed Reactors – Protocol is Critical to Correct Performance Ranking. Int. Conf. Ref. Proc. AIChE. 1998, 237.

63e Modelling and Model Validation of a FCCU Regenerator

Jober C Penteado, Luciano F S Rossi

This work aims the numerical modeling and analysis of a FCCU (Fluid Catalytic Cracking Unit) regenerator. In the modeling, the regeneration is divided into two regions: dilute and dense. The latter comprises a bubble and an emulsion phase. The dense region, both the bubble and the emulsion phases, is model as a CSTR (Continuous Stirred Tank Reactor) in which ordinary differential equations represent conservation equations of mass, energy and species. In the dilute region, the conservation principles are written as partial differential equations to describe the change in space and time of the variables. The partial differential equations are solved by the finite difference method. Comparisons of the results of the current model with experimental data of a pilot plant operating in steady-state regime are made. A comparison with a dynamic model found in the literature is also conducted and the results are corroborated.

63f Laboratory Evaluation of Emission Control Additives and Their Deactivation Using In-Situ Coke Combustion

Darrell Rainer, Jorge A Gonzalez

Using additives to reduce NOx, SOx, and CO emissions from the FCC regenerator has been extensively applied in the refining industry, as this strategy offers the refinery reduction of environmental emissions without significant modifications to their regenerator design or unit operation. One primary challenge facing catalysts/additives providers is lack of realistic methods for lab-scale additive performance evaluation. Because the presence of an additive can impact the emissions of several gases in addition to those targeted, it is necessary to monitor COx, NOx, and SOx simultaneously for an accurate performance assessment.

Here we report our results of evaluating CO combustion promoters under in-situ coke combustion (Coke Combustion Test). In this test, the CO oxidation promoter is blended with a spent catalyst and the emission gases are monitored on-line during the coke combustion at typical regeneration temperatures of FCCU. The results are used to optimize the traditional Pt based CO combustion promoters as well as develop non-Pt containing products to minimize NOx while reducing CO emissions.

Additive deactivation is another important issue in additive development, since high activity of a fresh additive may not translate into high performance in the real unit. Many factors are involved in the deactivation of an FCC additive, such as collapse of the pore structure of the support, agglomerization of the metals, as well as poison of the active sites by SOx and /or contaminant metals in the FCC feedstock. Examples are given here, in which CO combustion promoters are deactivated together with an FCC catalyst under simulated FCC conditions (cyclic deactivation). The deactivated materials are then evaluated by Coke Combustion Test. The deactivation dramatically changes the ranking of CO oxidation capability of the additives and adds a new dimension for additive Development. Similar deactivation and testing protocol can be applied to SOx reduction and NOx reduction additives.

63g Catalytic Attributes of Self-Supported Hydrotalcite

Albert Vierheilig, Michael K Maholland

Hydrotalcites have long been known to have catalytic and adsorptive properties which make them suitable for applications in refining processes. One application of their use is in the removal of sulfur oxides (SOx) from the flue gas of Fluid Catalytic Cracking Units (FCCU). While hydrotalcites have long been proven in the laboratory for removing sulfur oxides from FCCU flue gas, their commercial use has been limited because they were not physically robust enough to withstand the severe hydrodynamic and hydrothermal conditions of the FCCU process without being diluted by a binding system. Recent advances in the manufacture of a self-supporting hydrotalcite material with the physical and chemical properties capable of withstanding the operating conditions of the FCCU have overcome this instability problem. Additionally, it has been found that significant improvements in SOx transfer performance could be achieved through careful control of the dehydroxylation conditions during the manufacturing process. The latest generation of commercial catalytic SOx reduction additives from Intercat is based on these physically strong, self-supporting hydrotalcites.

This paper discusses the development of self supporting hydrotalcite technology along with their application in reducing SOx from FCCU refinery flue gas streams. These self-supported hydrotalcite products are the only catalytic additive technology available which has been proven to consistently reduce SOx emissions to below 25 ppm in FCCU flue gas streams.

63h New Developments in Low-NOx CO Combustion Promoters

Darrell Rainer, Lin Luo

Carbon monoxide combustion promoters are commonly employed in FCC regenerators to minimize "after-burn" in the dilute phase of the catalyst bed, in order to avoid damage to cyclones and other overhead hardware from the elevated temperatures. This approach uses an additive (usually some type of supported Pt) to catalytically enhance the CO combustion in the dense phase of the bed, thereby depleting CO in the dilute phase. While this is an effective means of avoiding afterburn, there is the undesired side effect of increased NOx production, particularly associated with Pt-containing additives. This is due both to lowering the concentration of CO available for NOx reduction and to the activity of Pt for direct conversion of HCN (a common product released during coke combustion) to NO. Several non-Pt CO combustion promoters have been synthesized and tested in the lab using the smalle-scale Advanced Additives Testing Unit to monitor CO and NO concentrations during in-situ coke burn-off from a common spent catalyst with varying additive levels and types. These mono- and bimetallic non-Pt subject catalysts were tested and compared to current commercial Pt and non-Pt combustion promoter additives, exhibiting clearly superior NO yields (i.e. decreased NO yields) at the same or even higher CO oxidation activities. The best example showed over 60% less NO produced at a slightly higher CO oxidation activity relative to the commercial Pt benchmark. These novel, low-NOx non-Pt combustion promoters were also tested after the application of a proprietary deactivation technique, and showed the same advantages relative to the commercial benchmarks after deactivation as when tested fresh.

A Rolling Horizon Approach with Aggregation for Supply Chain Optimization of Refinery Industry

Jehoon Song, Minhwan Park, Sunwon Park* and Ignacio E. Grossmann***
LG CNS, Seoul finance building, 84, Taepyungro 1ga-dong, Jung-gu, Seoul 100-768, Republic of Korea
**Department of Chemical and Biomolecular Engineering, Korea Advanced Institute of Science and Technology, 373-1 Guseong-dong, Yuseong-gu, Daejeon 305-701, Republic of Korea*
***Department of Chemical Engineering, Carnegie Mellon University, 5000 Forbes Avenue, Pittsburgh, Pennsylvania 15213-3890, USA*

Abstract

Supply Chain Management (SCM) is a win-win strategy among enterprises. It pursues cost reductions and profit improvements by maximizing customer satisfaction. Supply Chain Optimization (SCO) on the other hand is a systematic and mathematical approach that supports optimal choice with a global visibility. Accordingly, SCO plays an important role for decision -makers in various industries in that its results can provide guidelines for the optimal planning or execution of supply chains. Not less important is the oil refinery industry because of its large-scale operation and complex supply chains. Therefore, there would be scope to maximize economical benefits through SCO. However, there are still few developments for the large-scale supply chains of refinery industry.

The previous works have focused on just a part of the whole supply chains of an oil company. In order to make more profit or reduce costs further the whole supply chains, e.g. source, production, and distribution, should be considered at the same time. In other words, in order to apply SCO to the whole supply chain of a real size oil industry, more specific multi-echelon production network and distribution network should be included in the optimization model. Therefore, it is necessary to build an extended and concrete whole supply chain model for finding optimal solutions through SCO in refinery industry.

In this paper, we develop an extended SCO model for refinery industry. The extended

SCO model consists of several echelons. Accordingly, it has a large number of integer decision variables, which cause computational complexity for solving an MILP problem. Additionally, we consider work transportation delays of crude oil that makes the MILP problem more difficult to solve because the time horizon must be extended. Since the proposed problem cannot be solved in reasonable time due to the many integer variables, we propose a technique to decompose the problem. We use a rolling horizon approach with aggregation concept. This approach has a merit compared to the normal rolling horizon approach because it considers not only the optimization horizon but also the whole scheduling horizon at each optimization.

The result of the proposed method shows better solutions than that of the rolling horizon approach without aggregation concept.

Key words: Supply Chain Management, Supply Chain Optimization, Rolling Horizon, Aggregation, MILP, Refinery Industry

1 Problem Statement

A typical SCO model for the refinery industry is shown in Figure 1. There are material flows which are transported from suppliers to customers and information flows customers to suppliers. The material flow in which crude oil is transported to customers as products from the oil field is as follows. Oil tankers transport different types of crude oil from different oil fields to the refinery plant. In this case, we assume there is no crude oil transportation through the pipelines. When the oil tanker arrives at the refinery plant, the crude oil is unloaded to the storage tanks under an unloading schedule. A blending tank plays a role of keeping the specification of the feed to CDU through mixing different types of crude oils in order to operate the CDU.

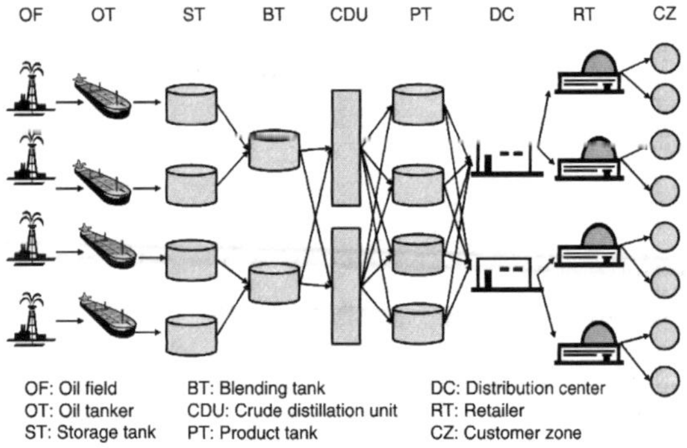

Figure 1. Graphical overview of a refinery industry

The CDU produces oil products at the different production ratios according to the feed specifications. Each product is stored in each product tank. The stored products are sent to distribution centers for providing to retailers. Distribution centers send all the products to the retailers, which in turn distribute the products to customers. In this supply chain, it takes much time for the oil tanker to transport crude oil to the refinery plant and it also takes time to transport between product tanks and distribution centers, and between distribution centers and retailers.

Owing to the development of IT using EPOS and internet, the information of customers' demands is processed immediately as if it is single company from material supplier to final retailer. In case that a product tank, a distribution center, and/or a retailer do not have enough inventories to satisfy customers' demands, this gives rise to shortages, which produce economic losses in the refinery plant.

2 Mathematical Formulation

2.1 Assumption

The assumptions which are used in supply chain model for the refinery industry are as follows (Lee et al. 1996, Park et al. 2001, Song et al. 2002).

(1) Each oil field produces different kinds of crude oil.

(2) The concentration of a specific component in crude oil determines the property of the crude oil.

(3) Each oil tanker transports different kinds of crude oil with full capacity.

(4) The crude oil in the oil tank can be unloaded Only if there is enough space in the storage tank.

(5) Two or more oil tankers cannot unload simultaneously.

(6) Each oil tanker delivers the specific crude oil from a specific oil field to the assigned storage tank.

(7) Each oil tanker has a constant transportation time.

(8) The concentrations of the specific component in blending tanks are constant.

(9) There is perfect mixing in the blending tanks.

(10) The crude oil flows from a storage tank to a blending tank at the maximum flow rate.

(11) The blended oil flows from a blending tank to a CDU at a flow rate between maximum flow rate and minimum flow rate.

(12) The crude oil cannot be fed into a blending tank while the blending tank is charging the CDU.

(13) A CDU cannot be charged simultaneously from different blending tanks.

(14) The changeover time in the CDU is negligible.
(15) The production ratio varies according to the feed into the CDU.
(16) Each DC and RT has independent tanks for the all products.
(17) There are transportation times between PT and DC, and DC and RT.
(18) The transportation time between RT and CZ is negligible.

2.2 Objective function and Constraints

The objective of this problem is to maximize the total profit. The objective function, total profit, is expressed as follows:

Total profit = Revenue – Crude oil transportation cost – Tank inventory cost – Change over cost – Operating cost – Distribution inventory cost – Product transportation cost – Shortage penalty cost.

There are constraints which balance the flow rates between storage tank, blending tank, product tank, distribution center and retailer. This problem has the inventory limitation of each tank and the demand limitation.

2.3 Decision Variables

The decision variables in this model are as follows:

(1) Inventories of crude oil in storage tanks and blending tanks
(2) Inventories of product in product tanks, distribution centers, and retailers
(3) Crude oil flow rates from oil tankers to storage tanks
(4) Product flow rates from product tanks to distribution centers, from distribution centers to retailers, and from retailers to customers
(5) 0-1 binary variables denoting if an oil tanker charges a storage tank, a storage tank charges a blending tank, and a blending tank charges a CDU
(6) Variables denoting if a changeover of blended oil occurs in a CDU

3 Solution Strategies

First, we start by reviewing simple approaches for handling large scheduling problems. Then, wc propose an enhanced rolling horizon approach with aggregation in order to obtain better solutions and in reasonable computation time.

3.1 Simple approach

Suppose there is a large-scale MILP problem, which has a long scheduling horizon, e.g. 30 days, like Figure 2. It is not easy to solve this kind of optimization problems in reasonable time if we use uniform discretization time model.

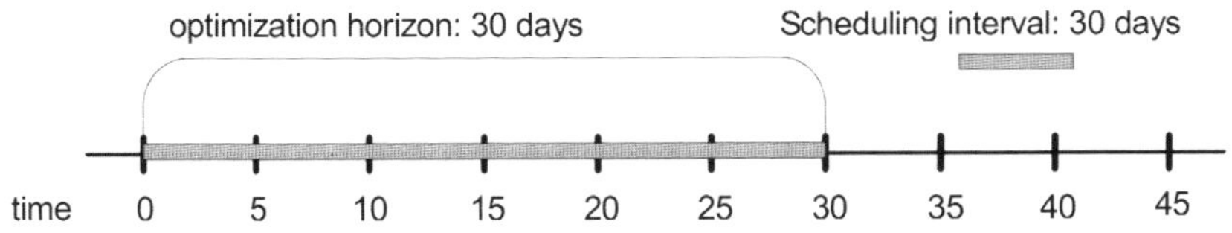

Figure 2. 30-day scheduling with single 30-day optimization horizon.

The easiest and simplest approach to solve this optimization problem is to solve the model over the whole optimization horizon at just one time. This approach is the most desirable as long as there is no computational problem because it provides the optimal solution for the given horizon. However, it is not easy to solve it in reasonable time because many binary variables are needed in order to consider the long scheduling horizon. In this case, we should seek better solutions with heuristics even though the optimality is not guaranteed. A strategy to overcome the computational problem is to solve a sequence of several short-term optimization problems, which can be solved in relatively short time.

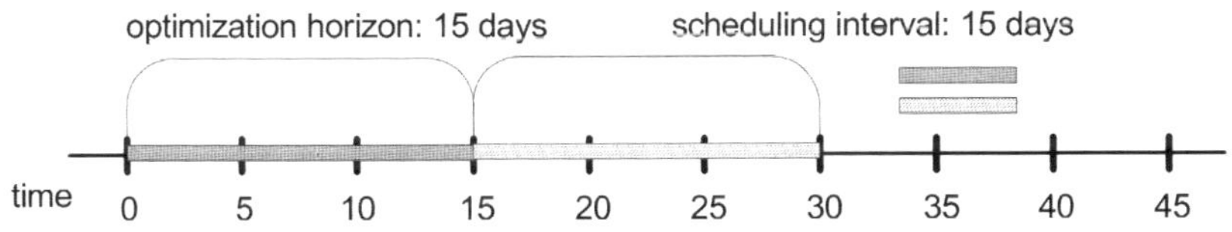

Figure 3. 30-day scheduling with two 15-day optimization horizons.

Figure 3 shows an approach to implement two successive 15-day optimizations when solving a 30-day scheduling problem. Solving two 15-day optimization problems takes less time than solving one 30-day problem. However, this approach cannot always give feasible solutions because the first 15 day optimization problem does not take into account the demand and constraints during the second 15 day period. The reason is, as shown in Figure 4, that the results of the last day in the first optimization problem are used as initial parameter values in the second optimization problem.

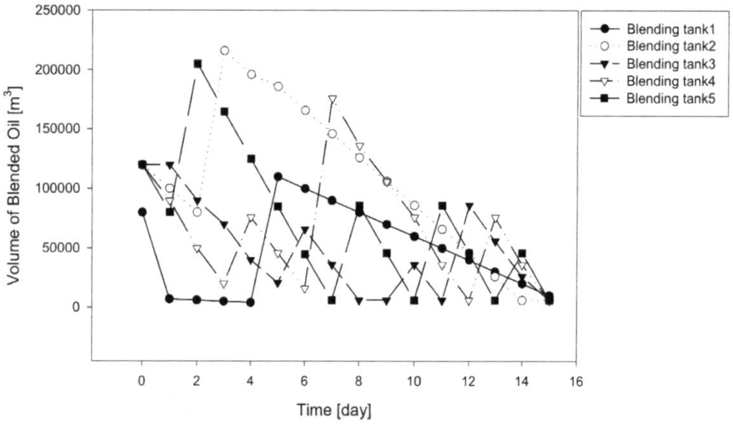

Figure 4. Typical example showing optimal inventory operation by short-term scheduling (Song et al. 2002).

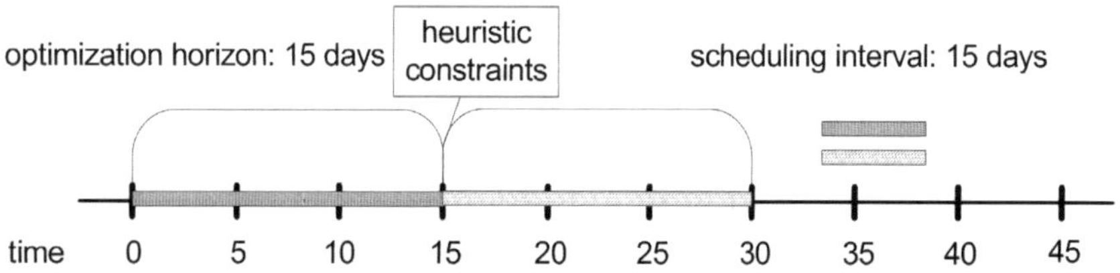

Figure 5. 30-day scheduling with two 15-day optimization horizons using heuristic constraints.

In order to avoid this infeasibility problem, we can think of imposing additional constraints that can provide at least minimal parameters to solve the next optimization problem when solving the first optimization problem.

3.2 Rolling horizon approach

The Simple decomposition approach such as the one in Figure 3 may have some problem related with the results of the last day. It can be an approach to implement optimization again before using the information of the last day. To solve a 30-day scheduling problem, we solve the first 15-day short-term scheduling problem and accept the results of the first five days, and solve another 15-day scheduling problem to find 5-day schedule and repeat this procedure until the whole 30-day schedule is found as shown in Figure 6. After all, it means determining 5-day schedule every time we solve a 15-day rolling horizon optimization problem six times. In this way, by solving 15-day rolling horizon optimization problem, we ultimately obtain the results for 30 days. This approach does not guarantee the optimality of the solution, but it has the advantage of being more likely to find a feasible solution.

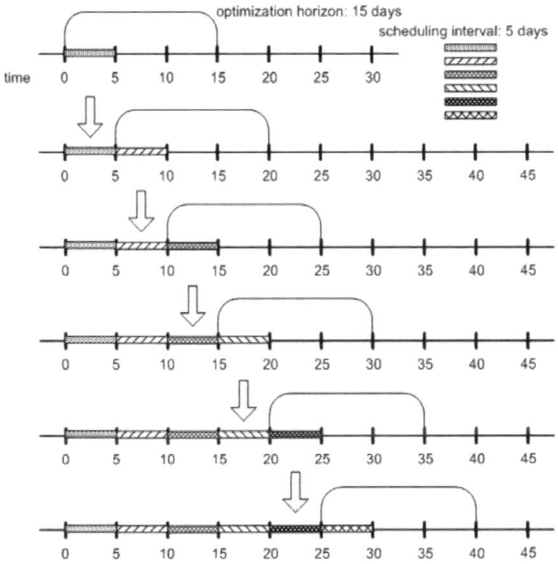

Figure 6. 30-day scheduling by a rolling horizon approach.

3.3 Rolling horizon approach with aggregation

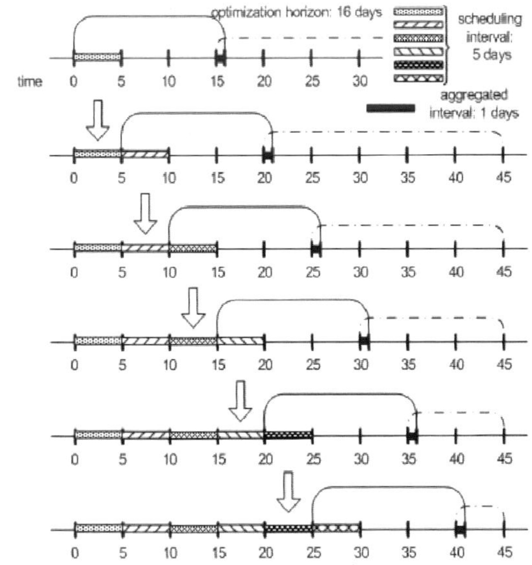

Figure 7. 30-day scheduling by a rolling horizon approach with aggregation.

The rolling horizon approach described in section 4.2, for example, solves six 15-day problems successively when solving a 30-day optimization problem. When it solves a 15-day optimization problem, it does not use the information after the period. If this information is used and reflected into the optimization, better solution can be obtained. However, considering the remaining information increases the problem size. We propose a rolling horizon with aggregation concept, which reflects the information after the optimization horizon without increasing the problem size too much. In order to consider the rest of the scheduling horizon after the 15th day, we aggregate the excess demand during the remaining days into one period, which is the value of demand minus capacity during the remaining period. This aggregation means that we can consider the whole horizon without much additional computational burden.

4 Example

In order to provide insight into this solution strategy for a large scale refinery industry, we consider the following example. Figure 8 shows the system overview for this example. The whole supply chain include four oil fields, four oil tankers which transport four different crude oils respectively, four storage tanks, two blending tanks, one CDU, three product tanks, two distribution center, four retailers, eight customer zones. The parameters used in this model are shown in Tables 1-9.

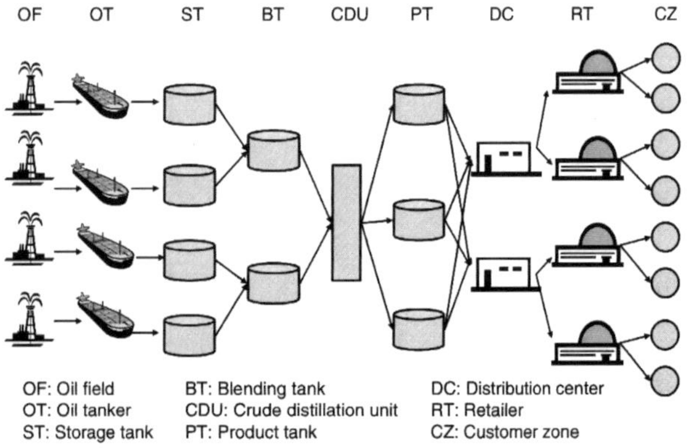

Figure 8. Supply chains for an example problem.

The scheduling horizon of this problem is 30 days. We assume that the demand of 60 days can be predicted. The model involves 780 discrete variables, 3021 continuous variable, and 4431 equality/inequality equations. The number of products is three. This MILP model is solved with CPLEX 7.0 in GAMS 2.5 on a PC operated with Pentium 4 2.4GHz RAM 1GB.

4.1 Optimization with the whole scheduling horizon

Optimizing simultaneously the entire schedule, the objective value is 20101.567 requiring 2012.109 s of CPU time. The inventory operation of each echelon is shown in Figures 9, and the flow rate of crude oil is shown in Tables 1.

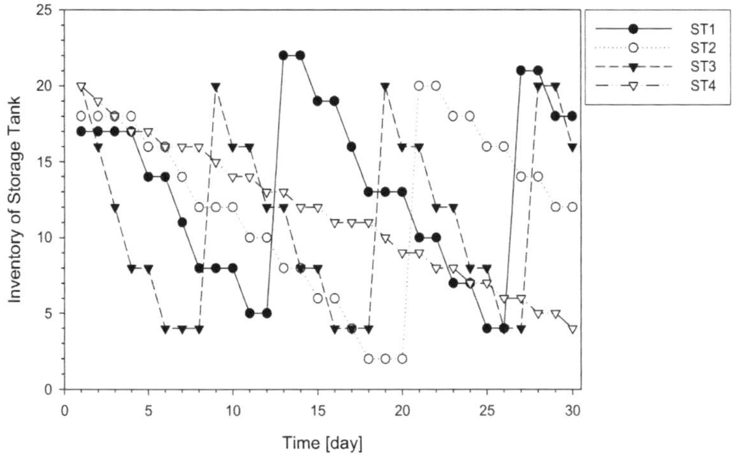

Figure 9. Optimal inventory schedule of storage tanks.

Table 1. Crude Oil Flow Rates from Oil Tanker to Storage Tanks

OT→ST	1	2	3	4	5	6	7	8	9	10	11	12	13	14	15	16	17	18	19	20	21	22	23	24	25	26	27	28	29	30
1→1													20														20			
2→2																					20									
3→3									20										20									20		

4.2 Optimization with rolling horizon approach

As mentioned above, although there is no guarantee of optimality, a rolling horizon approach is introduced as decomposition heuristic for reducing the computational burden when solving a large scale scheduling problem. When using rolling horizon approach, there are selection problems about the optimization horizon and the scheduling interval. For this, the problem is solved repeatedly by varying both optimization horizon and scheduling interval in order to obtain the objective values and CPU time. The results are shown in Figures 18-19. As an example selecting a horizon time of 16 days and an interval of 14 days the predicted profit is 20101.520 versus the value of 20101.567 from the simultaneous solution. Not only is this value close to the optimum but the time required was also much smaller, 1.7 sec of CPU time versus the 2012 secs of the simultaneous solution.

From Figs. 10-11 it can be seen that the CPU time increases as the optimization horizon increases and the scheduling interval decreases. Regardless of the scheduling interval size, if the optimization horizon is less than 20 days, it takes little time to calculate the optimization problem with the rolling horizon approach.

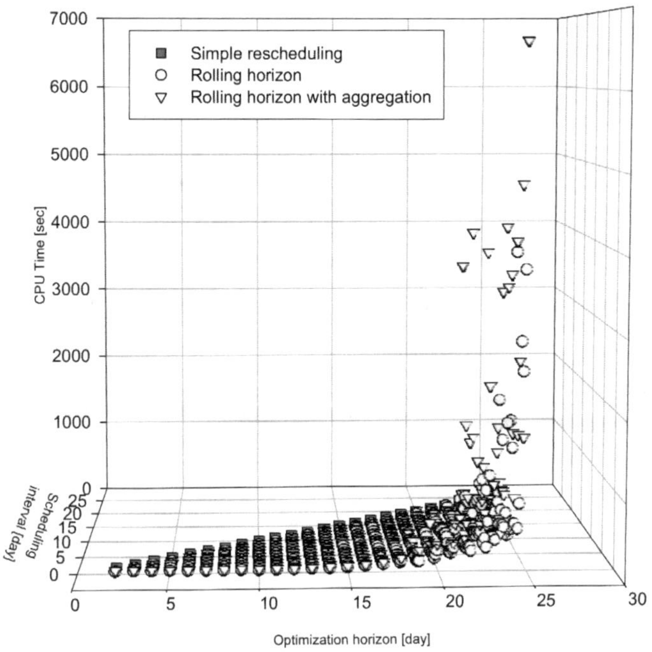

Figure 10. CPU time taken in various cases.

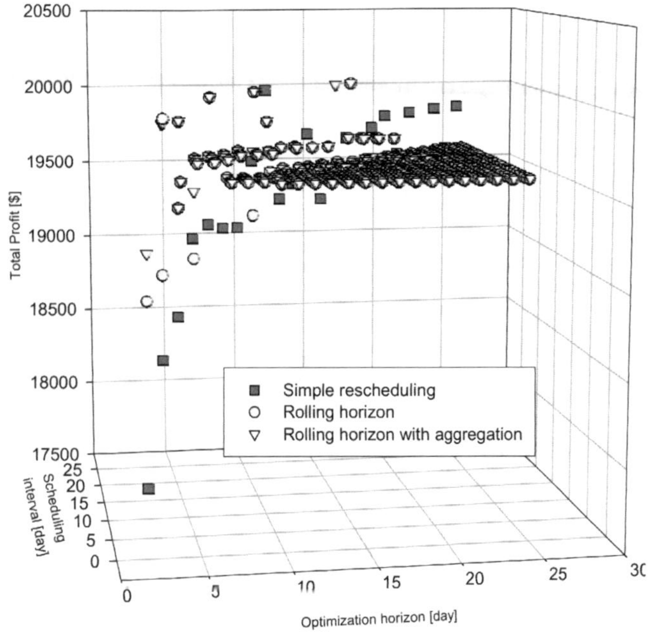

Figure 11. Total profit made in various cases.

As proposed above, among the results less than 20 days, the objective values of the optimization by rolling horizon approach are better than those of simple rescheduling. Among the result by rolling horizon approach, using aggregation concept shows better objective values than using no aggregation concept. There is only about 2% of gap between the objective value of 30-day optimization problem and the objective value of optimization problem with rolling horizon approach. However, the CPU time of rolling horizon approach is less than 1/1000 of the 30-day optimization problem.

5 Conclusions

In this paper, an extended supply chain optimization model was developed for a large-scale refinery industry. An MILP model was used to implement optimization. The optimal scheduling can be obtained for the entire scheduling horizon by solving the model. But it may take forever to solve it. In order to overcome this problem, optimization horizon is chosen to be shorter than the scheduling horizon and a rolling horizon approach is introduced to determine scheduling for the entire scheduling horizon by repeatedly determining the schedule within the rolling optimization horizon. Using the rolling horizon approach, the decrease of objective value from the optimum is just 2%, but the CPU time was dramatically reduced. For better solutions, a rolling horizon approach with aggregation was proposed. The result of the proposed method shows better solutions than that of the rolling horizon approach without aggregation concept.

References

Lee, H.; Pinto, J. M.; Grossmann, I. E.; Park, S. Mixed-integer linear programming model for refinery short-term scheduling of crude oil unloading with inventory management. *Ind. Eng. Chem. Res.* 1996, *35*, 1630.

Park, H.; Bok, J. -K.; Park, S. Scheduling of refinery processes with optimal control approach. *J. of Chem. Eng. Japan*. 2001, *34*, 411.

Song, J.; Park, H.; Lee, D. -Y.; Park, S. Scheduling of actual size refinery processes considering environmental impacts with multiobjective optimization. *Ind. Eng. Chem. Res.* 2002, *41*, 4794.

64b Evaluating Materials for High Temperature Hydrogen Separation Using Grand Canonical Monte Carlo and Molecular Dynamics Simulations

Marco Gallo-Estrada, Martha C Mitchell, Tina M Nenoff, V K K Upadhyayula

Finding more economical and less energy-demanding ways of separating hydrogen from methane in refinery waste streams and from the output streams of steam reforming is very important. Ideally, these separations should take place at high temperature to avoid excessive cooling costs. Separation of hydrogen from other gas phase species can be carried out using inorganic molecular sieve membranes. We have used computer simulation techniques to evaluate the ability of several molecular sieve materials to separate hydrogen from methane, carbon dioxide and carbon monoxide. The materials evaluated were three titanosilicate materials: zorite, ETS-4, and ETS-10; as well as zeolite ZSM-5 and its siliceous analog, silicalite. Adsorption isotherms were generated for each material using Grand Canonical Monte Carlo (GCMC) simulations. High temperature molecular dynamics (MD) simulations were also carried out for single components, binary mixtures of hydrogen/methane, hydrogen/carbon dioxide and a quaternary mixture of hydrogen/methane/carbon dioxide/carbon monoxide. These simulations were used to determine self-diffusion coefficients. Separation factors were calculated to determine the separation capabilities of each material. It was concluded ETS-10, with a three-dimensional pore structure, could provide kinetic separations of hydrogen. The other two titanosilicate materials, zorite and ETS-4, can serve as true molecular sieves for hydrogen in the mixtures studied, as the hydrogen is the only species with significant mobility in these pore structures. ZSM-5 and its analog silicalite are effective materials for separating hydrogen from other gas-phase species.

Coriolis Meters Improve the Efficiency of Hydrogen Production Process

Win Hoglen, Senior Principal Process Controls Engineer, Air Products
Julie Valentine, Refining Industry Marketing Manager, Micro Motion Inc.

Introduction

Cost-effective production of hydrogen has become a significant concern in the refining industry. The need for deeper hydrotreating and for the processing of heavier and higher sulfur crude slates is driving the demand for reliable hydrogen and steam availability.

Precise control of the steam-to-carbon (S/C) ratio in the hydrogen production process can have a major impact on the efficiency of the facility. Running the S/C ratio closer to target and running the facility consistently can significantly reduce operating costs, with a potential payback period of less than a week. Possible results from replacing the existing hydrocarbon measurement technology with Micro Motion include:

- Decreased capital investment for new units
- Decreased maintenance costs
- Decreased energy costs
- Increased safety
- Increased throughput

Background

Steam Methane Reforming (SMR) technology is the primary process for high-purity hydrogen production. This process involves introducing preheated natural gas (or another hydrocarbon stream such as Refinery Off Gas) with steam across a catalyst to produce a 75–80% hydrogen stream composed primarily of hydrogen and carbon monoxide, then purifying this effluent, using either an MEA scrubber or a PSA unit, to produce 99%+ hydrogen. See Figure 1.

Figure 1 Hydrogen Production in SMR Unit

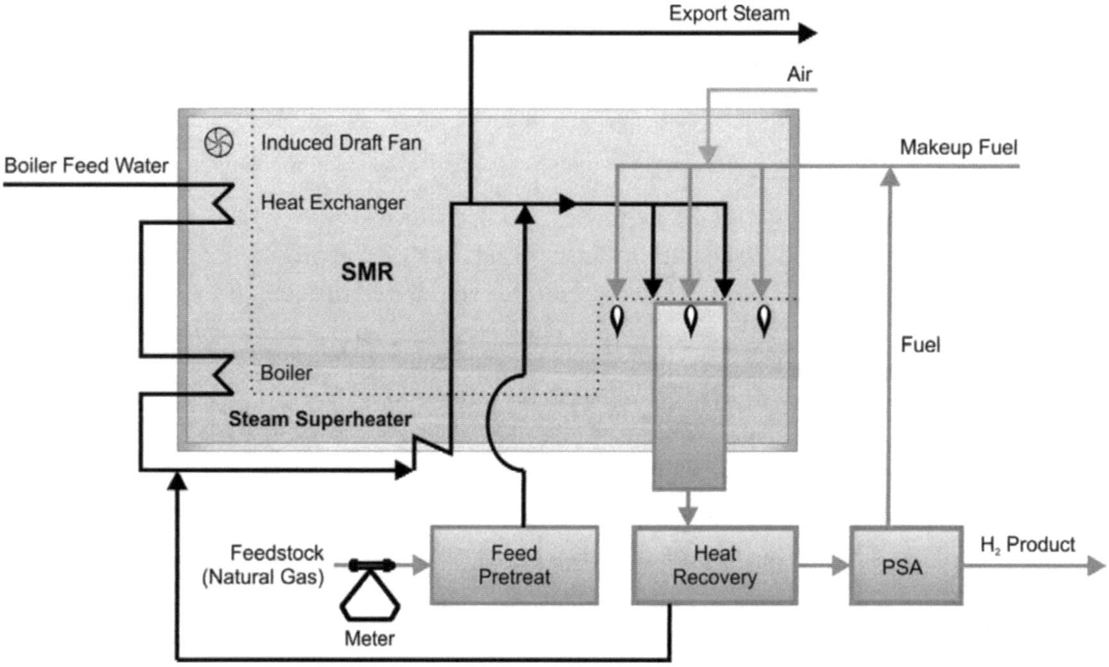

Two reactions occur in the SMR:

- First, the reforming reaction that converts hydrocarbon feedstock to carbon oxides and H_2

$$C_nH_m + nH_2O + \text{heat} \rightarrow n\,CO + (n/2 + n)\,H_2$$

- Second, the shift reaction that converts CO produced during the reforming reaction to carbon dioxide and additional H_2

$$CO + H_2O \rightarrow CO_2 + H_2 + \text{heat}$$

The S/C ratio is one of the most important process measurement areas for an SMR. The minimum S/C ratio is dependent on the feedback, but must typically be above 3:1 to prevent carbon deposit formation on the catalyst. For insurance, most hydrogen producers run the S/C ratio between 3.2:1 and 3.7:1. Any change in the composition of the feedstock requires a corresponding change in the amount of steam, and this is where the difficulty lies.

In most SMR units, the S/C ratio is controlled using differential pressure measurement, corrected for temperature and pressure, plus an analyzer (typically a gas chromatograph or a mass spectrometer) to calculate the molecular weight (MW) and determine the carbon weight contribution of the feed stream. This approach is expensive both in terms of the initial capital cost of the instruments and analyzer and in terms of the operations and maintenance required for the analytical devices.

This technology has additional drawbacks:

- Because calibration of the dP meter is highly dependent on fluid properties, changes in the composition of the feed stream will be a significant source of error.
- If the composition of the feed stream changes rapidly, response time problems can occur.

As a result, many operators run their SMR units at S/C ratios considerably above target, to provide a margin of safety that can compensate for changes in composition.

An alternative to the volumetric flow measurement and analytical devices is to use a Micro Motion Coriolis meter to measure the mass flow of the gas directly and to ratio the steam accordingly.

Accuracy of Orifice / dP Technology

Orifice plate differential pressure technology is well understood and has an expected error of 1–2% (the orifice plate itself has an inherent error of 0.5% as a starting point) if the measurement is compensated for temperature and pressure, as is typical in custody transfer applications. Typical orifice plate installations for process quality metering will have even higher errors, as these systems are normally not temperature- or pressure-compensated. Composition changes in the fluid being measured will lead to further errors because the calibration is strongly dependent on the fluid properties.

If the hydrocarbon stream has a fairly constant MW, the mass flow can be estimated by assuming a fixed MW or density and applying temperature and pressure compensation to the orifice plate/dP transmitter installation.

If the MW of the hydrocarbon stream is not constant, an analyzer is required to determine the stream composition and MW. Minimum error of the volumetric meter, for changing composition of the process fluid, is 2%. The required S/C ratio must be calculated by extracting the carbon mole weight value from an analyzer. Mass spectrometers may be able to update this value every few seconds, while gas chromatographs can require up to 15 minutes to update.

Accuracy of Coriolis Technology

The new technology utilized in this application is a Coriolis mass flow meter from Micro Motion, who holds a patent on this application, US Patent No.: US 6,758,101 B2, titled "Steam to Carbon Ratio Control in Steam Reforming of Hydrocarbons."

Coriolis meters measure mass directly: the measurement is not affected by density or composition changes, and the calibration holds across all process fluids – liquid or gas. Micro Motion's ELITE Coriolis meter has a rated flow accuracy of 0.35% on gases. Because the hydrocarbon string typically changes by 1 carbon per 2 hydrogen, the mass measurement from the ELITE meter can provide a very good estimate of the carbon weight for any hydrocarbon stream.

Accuracy of the ELITE meter was tested at a site where the composition of the natural gas stream changed significantly during the testing period:

- Methane range: 77.5% to 89%
- Ethane range: 6.8% to 15.23%
- Inerts:
  - Nitrogen fluctuated by +/- 0.185%
  - Carbon dioxide fluctuated by +/- 0.567%

An ELITE meter with a gas chromatograph was assumed to provide the most accurate data, so this technology was used as a reference for two test technologies: an orifice plate with a gas chromatograph, and an ELITE meter configured with a fixed MW value. The gas chromatograph was set to look for most hydrocarbons from methane to C_{6+}, plus CO_2 and N_2. The objective was to control the S/C ratio to within +/- 0.1.

During the two-month testing period:

- The maximum error of the ELITE meter with a fixed MW value was 0.02, compared with the reference technology.
- The maximum error of the orifice / dP system with the gas chromatograph was 0.2 compared with the reference technology – 10 times greater than the ELITE meter with a fixed MW value.

In these tests, no attempt was made to compensate for changes in the inerts. In production installations, testing is recommended to determine whether or not compensation for inerts must be implemented. For example, calculations performed during this study indicate that if the concentration of nitrogen changes by 3%, with a corresponding change in the methane concentration, the S/C ratio in this process will change by 0.1 points, and a nitrogen analyzer might be recommended.

Benefits

The economic benefits of tighter control of the S/C ratio are significant, both in terms of reduced fuel consumption and in terms of increased export steam credit. Reducing the variability in the S/C ratio by 0.2 points could increase plant efficiency by as much as 8 BTU/SCF of H_2 produced. With natural gas at $5.50 per MMBTU, an H_2 plant with a capacity of 80 MMSCFD would realize savings of close to $1.3 million a year.

Additional economic benefits of utilizing the Micro Motion meter instead of traditional measurement technologies include lower capital costs for new unit designs:

- Gas chromatograph analyzer and orifice plate technology: $50,000
- Mass spectrometer and orifice plate technology: $130,000

- Coriolis meter: less than $20,000

It is well recognized that gas chromatographs and mass spectrometers are expensive to maintain. In addition, when the analyzers are down for maintenance, the unit must run in a conservative, less efficient mode.

Conclusions

Tighter control of the steam-to-carbon ratio in the Steam Methane Reforming unit will result in significant increases in the efficiency of the unit. Coriolis direct mass flow technology allows the user to control this ratio more accurately, especially under changing conditions of the hydrocarbon feed stream. This significantly reduces fuel costs, equipment costs, and maintenance costs.

64d Structure-Based Generalizations for Multiple Temperature Infinite-Dilution Activity Coefficients Data for Aqueous Systems

Brian J Neely, Khaled A M Gasem, Robert L Robinson Jr., Srinivasa S Godavarthy

Often phase equilibrium data are necessary for the proper design and operation of most chemical processes. When experimental data are not available, generalized thermodynamic models, including group contribution methods, are used to predict the equilibrium phase behavior. Recently, we have developed structure-based quantitative structure-property relationship (QSPR) models to correlate the infinite-dilution activity coefficients (γ∞) of aqueous systems. These QSPR models were successful in correlating γ∞ at a single temperature; however, they failed to account accurately for the temperature dependence in this property. As such, a suitable theoretical framework is required to provide a sound basis for generalized γ∞ predictions over a wide range of temperature.

Previously, we have employed the Soave-Redlich-Kwong equation of state to correlate γ∞ of aqueous systems [Bader and Gasem, 1993]. Generalized EOS interaction parameter estimates were presented for various homologous subsets of molecules. In this work, we integrate this EOS approach with our recently developed QSPR algorithms to provide structure-based generalizations for the EOS interaction parameters. The database employed in this work includes data for 287 chemical species at temperatures ranging from 283.2 to 373.2 K.

The results obtained indicate that this integrated approach of using the SRK EOS and the structure-based QSPR parameter generalization provide more accurate estimates for γ∞ of aqueous systems.

65a Recent Advances in Residue Upgrading Fundamentals

Murray R Gray

The tightening supply of light crude oils and the increased supply of bitumens and heavy oils has led to a renewed interest in the conversion of vacuum residues and heavy oils. At the same time, renewed attention to the chemistry of asphaltenes and the emerging field of supramolecular chemistry has rekindled the debate on the structure of the heavy fractions of crude oil. This tutorial will cover three main topics:

1) the chemistry of asphaltenes and vacuum residue and how it dictates yield and liquid-phase behavior

2) the connection between cracking in the liquid phase and the evolution of products and how this affects reactor design

3) how cracking and hydrogenation processes are being combined with deasphalting and gasification to give new flowsheets for processing of residues.

65b Tutorial: Delayed Coking Fundamentals

Christopher A Paul, Paul J Ellis

Great Lakes Carbon Corporation has worked closely with refineries producing delayed coke in all forms, fuel grade (shot or sponge), anode grade (sponge), and electrode grade (needle) since start-up of the company's first calcining operation in 1935. With on-going research in the area of delayed coking since 1942, Great Lakes Carbon (CLC) has operated delayed coking pilot units including an excellent large-scale pilot unit with a coke drum 0.3 meter (1 ft) diameter by 2.1 meters (7 ft) long and has developed physical models which explain coke formation in coke drums.

Knowledge of commercial delayed coking units as well as that of the GLC Pilot Delayed Coker is used in this tutorial paper to describe the formation and uses of the three types of structures of delayed petroleum coke: needle, sponge, and shot. Troubleshooting tips are included on many aspects of the delayed coking drum cycle including: steam stripping, water quenching, coke cutting, drum warm-up, and drum switching technique. Suggestions and descriptions of delayed coking unit hardware are included.

The objective of this tutorial paper is to acquaint the non-refinery technologist and further the knowledge of refinery personnel with the delayed coking operation, delayed coking unit hardware, types of coke that can be produced, coke formation models, and the uses of petroleum coke. Also, by discussing some of the delayed coking process problems encountered in the industry, we hope to encourage new advances, solutions, and improvements for the future.

Resid Upgrading: Technology, Limitations, and Potential

Irwin A. Wiehe
Soluble Solutions
3 Louise Lane, Gladstone, NJ 07934

Prepared for Presentation at Spring AIChE Meeting on April 13, 2005
Session on Resid Upgrading Tutorial
Unpublished

Introduction

Petroleum resids are the heavy fraction remaining after distilling petroleum crudes at atmospheric pressure (atmospheric resid) or at reduced pressure of 25 to 100 mm Hg (vacuum resid). Not only do resids have high molecular weight (≈1000 amu number average for vacuum resids) but they contain polynuclear aromatics (PNA's). The PNA's greater than 3-4 rings provide the greatest limitation to the conversion of resids. This is because of their high thermal stability and because desirable aromatics in transportation fuels are limited to alkylated single rings. In addition, the high concentrations of heteroatoms (sulfur, nitrogen, vanadium, and nickel) in petroleum resids act along with PNA's to poison catalysts. Therefore, care must be taken to protect catalysts, to replace and regenerate catalysts, to use throwaway catalysts, or to use purely thermal processes. No matter which type of process is used, a substantial fraction of resid molecules can be cracked off as fragments to become liquids in the transportation fuel and vacuum gas oil boiling ranges. However, one should not try to overly convert resids or the PNA limit will force the selectivity to go to the thermodynamically favored[1] but lower valued products: coke and/or hydrocarbon gases. As a result, all economical resid conversion processes have an even heavier carbonaceous byproduct. It is critical to have an economical and environmentally acceptable use for this byproduct for the resid conversion process to be viable.

Characterization of Resids

The understanding of the chemical transformations of the macromolecules during resid conversion is greatly limited by the wide diversity (over a million chemical structures) of these complex macromolecules. One way to sample this molecular diversity is to separate the resid and its conversion products into diverse fractions using solubility/insolubility in volatile liquids as well as adsorption/desorption on solids. In this way a number of resids and resid conversion products were separated into coke (toluene insoluble), asphaltenes (toluene soluble/n-heptane insoluble), resins (n-heptane soluble/adsorbs on Attapulgus clay), small ring aromatics (n-heptane and cold methylethyl ketone soluble/does not adsorb on Attapulgus clay), and saturates (n-heptane soluble/ cold methylethyl ketone insoluble/does not adsorb on Attapulgus clay). A plot of molecular weight versus hydrogen content, the solvent-resid phase diagram in Figure 1, shows that each of these fractions occupies an unique area that better defines the fraction than the separation method[2]. Thus, to a first approximation resids differ in the relative amounts of similar fractions: saturates, small ring aromatics, resins, and asphaltenes (unconverted resid contains no coke).

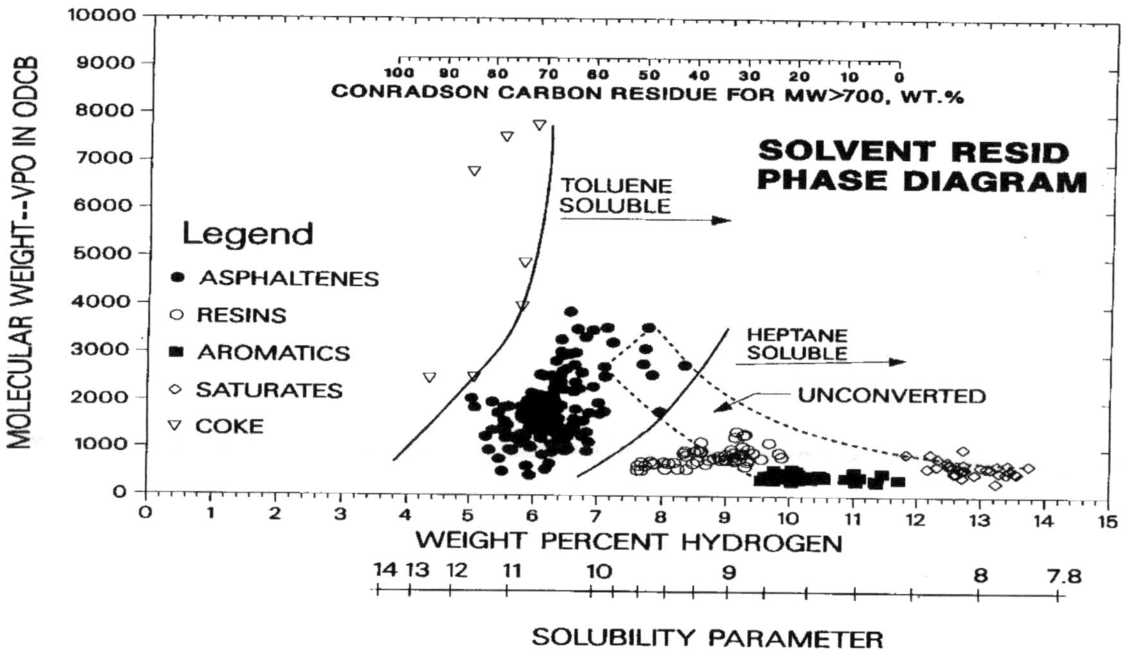

Figure 1. The Solvent - Resid Phase Diagram

While saturates and small ring aromatics contain sulfur, they contain little or no nitrogen, oxygen, vanadium, nickel, or PNA's of 3 rings or larger[2]. Thus, they are relatively catalyst friendly. During conversion, the saturates can be completely converted to lower molecular products without forming a carbonaceous byproduct. However, the small ring aromatics combine during thermal cracking to form resins. The resins crack off side chains from PNA's to form lower molecular weight, lower hydrogen content byproducts. When these PNA containing molecules combine, they form higher molecular weight asphaltenes. The asphaltenes crack off side chains from PNA's to form lower molecular weight, lower hydrogen content byproducts until they combine to form higher molecular weight coke[2]. The addition of hydrogen and the presence of catalysts can alter the rates of these aromatic combination reactions and thus the selectivity to products. However, at high enough conversion, coke and gas are the principal products, independent of conversion chemistry as the remaining molecules approach partially saturated PNA's with very short side chains.

Hydrocarbon Solubility

Above the wax melting point (40 to 60°C) where resids are processed, all fractions become less soluble in hydrocarbons with lower hydrogen content and higher molecular weight[3]. Figure 1 shows this with the solubility parameter increasing with lower hydrogen content (higher aromaticity) because the attractive force between PNA's is the strongest intermolecular force in resids and resid conversion products. This is very important in the mechanism of coke and sediment formation that often plagues resid conversion processes.

Pendant-Core Building Block Model and Conradson Carbon Residue

Models of petroleum with detailed chemical structures are misleading with the present characterization uncertainty. However, one does not approximate reality if the wide diversity of molecular types is not included. A compromise that satisfies both of these requirements is to represent molecules as combinations of pseudo building blocks. The simplest model has two building blocks[4]. Since thermally stable polynuclear aromatics are such an important part of the mechanism of coke formation, they are represented by the aromatic core building block. The other building block represents the fragments that become volatile liquids when cracked off the core, called pendants. Then a vacuum resid can be represented by a wide distribution of species made up of various combinations of these two building blocks (Figure 2). If each building block is assigned a hydrogen content and a molecular weight, the diversity of the solvent-resid phase diagram (Figure 1) will be represented.

Saturates	**Aromatics**	**Resins**	**Asphaltenes**
P-P	P-•-P	P-•-P-•	P-•-•-P-•-P-P
P-P-P	P-•-P-P	P-P-•-P-•	P-•-P-•-P-•-P-•
P-P-P-P		P-•-•-P-P-P-P	P-•-•-P-P
			•-•-•-P

Distillable Liquids	**Toluene Insoluble Coke**
P	•-•-•-•-•-P
	•-• •-•-•

Building Blocks

P = Pendant

• = Core

Figure 2. The Pendant - Core Building Block Model Allows for a Wide Distribution of Specie

Conradson carbon residue (CCR) is a standard petroleum industrial coking test for characterizing the coke forming tendency of petroleum liquids that today is measured by the Microcarbon method[5] (ASTM D4530). Since this test is widely used, it is extremely important for those working in resid conversion to understand what this test does or does not reveal. In the terms of the Pendant-core Building Block Model, CCR measures the percent fraction of the resid that is cores. A hydrogen balance around the Microcarbon test (Figure 3), shows that the CCR is a linear function of hydrogen content. This enables determining that the hydrogen content of the core is 3.8 wt.% and the hydrogen content of the pendant is 11.6 wt.% (Figure 4). These values were independently verified by directly measuring the hydrogen content of the coke in the microcarbon test and the conversion independent hydrogen content of distillable liquid product of vacuum resid thermolysis[4]. In addition, the pendant-core model explains why CCR is conserved for physical separations (hydrogen is conserved), increases for thermal conversion (hydrogen escapes as part of hydrocarbon gas byproduct), and decreases for hydroconversion (hydrogen is added).

Selection of Resid Conversion Processes

A world shortage of resid conversion capacity has resulted in heavy crudes with a high fraction of vacuum resid being priced much lower than light crudes. This is encouraging refineries to increase resid conversion capacity and requiring producers of extra heavy oil to upgrade near the production site before marketing the oil. Most petroleum resid processes

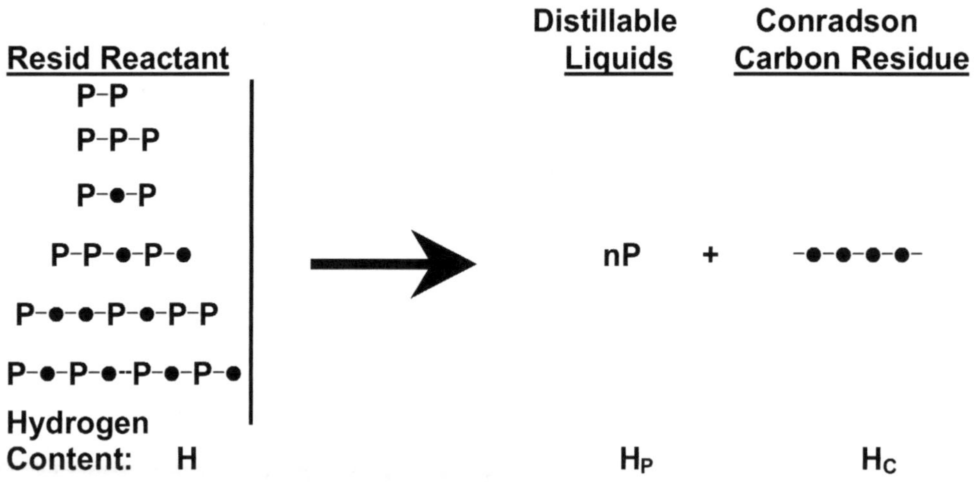

H Balance: $100\ H = (100\text{-}CCR)\ H_P + CCR\ H_C$

$$CCR = 100\ (H_P - H)\ /\ (H_P - H_C)$$

Figure 3. Model shows Conradson carbon residue (CCR) is a linear function of hydrogen content

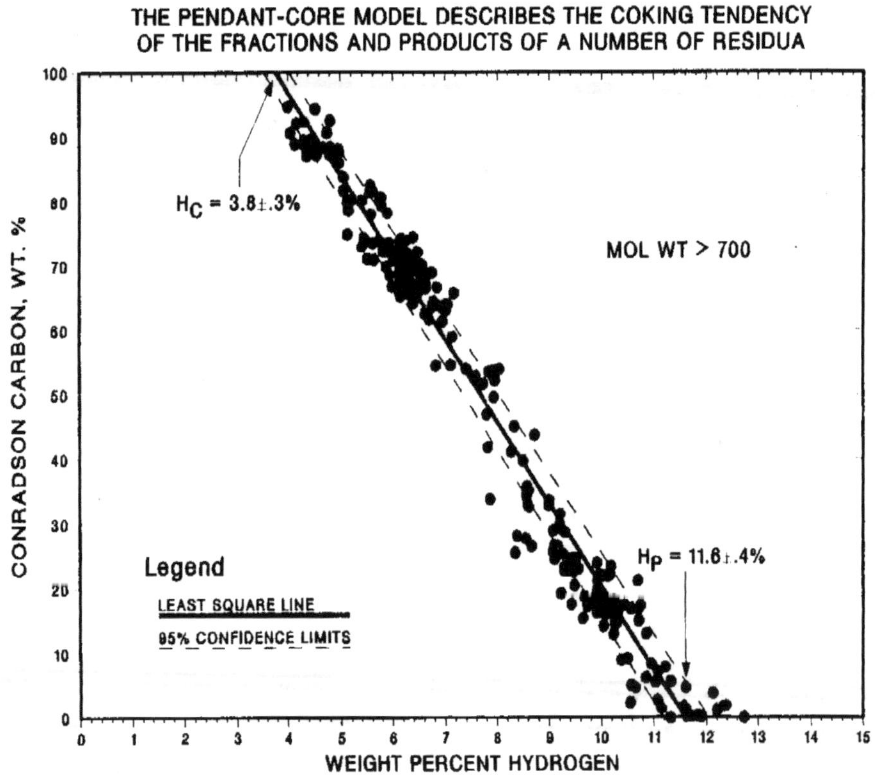

Figure 4. Conradson Carbon is a Linear Function of Hydrogen Content

can be made up of combinations of separations, thermal conversion, hydroconversion or hydrotreating, and fluid catalytic cracking. Distillation is the most common separation choice because of its low cost. However, fuels deasphalting is becoming more popular because it can separate saturates and small ring aromatics out of resids. Nevertheless, the selection of the conversion process depends on the quality of the resid and the selection of the crude for the refinery depends on the resid conversion process previously selected. Resids high in saturates and small ring aromatics and low in resins and asphaltenes are more appropriate for catalytic processing. On the other hand, resids high in resins and asphaltenes are more suited for coking.

Resid Catalytic Cracking[6]

Resid catalytic cracking has much better selectivity to desired products (high gasoline and low gas yields) than for coking or hydroconversion. In fluid catalytic cracking feed is sprayed on zeolite catalyst within a porous silica alumina support in a short contact time, riser reactor. The vaporized product flows to a fractionator while the catalyst with coke and adsorbed hydrocarbons flow to a fluidized bed regenerator where the coke and hydrocarbons are burned off the catalyst. Fluid catalytic cracking requires much higher quality feeds than coking or hydroconversion. This is because of expensive zeolite catalysts, intolerance to sodium, nickel, vanadium, and basic nitrogen as well as limitations on the amount of coke that can be burned in the regeneration step by cooling capacity. As a result, the feed in resid catalytic cracking is at worst an excellent quality atmospheric resid but mixtures of vacuum gas oil and atmospheric resids are more common. The total feed is limited to Conradson Carbon Residues of 3 to 8 wt.% , depending on the cooling capacity. This and other typical feed limitations are found in Table 1.

Table 1
Resid Fluid Catalytic Cracking Feed Guidelines[6]

Feed Property	Limit
Conradson Carbon Residue	3 - 8 wt%
Nickel Content	20 - 25 ppm
Vanadium Content	10 - 15 ppm
Sodium Content	5 - 10 ppm
Basic Nitrogen	800 ppm
Sulfur Content	3 wt%
Maximum Density	0.940 g/cc
Minimum Hydrogen	11.5 wt%

Deasphalting[7,8] and Separation Processes

Deasphalting can be used to separate the higher quality oil from vacuum resids for making a resid feed for fluid catalytic cracking (FCC). The deasphalted oil is typically blended with the vacuum gas oil and hydrotreated before catalytic cracking. This combination of deasphalting and hydrotreating can enable producing a good FCC feed from a low cost, poor quality atmospheric resid. Disposing of the asphalt can be problem because asphalt products and heavy fuel oil usually cannot absorb such high volumes. The most common choice is to add the fuels asphalt to coker feed. However, since the resulting coke is becoming difficult to sell, gasification of asphalt and/or coke is probably the better long term choice in many locations.

Deasphalting is done in a countercurrent treater (or separator) using n-propane, n-butane, isobutane, or n-pentane as the precipitating liquid. For fuels deasphalting the uniform quality is not as important as for lubes. What is more important for fuels is processing large volumes at low cost, both capital and expense. Hydrotreating further cleans up the deasphalted oil and FCC can tolerate some impurities. Therefore, fuels deasphalter treating towers are shorter (order of 40 ft. high by 13 ft. diameter) and contain packing instead of trays or a combination of packing and trays. Fuels deasphalters typically use butane or pentane at a low solvent to oil ratio (4-8) and do not heat the top of the tower.

Separation of Vacuum Resid Curves representing the separation of several vacuum resids are shown in terms of Conradson carbon residue (Figure 5) and vanadium (Figure 6) in the deasphalted oil as a function of yield of deasphalted oil[7]. The lowest concentration of

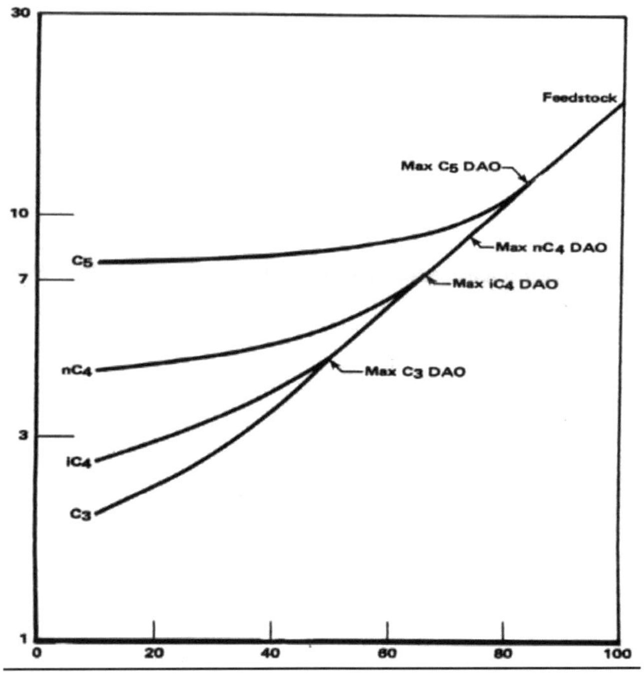

Yield of Deasphalted Oil, wt%

Figure 5. Conradson Carbon Residue (wt%) vs. Yield of DAO for Varying Solvent[7]

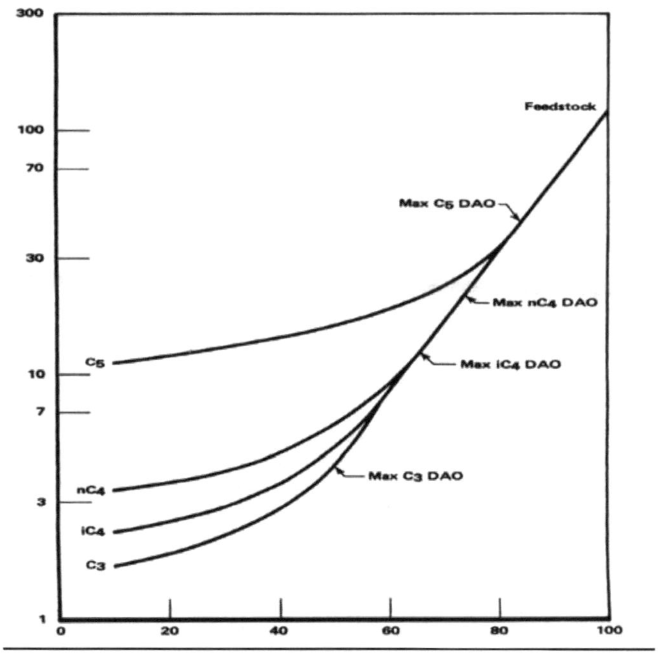

Yield of Deasphalted Oil, wt%

Figure 6. Vanadium Content (ppm) vs.Yield of DAO for Varying Solvent[7]

Conradson carbon and metals are in the propane deasphalted oil but the maximum yield is about 50 wt%. As the carbon number of the alkane solvent is increased, the possible yield of deasphalted oil is increased but so are the Conradson carbon and the metals in the deasphalted oil. The overlapping curves for the lowest concentration of each contaminant as a function of yield of deasphalted oil approximates the separability of vacuum resid. There are limits imposed by the molecules in the resid of how well we can separate according to Conradson carbon residue and this overlapping curve approaches this limit.

Molecular Limitation to Separation[9] The limit for separation on the basis of Conradson carbon is a molecular limit as demonstrated by the pendant-core building block model. While there is a large fraction of vacuum resids with no cores (zero Conradson carbon), there is little or no fraction with Conradson carbon above 50 wt%. Thus, since there is a large fraction with pendants and cores covalently bonded in the same molecules, there is a molecular limit of how well heavy oils can be separated on the basis of Conradson carbon. Thus, it makes no sense to research other ways to obtain a better separation than by deasphalting. Instead, one needs to crack bonds between pendants and cores by visbreaking or by hydroconversion prior to deasphalting. Alternatively, one could separate the low Conradson carbon fraction by deasphalting and rely on coking of the asphalt to reactively separate the pendants and cores. This alternative was already discussed.

Although the metals also appear to have a separation limit, an analysis of the concentrations of metals and the molecular weights present shows there cannot be a molecular limitation. Typically, all the vanadium and nickel in a vacuum resid can be contained in less than 1 wt% of the resid. However, the molecules containing the metals are in large polynuclear aromatics (porphrins) that separate much like other large polynuclear aromatics. It is this similarity that has limited the selective separation of metals.

Coking

In coking processes the resid is heated to high temperatures (450-500°C) with the light boiling products distilling overhead and leaving the solid coke behind. Coking is the most popular conversion choice for resids high in cores (low hydrogen, high heteroatoms, and high Conradson carbon) because it reactively separates the cores, the catalyst poisons, into a low value byproduct (coke) leaving the pendants for catalytic upgrading. Therefore, coking should be called a "catalyst poison rejection" process because a principal reason for coking is to improve the liquids for further upgrading by catalytic processing[10]. Unfortunately, coking has been mistakenly called a "carbon rejection" process, a term that continues to be repeated without critical examination. For example, Table 2 shows that the coke product in a microcarbon tester has a much higher concentration of sulfur, nitrogen, and metals than the vacuum resid feed and by atomic balance, the volatile liquid products. However, the concentration of carbon in the vacuum resid, in the coke, and in the liquid product is nearly the same. The rejection of heteroatoms is counter balanced by an increase of hydrogen content in the liquid product. This highly desirable result is because most of the polynuclear aromatic building blocks in the petroleum macromolecules contain heteroatoms that are reactively separated with the polynuclear aromatics into the coke.

Kinetics of Coke Formation and Mesophase The thermal cracking of resid is a first order reaction with a constant activation energy over a large range of temperatures[11]. There is an induction period before coke begins to form (Figure 7). It is now understood[12] that coke formation is triggered by the phase separation of a second liquid phase formed from partially converted asphaltene cores. Evidence of the liquid - liquid phase separation of asphaltene cores, promoted by the stacking of polynuclear aromatic structures, is the presence of spherical liquid crystalline coke[12], called the carbonaceous mesophase. Mesophase in a coke sample can be used as a diagnostic tool to show that it was caused by thermal cracking and insolubility at high temperatures[13].

Table 2
Elemental Balance Around Microcarbon Test

Element	Arabian Heavy Vac. Resid	22.3 wt% Microcarbon Coke	77.7 wt% Liquid Balance	Resid to Liquid % Change
Carbon	83.5%	85.8%	82.9%	-0.8
Hydrogen	9.9%	3.7%	11.7%	+18
Sulfur	5.8%	8.6%	5.0%	-14
Nitrogen	0.45%	1.3%	0.21%	-53
Vanadium	165 ppm	808 ppm	0 ppm	-100
Nickel	40 ppm	165 ppm	4 ppm	-90

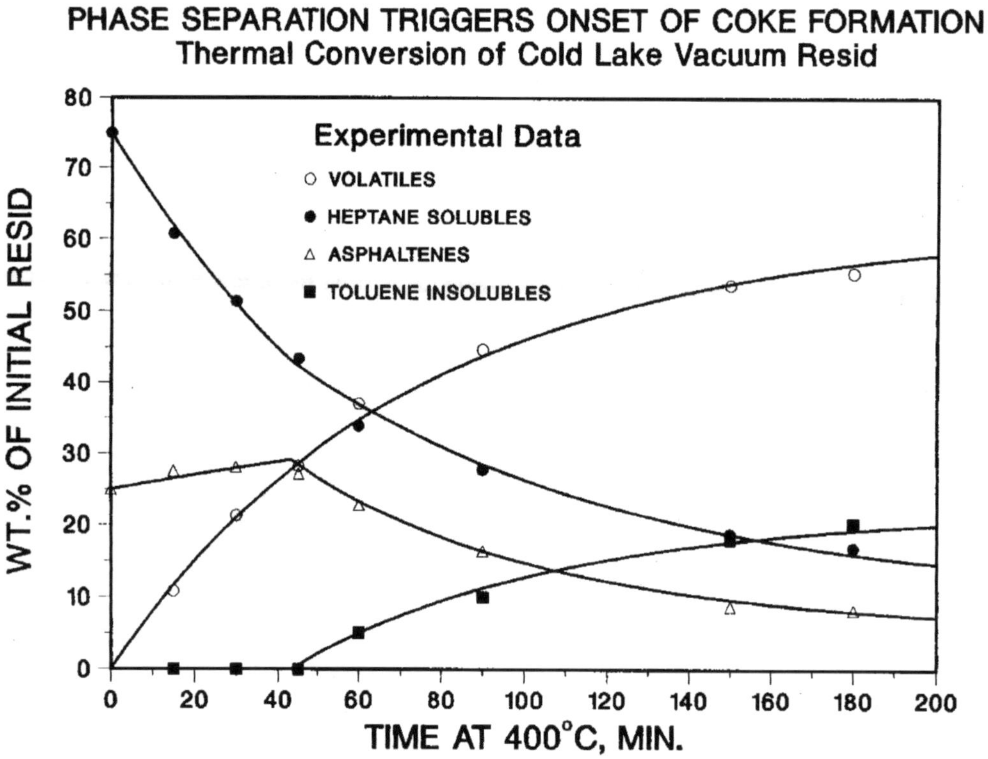

Figure 7. Kinetic Data Has a Coke Induction Period when Asphaltenes are Soluble

Delayed Coking[14,15] The delayed coking process is the oldest and most popular choice. The resid is heated by flowing through a long tube in a furnace and then reacted by flowing into the bottom of a high, cylindrical, insulated drum. The volatile products go overhead into a fractionator and coke accumulates in the drum. The heavy liquid product from the fractionator is recycled back to the furnace. Meanwhile, when the drum fills up with coke, the feed is switched to a second drum. The coke is mined out of the drum by drilling a hole down the center with high pressure water and cutting out the remainder also with high-pressure water to get the drum ready for the next 16 hour reaction cycle. Anode grade coke for making electrodes for manufacturing aluminum commands a higher price than fuels grade coke. However, the low specification on sulfur, vanadium, and nickel content for the coke requires a higher quality (high in saturates and small ring aromatics) and more expensive resid feed. Needle coke for making electrodes for steel manufacturing is priced even higher ($500/ton) than anode grade coke but it requires low sulfur, fluid catalytic cracker bottoms, not resid, as a feed.

Fluid Coking and FLEXICOKING[16] These processes are licensed by ExxonMobil. In Fluid Coking the resid is sprayed on a hot, fluidized bed of coke particles in a vessel, the reactor. The volatile products go overhead to a fractionator while the coke particle are removed out of the bottom and transferred to another vessel, the burner, where the coke is partially burned with air to provide the heat for the process. This coke then is recirculated back to the reactor. Since this process produces much more coke than is required for heat, fluid coke is withdrawn at the bottom of the reactor. This is a high sulfur, fuels grade coke of low value. A heavy liquid product is recycled back to the reactor from the fractionator. This is used to scrub coke fines from the reactor vapors and to improve the quality of the liquid product. In FLEXICOKING a third vessel, the gasifier, is added to the Fluid Coker. In the gasifier coke is gasified with steam and air in net reducing conditions to produce a low BTU gas containing hydrogen, carbon monoxide, nitrogen, and hydrogen sulfide. After adsorbing out the hydrogen sulfide, this low BTU gas is burned as a clean fuel within the refinery and/or in a nearby power plant. Only small amount of coke (about 3%) needs to be withdrawn in FLEXICOKING to keep vanadium and nickel from accumulating in the gasifier. With the trend of decreasing high sulfur coke prices, gasification of coke to form fuel and syngas to be used within the refinery may be the only economical and environmental option for refineries in the future.

Reactive Separation by Coking Coking has no limitation on separation due to the bonds between pendants and cores because these bonds are thermally broken during the coking process to reject the cores at 100% CCR. However, previously[17] we discussed three limitations to coking chemistry:

1. Secondary cracking of volatile liquid products
2. Formation of 5+ ring polynuclear aromatics by combining of reaction fragments containing smaller ring aromatics
3. Aromization of hydroaromatics to form 5+ ring polynuclear aromatics

Any significant improvement in the yield of high quality liquids depends on one devising ways to overcome these limitations. Secondary cracking degrades the quality of the most desirable product while forming hydrocarbon gases, an undesirable byproduct. Previously[17], we showed that the secondary cracking of volatile liquid products can be minimized by using a coking reactor with a very short vapor residence time but with a long resid residence time. This latter feature is necessary to go to complete conversion to coke of 100% CCR. Figure 9 shows the result using four different ideal laboratory-coking reactors with these two features and a feed of Arabian Heavy vacuum resid[17]. Hydrogen content of the volatile liquids is used as the measure of quality and the resulting yield-quality from the different reactors is compared with a coking model prediction for a commercial coking process. Either the quality can be improved at the same yield or the yield can be increased at the same quality. While this 7.9 wt.% increase in yield (69.3 wt.% total yield) provides incentives to improve existing processes, it falls short of the requirement for a new coking process. Of course, one must realize that no commercial coker is going to match the yields obtained from a small, ideal laboratory reactor. Therefore, to justify developing a new coking process, one needs to both reduce secondary cracking and minimize the remaining limiting chemistries.

Coking-Separation Synergy The chemical combination of aromatics and aromization of hydroaromatics are two ways that coking degrades the resid feed. Actually, both of these chemistries occur during the microcarbon test for measuring Conradson carbon residue (CCR). Therefore, part of the CCR (5+ ring polynuclear aromatics) is formed during the measurement. As a result, if these chemistries are minimized, one should form less coke than CCR while commercial cokers form 1.2-1.6 times as much coke as CCR[14, 15,16]. Since coking degrades the feed, it makes sense to separate the portion of the resid that already meets the quality requirement before coking. At least this portion will be spared from being degraded. In the laboratory we decided to use high cut-point vacuum (Distact) distillation as the only separation process. Deasphalting could also be used on the feed. Thus, the overhead of high cut-point distillation to an atmospheric equivalent boiling point of 1300°F is combined with the overhead product of coking the 1300°F+ resid in an enhanced microcarbon tester[16], that is capable of short vapor and long resid residence times while collecting the volatile liquid product. Since the microcarbon tester measures CCR, the yield of coke is the CCR of the 1300°F+ resid or on a resid basis it is this value times the fraction of the resid that is 1300°F+. A schematic of the process and the calculation of the coke yield for three different vacuum resids are in Figure 8. For each resid the coke yield is less than the CCR of the resid feed (average of 0.8 times CCR) but since it also depends on the fraction of 1300°F+, the coke yield is not proportional to resid CCR. Therefore, Syrian Light resid, that has a higher CCR than Arabian Heavy resid, has a lower coke yield.

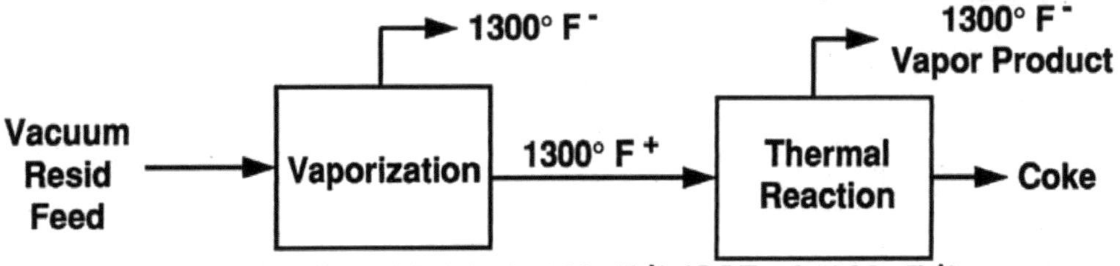

Minimum Coke = (Yield of 1300° F$^+$) (CCR of 1300° F$^+$)

Berri Vacuum Resid (CCR = 18%): Coke = (.382)(35.1) = 13.4 wt%

Arabian Heavy Vacuum Resid (CCR = 22%): Coke = (.546)(35.9) = 19.6 wt%

Syrian Light Vacuum Resid (CCR = 23%): Coke = (.541)(31.9) = 17.3 wt%

Figure 8. Separation before low vapor residence time coking reduces coke yield[18]

In order to measure the yield and quality of the liquid product, the 1300°F$^-$ fraction is added to the liquids collected in the enhanced microcarbon tester as shown in Table 3. The

Table 3
Yields and analyses of products of separation plus low vapor residence time coking on Arabian Heavy vacuum resid

	Yield (wt%)	C (wt%)	H (wt%)	S (wt%)	N (wt%)	V (ppm)	Ni (ppm	CCR (wt%)
1300°F- Feed	45.4	84.33	11.11	4.12	0.23	45	7	7.4
Coking Liquids	31.4	84.33	11.32	3.99	0.24	0	0	0
Total Liquids	76.8	84.33	11.20	4.07	0.24	27	4	4.4
Coke	19.6	84.99	3.85	8.85	1.5	835	260	100
Gas (By Diff.)	3.6							
Arab Heavy Resid	100	83.67	10.18	5.13	0.42	190	55	22.4
% Reduction				20.7	43.8	85.9	92.6	80.5

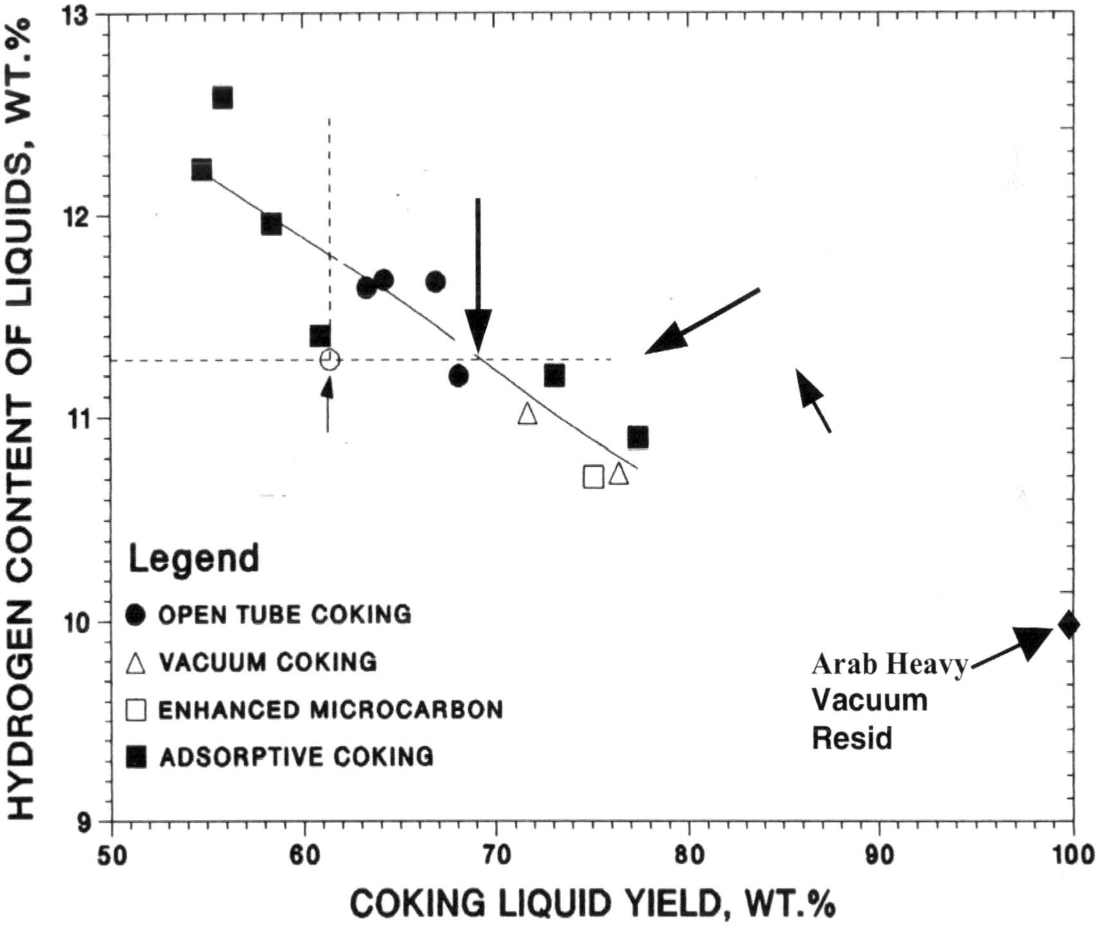

Figure 9. **Separation prior to low vapor, high resid residence time coking gives high liquid yield**

result is that the liquid yield is increased by 14.8 wt% over the commercial coker at only a slightly lower quality (11.20 wt% hydrogen as compared with 11.28 wt% hydrogen predicted for a commercial coker). Even while separating part of the resid without coking, the process removes 21% of the sulfur, 44% of the nitrogen, 86% of the vanadium, 93% of the nickel, and 80% of the Conradson carbon residue. Since the gas yield is 3.6 wt.%, there is still room for further improvement towards the upper bound of a liquid yield of 85.3 wt.% for no gas using hydrogen balance (Figure 9). Therefore, this result shows that the combination of the separation of the higher quality fraction of resid without coking and coking of the lower quality fraction of the resid at long residence time, but with short vapor residence time, can achieve a substantially higher liquid yield over commercial cokers at nearly the same liquid quality. Thus, despite the current perception, the coking of petroleum resids has the potential for significant improvement.

Visbreaking

This is a low conversion thermal process used originally to reduce the resid viscosity for heavy fuel oil applications. However, today it uses a resid that exceeds minimum heavy fuel oil specifications and converts just enough to obtain 15-30% transportation boiling range liquids and still have the heavy product meet heavy fuel oil specifications. Since this process cannot tolerate coke formation, it is required to be within the coke induction period that may limit conversion. Increasing the solubility of converted asphaltenes in the reacting oil can increase the induction period and thus, the conversion. An example is to strip out the volatile liquid product, a nonsolvent for converted asphaltenes[13]. A visbreaker reactor may be a furnace tube followed by a soaker drum or a long tube coiled within a furnace. Upsets cause coke to form and accumulate on visbreaker walls so that they need to be decoked periodically.

Hydroconversion

Hydroconverion combines thermal cracking with hydrogenation. The active catalyst needs to be a transition metal sulfide because the high level of sulfur in resid feeds will poison other hydrogenation catalysts. The addition of hydrogen increases the coke induction period, primarily by terminating free radicals and reducing the frequency of aromatics from combining to form larger PNA's. Thus, hydroconversion of vacuum resids to volatile liquids can be over 85% as opposed to 50-60% for commercial coking. However, one has to deal with the cost of hydrogen and catalyst, high-pressure vessels, poisoning of catalysts, the difficulty of asphaltenes to diffuse through small pores, and the intolerance to coke and sediment formation. Since metals removal is one of the fastest reactions and since the metals accumulate in the pores of supported catalysts, it is common to have a guard bed in front of the fixed bed. In order to hydrogenate the largest macromolecules in the resid, the asphaltenes, some or all of the catalysts need to have pores 50 to 100 microns in diameter. Even with these precautions it is difficult to get longer than one year run lengths on fixed bed hydrocoversion units with vacuum resid feeds and conversions to volatile liquids of 50% or more. This is because of catalyst deactivation with coke or by coke and sediment formation downstream of the reactor. The LC Finer and H-Oil units mechanically ebulate the catalyst so that it can be mixed and replaced onstream. The conversion is greatly dependent on the feed but conversions of vacuum resids to volatile liquids of the order of 70% are possible. Often these units are limited by the deposition of coke and sediment downstream of the reactor in hot and cold separators[19].

One way to avoiding pore plugging of catalysts is to use finely dispersed catalysts[20]. Dispersed catalysts also can be continuously added in low enough amounts (i.e. 100 ppm) to consider them throwaway catalysts with the carbonaceous by-product. However, economics usually dictate some form of catalyst recycle to minimize catalyst cost. Nevertheless, by designing the reactor to maximize the solubility of the converted asphaltenes[13], the conversion of vacuum resids to gas and volatile liquids can be above 95% with greater than 85% volatile liquids. However, the last 5 to 10% conversion may not be worth the cost of hydrogen and reactor volume to produce hydrocarbon gases and very aromatic liquids from this incremental conversion. The answer depends on the value and use of the unconverted carbonaceous liquid byproduct (essentially all PNA cores). Recently, ENI[21] is developing a dispersed catalyst, hydroconversion process using deasphalting to remove the deasphalted oil product and recycling a major part of the asphalt with the dispersed catalyst. This satisfies the need to recycle catalyst and to maintain solubility of converted asphaltenes. By not requiring hydrogenated rings to be cracked, the reactor can be run at lower temperatures and still obtain reasonable kinetic rates.

Future Improvements of Resid Conversion Processes

We are still on the steep part of the learning curve as to the characterization, phase behavior, and conversion chemistry of petroleum resids. As a result, much room remains to improve resid conversion processes. However, the rate of construction of new resid conversion units in refineries has been decreasing. The future growth appears to be at or near heavy crude production sites to decrease heavy crude viscosity and improve the quality to ease transportation and open markets for crude oils of otherwise marginal value. As fuel grade coke continues to reduce in value, gasification of coke, deasphalted bottoms, or even poor quality vacuum resid to produce electrical power, refinery fuel, and syngas for chemicals will become a more popular alternative[22].

References

1. Fitzer, E., Mueller, K., Schaefer, W., in "Chemistry and Physics of Carbon", P. L. Walker, ed., Marcel Dekker, 7, 237 (1971).
2. Wiehe, I.A.,Ind. Eng. Chem. Res., 31, 530 (1992).
3. Wiehe, I.A., Liang, K.S., Fluid Phase Equilibria, 117, 201 (1996) and Fuel Sci. & Tech. Int'l, 14, 289 (1996).
4. Wiehe, I.A., Energy and Fuels, 8, 536 (1994).
5. Noel, F., Fuel, 63, 931 (1984).
6. Barnes, P.H., Proceedings of 1st International Conference on Refinery Processes, AIChE, 227 (1998).
7. Sprague, S.B., NPRA Annual Meeting, Paper # AM-86-36 (1986).
8. Low, J.Y., Hood, R.L. and Lynch, K.Z., PREPRINTS ACS Div. Pet. Chem., 40, 780-784 (1995).
9. Wiehe, I.A., Proceedings of Conference on Energy and the Environment, AIChE, 34 (2000).
10. Wiehe, I. A., ACS Div. Pet. Chem. Meeting, Chicago (1995).
11. Olmstead, W. N., Freund, H., AIChE Spring National Meeting, New Orleans, 1998.
12. Wiehe, I.A., Ind. Eng. Chem. Res., 32, 2447 (1993).
13. Wiehe, I. A., ACS Div. Pet. Chem. Meeting, San Francisco (1997).
14. Ellis, P. J. and Paul, C.A., Proceedings of 1st International Conference on Refinery Processes, AIChE, 151 (1998).
15. G. C. Hughes, B. J. Doerksen, A. M. Englehorn and S. Romero, Proceedings of the 3rd International Conference on Refinery Processing, AIChE, 66 (2000).
16. Hammond, D.G., Proceedings of 1st International Conference on Refinery Processes, AIChE, 170 (1998).

17. I. A. Wiehe, "The Enhanced Microcarbon Tester as a Laboratory Coker", Presentation at AIChE Spring Meeting, Atlanta (2000).
18. Wiehe, I.A., Proceedings of 4th International Conference on Refinery Processes, AIChE, 245 (2001).
19. Nowlan, V.J. and Srinivasan, N.S., Fuel Sci. & Tech. Int'l, 14, 41 (1996).
20. Bearden, R. and Aldridge, C.L., US Patent 4,134,825 (1979) and ACS Div. Pet. Chem. Meeting, ACS Award in Petroleum Chemistry Address, San Francisco (1997).
21. Delbianco, A., Panariti, N., et al., Proceedings of 3rd International Conference on Petroleum Phase Behavior and Fouling, AIChE, 520 (2002).
22. Penrose, C.F., Rodarte, A., Wallace, P.S., and Preston, W.E., Proceedings of 3rd International Conference on RefiningProcessing, AIChE, 385 (2000).

Tailoring RTDs in Fluidized Beds for Improved Coker Yields - The Envision Technologies Corp. Cross-Flow Coker

ROBERT PINCHUK[1], WAYNE BROWN[1], GERARD MONAGHAN[1], WILLIAM McCAFFREY[2], OSWALDO ASPRINO[2], PEDRO GONZALEZ[2]*

[1] *Envision Technologies Corp., Calgary, Alberta T2C 4T3, Canada*

[2] *University of Alberta., Edmonton, Alberta, Canada*

Abstract

The Envision Technologies Cross-Flow (ETX) Coker has been engineered to take advantage of the increased understanding of heavy oil processing gathered over the past 50 years. Based on a cross-flow fluidized bed, the ETX design provides for a particulate phase residence time distribution that approaches plug-flow; this allows for a smaller reactor than other industrial processes, and eliminates feed losses due to short circuiting. The ETX fluidized bed is also shallow, which reduces gas phase hold-up and product losses to over-cracking. Research is presented here which substantiates the solid and gas phase mixing patterns that lead to these advantages. Results of experiments on the effects of temperature on coking product yields and quality are also presented showing the potential for a 9% increase in product yield by reducing coking temperatures from 530°C to 470°C. The ETX design is able to capture 4% of this potential yield increase with no sacrifice of reactor capacity as would be seen with other industrial coking processes.

Introduction

In the current economic climate, the optimum coking process is one that maximizing the production of the condensable liquid products. To achieve this goal the process must ensure that all of the feed introduced into the reactor has reacted. Reaction times for coking are on the order of 1 minute at commercial processing conditions[1]. Once the feed has reacted and the liquid products have evolved into the gas phase it becomes important to remove them from the reactor quickly so that they can be quenched before they crack further. In addition, since vacuum toping can only remove a portion of the liquid products contained in the feed, the ideal coker should also allow for the vaporization of feed components up to 524°C with minimal reaction. These ideal attributes require the collection and quenching of the gas phase products within seconds of their evolution from the liquid phase. The large difference in optimal gas and solid phase residence times posses some serious constraints on the coking reactor design.

Although the impact of the gas phase residence time has been documented[2], only recently has evidence been produced suggesting that liquid phase mass transfer may also play an important role in determining coke yields[3]. In these studies, the coke yield was shown to increase as the reacting liquid film thickness was increased. Ideally, thin films allow for the rapid vaporization of light products, which minimizes the chance of losing these products to condensation reactions or physical trapping in a solidifying liquid phase.

The ETX is a new approach to heavy oil upgrading based on a cross-flow fluidized bed (Figure 1). Cross-Flow fluidization is an industrial process that is most commonly applied in the drying industry. The term cross-flow fluidization is used because the bulk flow of the fluidized solids is horizontal and perpendicular to the flow of the fluidizing gas. The mixing anisotropy associated with fluidized beds results in a residence time distribution for the flowing solids that is very uniform[4]. Similarly to other chemical processes, this approach to plug-flow in the particulate phase allows for a smaller reactor volume because short-circuiting is eliminated. In addition to increased capacity, the reactor configuration provides for a short gas phase residence time, which reduces the gas phase reactions that can lead to the loss of product value through over-cracking. The independent nature of the solid and gas fluxes allows for the residence times of the two phases to be optimized independently, which is vital in a coking system where the optimal residence times of the two phases can differ by an order of magnitude.

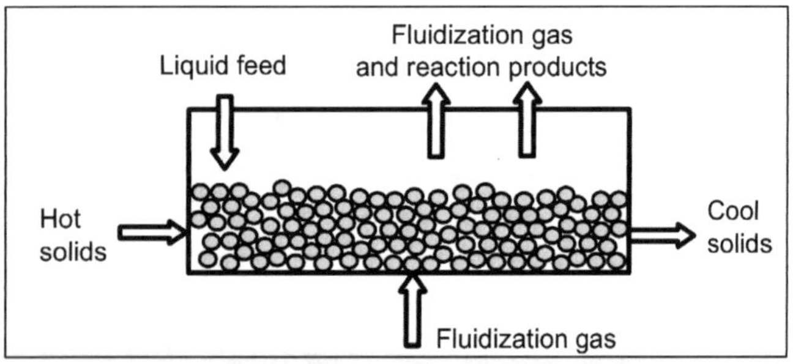

Figure 1. Schematic of the ETX coker showing the cross-flow fluidized bed.

Materials and Methods

Cold Flow Modeling

Cold flow modeling of the ETX design was used to investigate the solids and gas mixing characteristics of the bed. All experiments were carried out in a fluidized bed designed to emulate the performance of the commercial unit at reactor conditions. Scaling was based on the criteria put forth by Horio et al.[5], matching both the Froude number ($u_0/\sqrt{gd_p}$) and the velocity ratio (u_0/u_{mf}). Based on the results of the scaling exercise, glass beads with a density of 2500 kg/m^3 and a mean particle size of 0.180 mm were used in the study to maintain dynamic similitude with the commercial design

The bed was fluidized with air and could be operated with superficial fluidization velocities up to 0.51 m/s. The unit had a footprint of 0.75 m by 0.75 m, and a height of 2.0 m. The walls and bottom of the bed were made of acrylic, so that the characteristics of the bed could be visualized. Gas distribution was achieved by means of 16 bubble caps arranged in a parallel configuration. Slumped bed heights between 0.4 and 0.8 m were studied with this unit.

Gas tracer experiments

Experiments with a gas tracer were used to quantify the residence time distribution of the continuous phase in the system. Propylene was used as the gas tracer for all such experiments. Tracer gas was introduced directly into eight of the 16 bubble caps in the distributor. A photo ionization detector (PID) was used to quantify the concentration of propylene in the bed. The PID showed a linear response to the range of propylene concentrations used. Withdrawal of gas from the bed into the PID was accomplished by means of a tube 0.99 m in length and 0.794 mm in internal diameter. The sampling tube was located just above the surface of the expanded bed, at a distance 0.375 m from each wall. Based on the dimensions of the tube, this sampling method introduced a lag of 1.8 ms, at a nominal withdrawal rate of $1.1x10^{-3}$ m^3/min. Data was acquired at a rate of 225Hz from the PID.

Negative step change experiments were performed in which the flow of tracer gas was turned on until the concentration of tracer in the bed reached steady state; the flow of tracer gas was then stopped and PID measurements were recorded with time until no change in the output signal was observed. The raw data from the PID was first scaled and normalized linearly between the maximum and minimum recorded output values. To reduce the noise in the data, the normalized data from three experimental runs were combined then filtered using a centered 100 point moving average.

Solid tracer experiments

The horizontal mixing of the solid particles was assessed using glass tracer particles of the same density as the bed particles, but with a nominal size of 550 microns. The tracer particles were dyed black to assist in quantifying their concentration in withdrawn samples. The solid tracer experiments were commenced by releasing a column of tracer solids into the bed at a particular instant. This was accomplished by means of a tube 4.08 cm in diameter, with the same height as the expanded bed. The tube was located 5.84 cm from the nearest wall. The ends of the tuber were sealed with plugs. The plugs could be held stationary by means of a rod connecting the two, allowing the tube to be moved independently. The tracer was released as a column by quickly raising the tube while holding the plugs stationary.

After the tracer pulse was released samples were withdrawn from a point located half way up the bed, on the wall of the vessel furthest from where the tracer particles were introduced. Samples were taken through a 2.54 cm ball valve tapped into the wall of the vessel. Tracer particle concentrations were measured by separating the tracer from the bulk of the fluidized particles by means of a 0.33 mm (nominal) sieve and weighing the particles that remained on the sieve.

Hot coking experiments

The effect of temperature on coking product yields was investigated with the use of a small coking reactor (Figure 2). The reactor had a base 5.4 cm in diameter, while the neck was 1.2 cm, and the total height was 10cm. A molten salt bath was used to heat the reactor. Nitrogen sweep gas was flowed through the system at a rate of 2.1m^3/sec. Athabasca vacuum residue (524°C+) was cracked between temperatures of 450°C and 550°C. The thickness of the liquid film before cracking was 250 microns. The gas phase product stream was passed through a condenser, which was comprised of a glass U-tube filled with quartz batting immersed in liquid nitrogen. Reaction times were chosen conservatively to ensure complete reaction, based on literature values[1].

Coke was defined as all of the material that remained in the reactor after the reaction. Material collected in the condenser was considered condensable products. The remainder of the mass balance was attributed to non-condensable gasses. Closure of the mass balance to within 2% was confirmed in independent experiments designed for this purpose by adding n-hexadecane to the reactor drop-wise. Both the coke and the condensed product were analyzed for carbon, nitrogen and hydrogen content. The aromatic carbon and olefin content of the liquid products were assessed by a combination of ^{13}C and ^{1}H NMR.

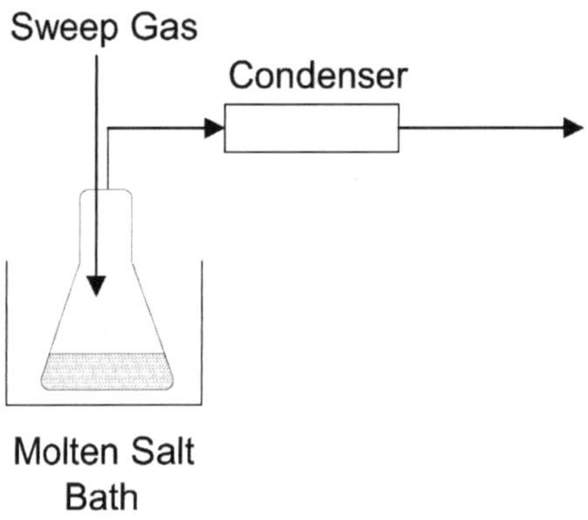

Figure 2. Coking reactor used for hot coking experiments

Results and Discussion

Gas and Solid phase residence time distribution

The fluidized solids provide both the heat for the coking reaction and a surface on which the reaction takes place. Because the solids carry the reacting liquid, knowing the residence time distribution of the solids is essential for proper reactor sizing. Commercial fluid coking reactors can be considered well mixed with respect to the solids[6]; this is expected for a fluidized bed system where the solids inlet and outlet points are separated vertically, since mixing in this direction is very good in fluidized systems. In the cross-flow design the solids inlet and outlet points are separated from each other horizontally. Typically, fluidized beds have solids mixing rates that are ten times less in the horizontal direction than in the vertical direction[4]. This leads to an expected drop in the Peclet number by an order of magnitude, and a solids residence time distribution that approaches plug-flow.

To confirm the approach to plug-flow of the particulate phase in the ETX design, the horizontal solids mixing rates were measured in the model bed. This was done by releasing a plug of a particle tracer from one side of the bed and measuring the concentration on the opposite side. The horizontal dispersion of the solids was extracted from the tracer studies using a two dimensional dispersion model. The measured dispersion coefficients varied between 2×10^{-2} and 2×10^{-3} m^2/s, and were shown to be sensitive to bed depth and fluidization velocity. Both the values and the sensitivities of the horizontal dispersion were in good agreement with the literature[4].

Once the horizontal solids dispersion was determined, a solids residence time distribution for the ETX design could be calculated by superimposing the solid mixing on a bulk flow. Figure 3 shows the solids residence time distributions estimated through this approach for a typical ETX reactor. Even in the case of maximum horizontal solids mixing, the ETX coker can be made 20 times smaller than a well-mixed reactor and still provide the same solids/feed hold-up.

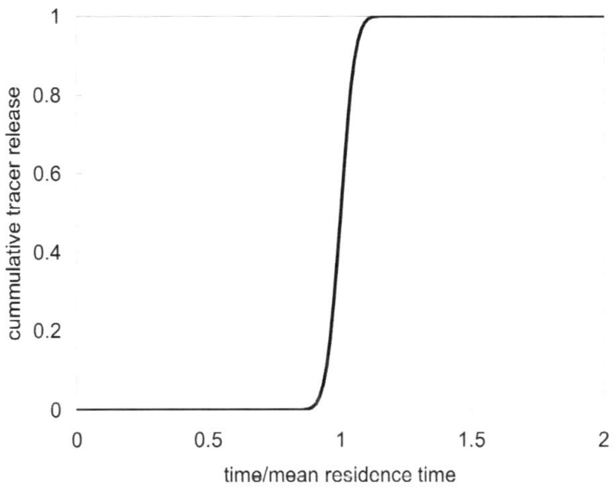

Figure 3. Predicted residence time distributions for the ETX coker design

The gas phase degradation of products can lead to a significant reduction in product yield and quality[2]. The shallow bed of the ETX design allows for a quicker release of gas phase products after they have evolved from the reacting liquid. This reduces the extent of gas phase over-cracking reactions, which can improve product yield and quality. The impact of reactor configuration on over-cracking was established by combining a gas phase mixing model[4] with an over-cracking reaction model[2]. The gas phase mixing model was validated experimentally in the cold flow bed using a gas tracer. Examples of data gathered from these experiments and the model predictions are shown in Figure 4. It should be noted that the model was not fit to the data, but was calculated based on empirical correlations from the literature. The effect of bed depth and fluidization velocity on overcracking losses as predicted by the coupled models is shown in Figure 5. The gas mixing model allowed for some interpolation and extrapolation on the effects of reactor configuration, specifically the effects of bed depths and fluidization velocity. Because the gas phase in fluidized beds demonstrates significant back-mixing, the effect of reducing bed depth on over-cracking losses is much more substantial than would be expected if the gas phase followed a more intuitive plug-flow mode of mixing.

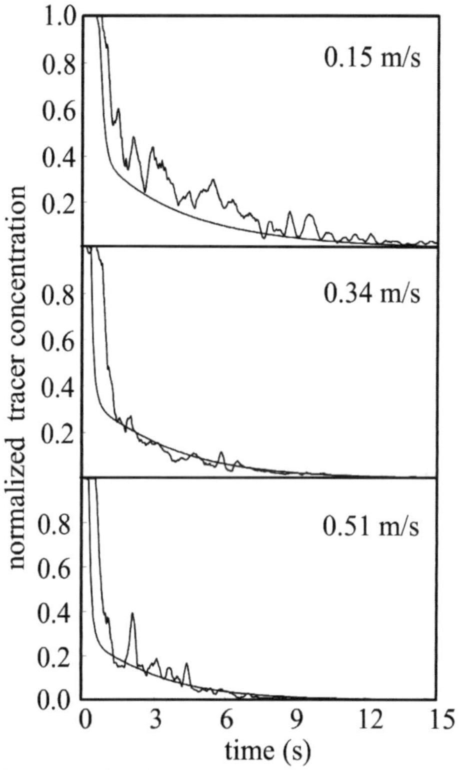

Figure 4. Gas tracer studies showing both the experimental data and the gas phase mixing model. Data shown is for a slumped bed height of 0.8m; fluidization velocities are shown on the figure.

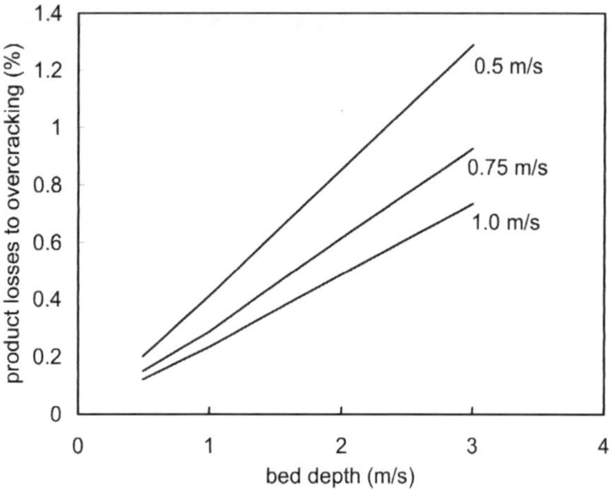

Figure 5. Losses of liquid product to over-cracking at varying bed height and fluidization velocity. Predictions are based on the combined gas phase mixing model and over-cracking kinetic model.

Temperature

A common approach implied by many researchers is to divide the coking process into two separate steps: A liquid-phase step, which includes the liquid phase reactions and associated mass transfer processes, and a gas-phase step, which describes the gas phase reactions leading to product over-cracking and potentially coke formation.

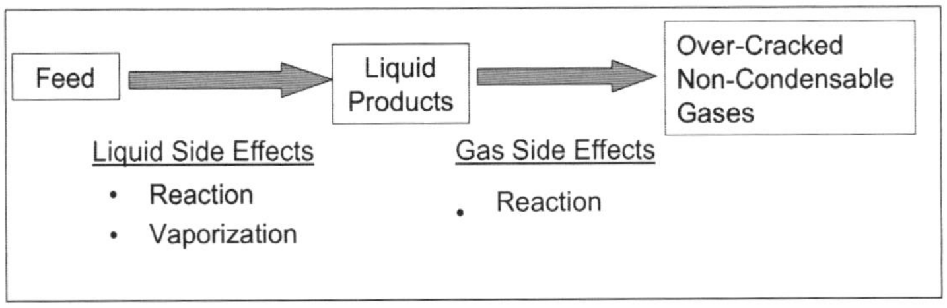

Figure 6. Sequential liquid and gas-side effects in the production of liquid products

Reducing the reactor temperature in coking has been shown to reduce overcracking losses by reducing the rate of overcracking. This is taken into account through the over-cracking kinetic model discussed above[2]. Using this reaction model it can be estimated that there is the opportunity in commercial fluid coking units to decrease product losses to overcracking by 5% (absolute) by reducing reactor operating temperatures from 530°C to 470°C.

In addition to the gas-side effect of over-cracking, recent work into the importance of mass transfer on liquid product yields has suggested that temperature may also play a role through a liquid-side mechanism, which was investigated with the use of the coking reactor described above (**Figure** 2)[3].

Reaction temperature was shown to have no effect on coke yield between temperatures of 450°C and 550°C (Figure 7). However, it was found that the hydrogen content of the coke decreased slightly as the temperature was increased. Other researchers have also observed an insensitivity of coke yield to reaction temperaure[1,8], but a small positive dependence of coke yield on temperature was seen in an industrial example[7]. Unlike the coke yield, reaction temperature was shown to have a significant impact on both the gas and liquid yields: Increasing reaction temperature was shown to decrease the condensable liquid yield and increase the non-condensable gas yield, which has also been observed in commercial reactors[7]. Liquid yields are shown in Figure 8. The decrease in liquid yield was accompanied by a decrease in liquid product quality as shown by a decrease in the hydrogen content and an increase in the aromatic carbon content (Figure 9). The olefin content in the liquid product was independent of reaction temperature.

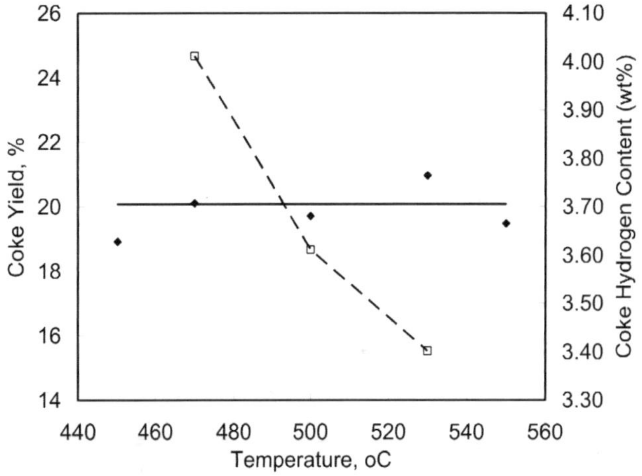

Figure 7. Effect of reaction temperature on coke yield.

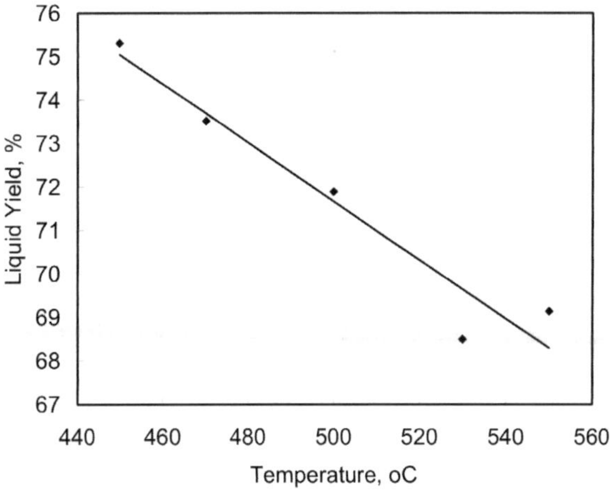

Figure 8. Effect of temperature on condensable liquid yield

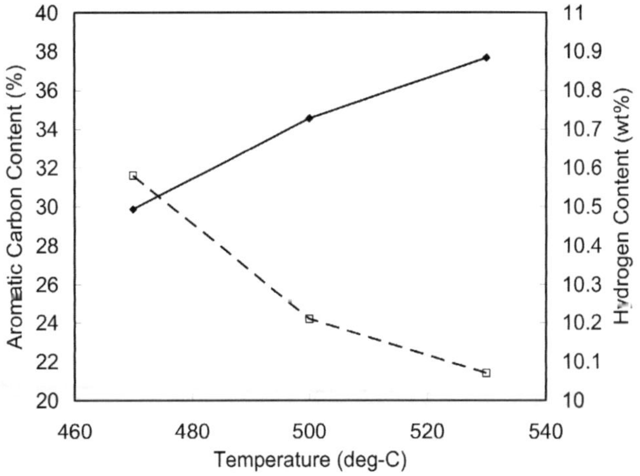

Figure 9. Quality of the condensable liquid product at different reaction temperatures. Quality was measured by hydrogen content and aromatic carbon content.

The mean residence time of the gas phase in the coking reactor was approximately 3 seconds, based only on the flow of sweep gas. The temperature of the gas phase exiting the reactor was measured during cracking experiments, and was shown to be a maximum of 200°C. Previous research into the liquid-side mechanisms of coking showed that gas-side effects where negligible when the gas-phase temperature was below 470°C for mean residence times of 3 seconds[3]. In addition, the data of Levy showed that the rates of gas phase reactions only become significant at a gas-phase temperature of 500°C, and that insignificant losses would occur at this temperature for stated mean residence time of 3 seconds[2].

The final distribution of hydrogen amongst the coking reaction products was found to be a strong function of temperature (Table 1). At the three reaction temperatures studied, the final hydrogen content of the coke remained approximately constant, while the hydrogen contained in the liquid products decreased by 9% (absolute) when the temperature was raised from 470°C to 530°C.

Table 1. Final distribution of hydrogen amongst reaction products (wt%)

Reaction Temperature	Liquid Products	Gas	Coke
470°C	82%	9%	8%
500°C	78%	13%	7%
530°C	73%	16%	8%

Impact on the ETX design

Demonstrating the benefits of the ETX design is most easily done by first establishing a base case design relative to an industrial benchmark. Fluid coking is an ideal benchmark since it is an industrially proven technology and is similar to the ETX coker. Typically, Fluid Cokers operate with a mean reactor temperature of 530°C [7]. Since the ETX design is also a fluidized bed coking process, operating at the same temperature should lead to a similar coke yield.

Taking 10,000 bbl/day as a feed basis, and a mean reaction temperature of 530°C, a reactor volume of 14 m^3 is required to provide the necessary hold-up for complete feed conversion. An additional 2m is added to the reactor length to compensate for the small amount of horizontal mixing and to provide space for solid feed and withdraw, and the reactor height is doubled to allow for solids disengagement. The final dimensions of the reactor vessel are 3m high, 2m wide and 6.5 m long. At this size, the ETX coker capacity is 65% larger than that of the fluid-coker operating at the same mean reaction temperature. In addition to the capacity increase, the lack of feed short-circuiting eliminated loses by this mechanism. The relatively shallow bed of the ETX unit reduces the gas phase residence time and the extent of over-cracking. The expected liquid yield benefit from the shallow bed is 2% even at this elevated temperature.

As shown, reducing the coking temperature has a beneficial impact on liquid yield through both gas and liquid side mechanisms. When both effects are considered a 9%

increase in liquid yield is achievable by reducing the reaction temperature from 530°C to 470°C. As with any reaction system, lower operating temperatures result in a slower rate of reaction, and reduced reactor capacity. The rate of feed conversion in coking has been shown to double with an increase in temperature of approximately 20°C[8]. Achieving the full 9% yield benefit would require an increase in reactor volume of 8 times. The lower operating temperatures would also likely require additional solids flux to reduce the chances of bogging.

Because the ETX coker already has an advantage in terms of capacity, it is worthwhile considering what fraction of the 9% yield benefit this translates into. At a mean reactor temperature of 515°C the capacity of the ETX reactor is equal to that of a fluid coker operating at 530°C. At this temperature the ETX design captures 4% of the potential 9% yield benefit with no loss of capacity compared to a commercial fluid coker. This includes both the yield benefit of temperature and the reduction in over-cracking from the reduced gas phase residence time.

Conclusions

The benefits of the ETX coker design are related to improved management of the residence times of the particulate and gas phases in the reactor. The residence time distribution of the solid particulate phase approaches that of a plug-flow reactor: in addition to eliminating feed losses through short-circuiting this also allows for a smaller reactor. The shallow bed of the ETX design reduces the hold-up of the gas-phase reaction products, which reduces the extent of gas-phase over-cracking reactions and product losses through this mechanism.

Temperature was shown to play a significant role in liquid product yields through both the gas-side mechanism of over-cracking and a liquid-side mechanism. It was estimated that a total combined increase in liquid yield of 9% can be achieved by lowering the coking reaction temperature from 530°C to 470°C. The increase in liquid yield was also shown to be accompanied by an increase in product quality. Because of the improved solids mixing, the ETX design is able to capture 4% of the potential yield increase with no loss of capacity when compared to similar industrial processes.

Acknowledgements

Envision Technologies would like to thank the National Research Council of Canada's Industrial Research Assistance Program (IRAP) and the Natural Sciences and Engineering Research Council of Canada for funding aspects of this research and Heather Dettman of the National Centre for Upgrading Technology (NCUT) for the liquid product quality analysis.

References

1. Gray MR, McCaffrey WC, Huq I, Le T. Kinetics of cracking and devolatilization during coking of athabasca residues. Ind. Eng. Chem. Res. 2003; 43: 5438-5445.

2. Levy JH, Mallon RG, Wall GC. Vapour phase cracking and coking of three Australian shale oils: kinetics in the presence and absence of shale ash. Fuel. 1987;66:358-364

3. Gray MR, Le T, McCaffrey WC. Coupling of mass transfer and reaction in coking of thin films of Athabasca vacuum residue. Ind Eng Chem Res. 2001;40:3317-3324.

4. Kunii D, Levenspiel O. Fluidization Engineering (2nd edition). Boston: Butterworth-Heinemann, 1991.

5. Horio M, Nonaka A, Sawa Y, Muchi I. A new similarity rule for fluidized bed scale-up. AIChE J. 1986; 32:1466-1482.

6. Knapper BA, Gray MR, Chan EW, Mikula R. Measurement of efficiency of distribution of liquid feed in a gas-solid fluidized bed reactor. Int J Chem Reactor Eng. 2003; 1:A35.

[7] Martin SW, Wills, LE. Coking petroleum residues. In: Kobe KA, McKetta JA, eds. Advances in petroleum chemistry and refining. Vol. 2. New York: Interscience Publishers; 1959.

8. Soundararajan S. Determination of thermal cracking kinetics of Athabasca bitumen vacuum residue [M Sc Thesis]. University of Saskatchewan: Dept. of Chemical Engineering; 2001.

SIMULATING THE ADSORPTION OF ASPHALTENES ON MACROPOROUS MATERIALS USING MODEL MOLECULES

Clementina Sosa*, Manuel González
Francisco López-Linares, Pedro Pereira-Almao

Department of Chemical and Petroleum Engineering,
University of Calgary, Calgary, Alberta, T2N 1N4 Canada

csosa@ucalgary.com
mfgonzal@ucalgary.com
flopezli@ucalgary.com
ppereira@ucalgary.com

Key words:
Asphaltenes, adsorption

Prepared for Presentation at the
2005 Spring National Meeting, Atlanta, GA, April 10-14
Unpublished

Introduction

Asphaltenes precipitation is an important problem faced at different levels in the petroleum industry. It has being found that asphaltenes are strongly adsorbed on certain mineral components of rock reservoir [1,2]. Also, it is known the capacity of asphaltenes to deposit (adsorb?) on production wells, where they are found in deposits of both organic and inorganic predominant nature [3].

Different studies relating asphaltenes adsorption on surface solids media, their characterization and precipitation capability are found in the literature [4, 5, 6].

Asphaltenes are defined as a fraction of a crude oil that dissolves in toluene and insoluble in n-alkanes (e.g. n-heptane). They contain polynuclear aromatics and are the heaviest components of crude oils with high heteroatom content [7]. Asphaltenes seem to contain a large heterogeneity of chemical functionalities whose variability depends upon the crude oil source and/or their dilution extent.

In order to understand the adsorption of asphaltenes from different media and origins on different surfaces we propose to use a set of structurally known model molecules and functionalize them in order to simulate their behaviour. These model molecules might contain chemical functionalities similar to the ones present on some averaged well-characterized asphaltenic fractions such as molecular weight, aromaticity, naphtenicity levels, and functional groups. Asphaltenes could also be compared in their stability at variable content of paraffins in the media with that of the model molecules, or in their tendencies to interact between them and with a substrate.

To explore this alternative, in this study we used two model molecules, originally water soluble, functionalized later to make them organo-soluble. The initial objective is to determine by comparison if those model molecules resemble the asphaltenes adsorption process on kaolin, a natural silica-alumina.

Materials and experimental method

Materials

Toluene (ACS or spctrophotometric grade), trioctylamine (95 %), concentrated in hydrochloric acid and anhydrous magnesium sulphate were obtained from Aldrich.

The adsorbent used was synthesized in our laboratory to obtain macro porous extrudates with cylindrical shape. The starting material was a natural silica-alumina, kaolin. The final solid obtained had a 20 m^2/g surface area and an average meso pore diameter between 20 and 30 nm determined by BET technique.

Asphaltenes from Athabasca bitumen were obtained via n-heptanes precipitation method (see references [8], and [9] for details) then re-dissolved in toluene. Their average molecular weight was 4000 g/mol as estimate by VPO. The asphaltenes used in these experiments were not washed so they might contain a small amount of resins.

The molecules selected were Xylene Cyanol FF from OmniPur – EDM (namely all the time in this work as MM1), and Xylenol Orange tretrasodium Salt GR ACS (namely all the time in this work as MM2), with a molecular weight in the range of 500-800 g/mol. Those model molecules were transformed in organosoluble by adapting procedure reported from literature [10]. The synthetic scheme with their chemical structures are shown in Fig. 1

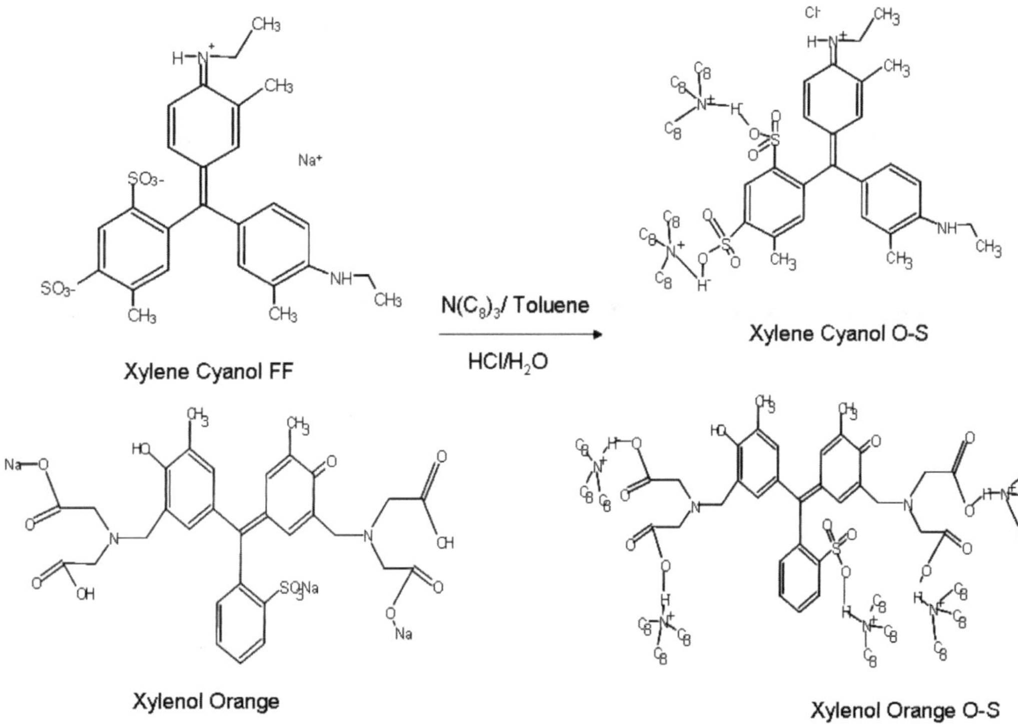

Fig 1. Synthetic scheme and chemical structure of the model molecules used in this study

Adsorption Method

Several solutions at different concentrations were contacted with a specific amount of adsorbent in a static adsorption experiment. The adsorptions were carried out at constant solution / adsorbent ratio of 10:1 (cc/gr).

The adsorption studies were performed using different concentrations of model molecules and asphaltenes solutions in toluene. Initial adsorption experiments of MM1 were carried out in water as reference.

The adsorption evolution was measured by means of the UV/VIS technique using a Cary 4E Spectrophotometer reading absorbance of λmax= 281 nm (for asphaltene) λmax= 619 nm (for MM1) and λmax= 440 nm (for MM2). A determination of the maximum wavelength was conducted following the a reported procedure [11].

An equilibration period of 48 hours at room temperature (22 C) was fixed. Dilutions of some samples in the case of the asphaltenes redissolved in toluene were done in order to be able to measure the absorbance.

Results and discussion

The adsorption studies were preformed over a natural silica-alumina (kaolin) at 22 C. Several solutions at different concentration from 1 to 60000 mg/L for asphaltene and 1 – 10000 mg/L for the model molecules were used.

In Figure 2 the adsorption isotherm corresponding to the model molecule Xylene Cyanol (MM1) performed in water is displayed. It is remarkable the lower adsorption values (0-0.03 mg/g) obtained for the MM1 dissolved in water in comparison with the ones obtained for the re-dissolved asphaltenes in toluene, with an adsorption of up to 50 mg/g of adsorbent for the monolayer (not shown).

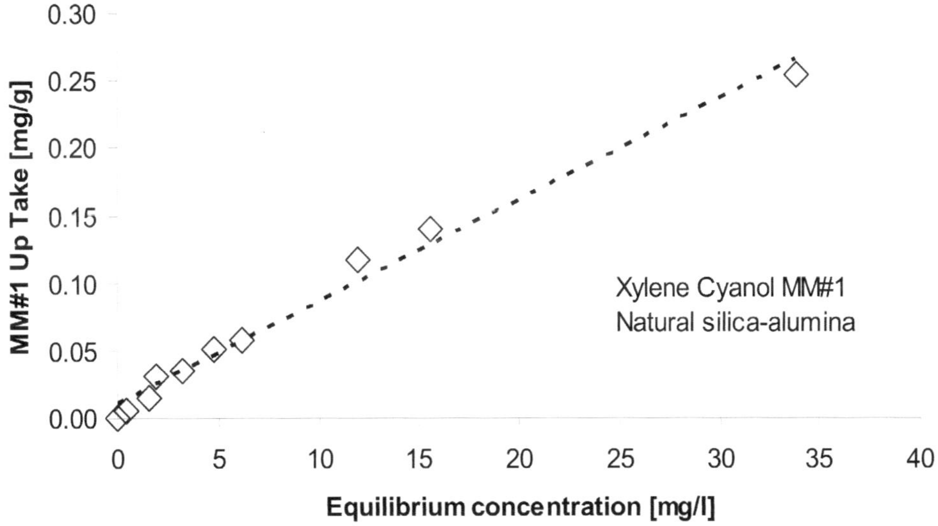

Figure 2 Adsorption Isotherm (25 C) of Xylene Cyanol / water solution on Natural Silica-Alumina

The lower adsorption of the MM1 dissolved in water suggests solvent-substrate competition for the adsorbent surface and eventually an additional effect of the ionic double layer formed around the polar adsorption sites of the adsorbate. The same performance is observed for the MM2 in water.

Adsorption studies of model molecules properly modified to make them soluble in toluene were performed. The solubility of the model molecules selected in toluene is extremely low, in the same way that asphaltenes deprived from alkyl paraffinic branches increase their tendency to segregate and become non soluble in organic media..

Figure 3 shows the adsorption isotherm for n-heptane asphaltenes, Xylene Cyanol (MM1), and Xylenol Orange (MM2) dissolved in toluene. For the three macromolecules evaluated Type I Langmuir isotherm behaviour was observed suggesting a typical monolayer adsorption.

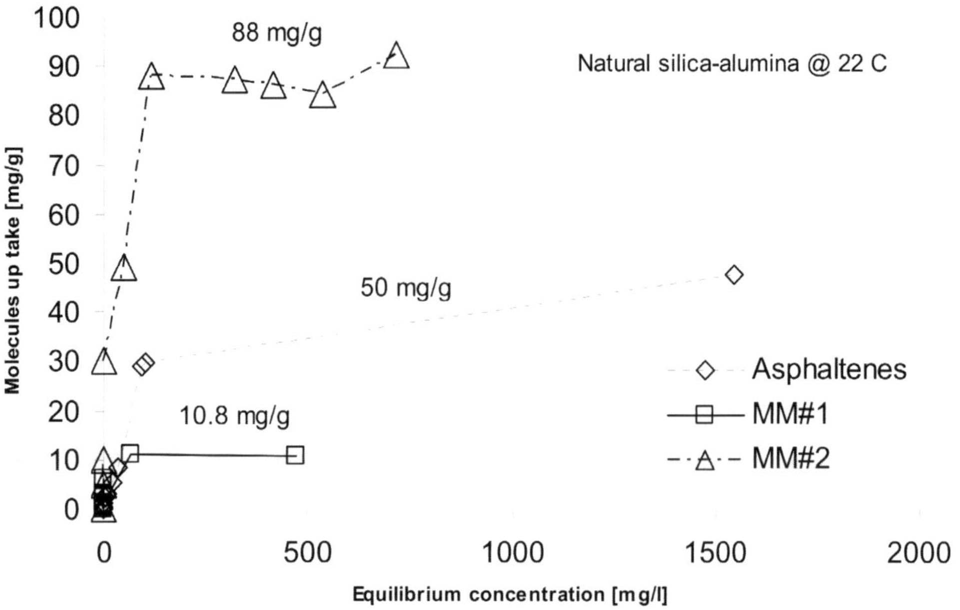

Figure 3 Adsorption Isotherm (25 C) of Macro Molecules/Toluene solutions on Natural Silica-Alumina

At 22 C, the maximum adsorption is observed for Xylenol Orange (80 mg/g) followed by the asphaltenes (50 mg/g) and Xylene Cyanol (10.8 mg/g). Different molecule uptake suggests different adsorption affinity of the macromolecules for the adsorbent surface. Several hypotheses can be presented in an initial attempt to explain this behaviour; several of them can act simultaneously, for we call them non exclusive hypotheses.

The first non exclusive hypothesis considers the presence of available heteroatoms in each molecule that may interact with the surface. For MM2, the oxygen and N atoms present (13 O, 7 N) are bigger than MM1 (4 O, 6 N). This fact allows this molecule to interact better with the surface perhaps through Si-OH-O-R or Si-OH-N-R bonds than MM1 [12]. Under this hypothesis our asphaltene sample would have and intermediate type of heteroatoms sufficiently exposed for adsorption.

A Second non exclusive hypothesis would be to consider the molecular size of the molecules being adsorbed as controlling the adsorption uptake at the equilibrium. In a very preliminary approach we can take the molecular weight as an indicator of molecular size. For MM1 and MM2, the corresponding molecular weights are 1274.45 g/mol and 2426.9 g/mol respectively while for asphaltene was considered 4000 g/mol as referenced. As it is observed, MM2 adsorbs more than the asphaltene and MM1 respectively. This result indicates that the molecular size is not controlling the adsorption uptake on Kaolin. It seems that the total adsorptive uptake at equilibrium is more determined by the presence of groups that can more effectively interact with the surface of the adsorbent.

The adsorption kinetics for all molecules was determined for each molecule at different initial concentrations. The results are displayed in Figure 4.

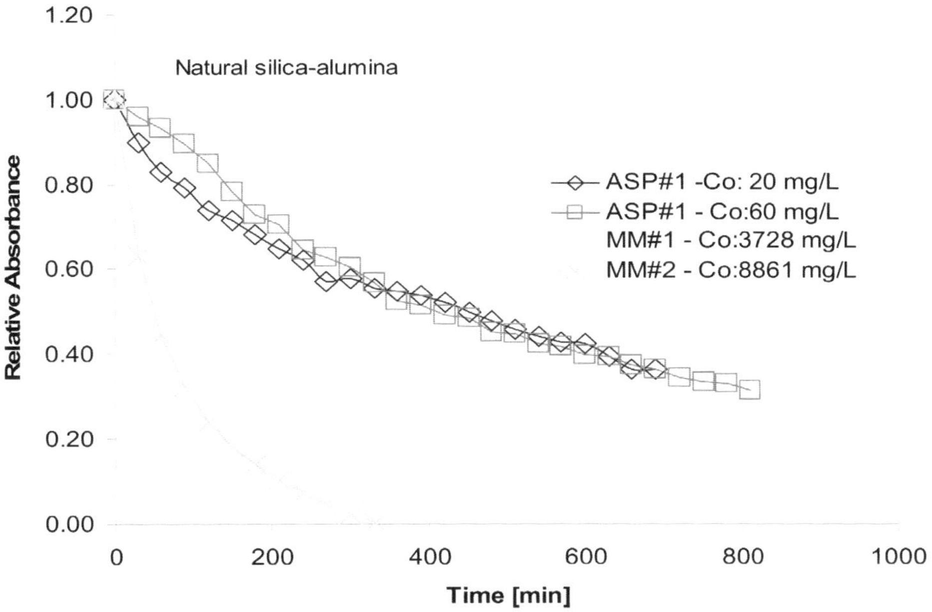

Figure 4 Kinetic of the adsorption of Macro Molecules/Toluene solutions on Natural Silica Alumina

As is observed therein, the model molecules have a faster initial adsorption rate (MM1: 8.13E-03 mg/LxMin^{-1}, MM2: 3.47E-03 mg/LxMin^{-1}) than the corresponding asphaltene (ASP 20 mg/L: 1.87E-03 mg/LxMin^{-1}: ASP 60 mg/L: 1.57E-03 mg/LxMin^{-1}) It is interesting to notice that MM1 has the maximum initial rate of adsorption. For the kinetics of asphaltene adsorption, it is found that the initial concentration does not affect considerably the adsorption rate. Regarding the dynamic of adsorption it seems to be confirmed that the lighter the molecule the faster the adsorption, even though the equilibrium uptake of the heavy molecules can rank in a different order.

Conclusions

The present study shows that the adsorption uptake at equilibrium of asphaltenes on a natural silica-alumina might depend more on the nature and number of the functional groups present on those molecules which allow them to interact stronger with the solid surface than on the molecular size of them, whereas the dynamic of the adsorption seems to be more determined by molecular size. This work opens the possibility of modeling the adsorption properties of asphaltenes via successive approximations both in molecular size and chemical functionalities starting with structurally known heavy molecules.

Acknowledgments

The authors acknowledge the substantial financial support provided by several institutions for the creation of a research group on topics related to the content of this paper at the University of Calgary, Alberta, Canada. In particular this work is being supported by the

Alberta Ingenuity Fund through the Ingenuity Scholar support provided to Professor Pedro R. Pereira-Almao. It is also acknowledged the financial support in the form of discovery and strategic research grants respectively, provided by the National Science and Engineering Research Committee (NSERC) of Canada and the Alberta Energy Research institute (AERI) of the province of Alberta, Canada.

References

1. Pernyeszi, T., Patzko, A., Berkesi, O., Dekany, I., Colloids and Surfaces A-Physicochemical and Engineering Aspects, 1998, 137, 373
2. Pernyeszi, T., Dekany, I., Colloids and Surfaces A-Physicochemical and Engineering Aspects, 2001, 194, 25
3. Acevedo, S., Ranaudo, M. A., Escobar, G., Gutierrez, L., Ortega, P., Fuel, 1995, 74, 595
4. Kokal, S., Tang, T., Schramm, L., Sayegh, S., Colloids and Surfaces A-Physicochemical and Engineering Aspects, 1995, 94, 253
5. Marczewski A., Szymula M. Colloids and Surfaces 208,2002, 259-266
6. Alboudwarej H. et al. The 5 Th International Conference on Petroleum Phase Behaviour and Fouling. Banff, Alberta Canada, June 13-17th, 2004
7. Speight, J.G., 'Petroleum Chemistry and Refining", Taylor & Francis, Washington, 1998.
8. Alboudwarej H. et al. Energy & Fuel, 2002, 16 (2), 462-469
9. Alboudwarej H. et al. AIChE Journal, 2003, 49 (11), 2948-2956
10. Herrmann W. A., Kohlpaintner C. W. Angew. Chem. Int. Ed. Engl., 1993 (32): 1524-1544
11. Marczewski, A. Szymula, M., Colloids and Surfaces A-Physicochemical and Engineering Aspects, 2002 (208), 259-266
12. For example of interaction of nitrogen group to the SiOH group: a) Müller, C. A.; Schneider; M.; Mallat; T.; Baiker., A. *Appl. Catal., A* 2000, (201) ,253 b) (b) Fey, T.; Fischer, H.; Bachmann, S.; Albert, K.; Bolm, C., *J. Org. Chem.* 2001, (*66*) 8154 c). McKittrick M. W and. Jones C.W., Chem. Mater. 2003, (15*),* 1132-113

ULTRA DISPERSED CATALYSTS FOR PROCESSING HEAVY HYDROCARBON FRACTIONS

Pedro Pereira-Almao*, Josephine M. Hill, Jie Wang, Alejandro Vasquez

Department of Chemical and Petroleum Engineering,
University of Calgary, Calgary, Alberta, T2N 1N4 Canada

ppereira@ucalgary.ca
jhill@ucalgary.ca
jwang@ucalgary.ca
avasquez@ucalgary.ca

Key words:
Molybdenum, Catalysts, Dispersion, Hydroprocessing

Prepared for Presentation at the
2005 Spring National Meeting, Atlanta, GA, April 10-14
Unpublished

Introduction

Heavy oils are starting to play a significant role in the world energy market. Due to a great deal of impurities in heavy oils, their production and refining requires more advanced technologies, such as more efficient catalysts for hydroprocessing[1]. Molybdenum based catalysts have been widely used in the petroleum industry for hydroprocessing of heavy oil fractions [2]. They can be used as both supported and unsupported catalysts. Supported catalysts suffer severe deactivation in the traditional hydrotreating process of heavy fractions coming from heavy oils [3], whereas unsupported molybdenum catalysts have higher catalytic activity with better metal dispersion, and can be recovered, recycled or discarded in the process [4]. In this study, micro/nano size (<1.5 μm) unsupported molybdenum catalysts were synthesized via water/oil emulsions. These catalysts were prepared in a continuous mode particularly designed for online application to hydroprocessing, or to in situ upgrading.

Materials and experimental method

Materials

Unsupported molybdenum catalysts were prepared via flashing and decomposition of micro-emulsions. The first step in the synthesis was to prepare a water-in-oil emulsion consisting of a surfactant (Brij 30 from Aldrich), vacuum gas oil (VGO, from PetroCanada, IBP 250°C, EP 550°C), ammonium heptamolybdate tetrahydrate (AHM, from Aldrich, ACS 99.99%), and distilled water. This emulsion was then pumped via a syringe pump through a ¼ inch tubular stainless steel reactor housed in a furnace (Lindberg) heated to temperatures between 250°C and 350°C with flow rates between 210 ml/h and 420 ml/h. The liquid product was collected at the exit of the tubing for analysis or for centrifuging and drying to obtain a solid product.

Another set-up also used for this study included the use of ISCO high pressure pumps, a heating tape coiled around a SS 3/8 inch stainless steel tubular reactor, heated between 200° and 300°C with the same flow range indicated for the set-up described above.

The liquid product of both systems was analyzed in a Malvern Zetasizer Nano series (Malvern, UK) equipped with a 4.0 mW He-Ne laser (633 nm) and operating at an angle of 173° and a temperature of 25°C to obtain the particle size distribution of the suspended particles. The solids were analyzed using nitrogen physisorption at 77 K (Autosorb 1c, QuantaChrome Instruments) to determine surface areas, X-ray diffraction (XRD, Rigaku) to determine crystal structure, and scanning electron microscopy (SEM, Philips FEI ESEM) to determine morphology. Reactivity tests were performed on a high pressure stainless steel kinetics apparatus connected to a gas chromatograph. Toluene hydrodearomatization was used as a test reaction to allow estimation of both cracking as well as hydrogenating capabilities of the catalysts.

Results and discussion

The average particle sizes obtained from the DLS measurements varied between 170 nm and 1600 nm depending on decomposition temperature, flow rate, concentration of

AHM, and water to oil ratio in the initial emulsion. Through a factorial design experiment set it was determined that decomposition temperature was the factor with the largest effect on the average particle size. As the temperature increased, the average particle size increased. The activation energy for the process of particle growth was calculated to be less than 50 kJ, suggesting that simple frequency of collisions more than chemical binding might be the determining process of agglomeration. A typical XRD spectrum is shown in Figure 1 and indicates that the particles are amorphous and mainly composed of MoO_2 and MoO_3 species. The average surface area of these particles is approximately 17 m^2/g, as determined by a multi-point BET analysis. Both SEM and DLS analyses show that there are a range of particle sizes produced. The SEM images indicate that the dried particles could be agglomerated from an original particle around 70 nm in diameter.

The results of the reactivity tests confirmed that the produced Mo particles are active catalysts. The conversion of toluene was over 94% and the distribution of reaction products is indicative of overcracking and hydrocracking.

Significance

Reactivity tests showed higher catalytic activity than commercial supported catalysts. This ultradispersed solid has potential application for both in-situ and ex-situ upgrading of heavy oil.

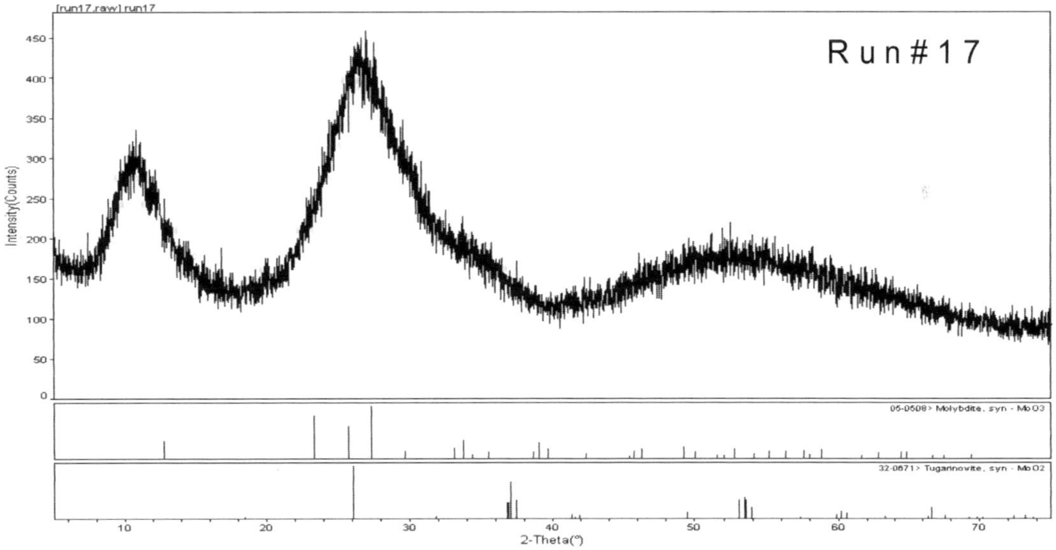

Figure 1. XRD spectra of Mo particles (run #17). The reference spectra are for MoO_3 (top) and MoO_2 (bottom).

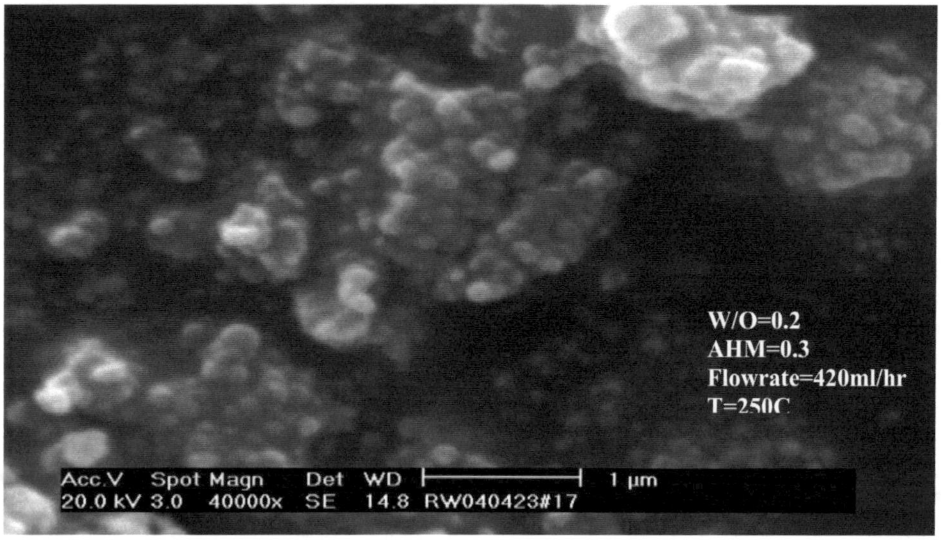

Figure 2. SEM image of Mo particles (run #17).

Figure 3. SEM image of Mo particles (run #17)

Observing the SEM image of figures 2 and 3, agglomeration of smaller particles to form the resulting agglomerates may become evident. Careful measurement showed that these original particles range between 50 and 80 nm.

A preliminary evaluation of the effect of temperature on the dynamics of formation of agglomerates was performed. It was assumed that the diameter of particles, which start at a level of 70 nm in average for the first particles formed, grows as a function of the concentration of particles, the residence time of the particles, the liquid fraction in the decomposition zone, the geometry of that zone, and the temperature of the media. It is also expectable that the viscosity of the media had an inverse effect on the particle growth.

To keep those variables constant to evaluate the single effect of temperature is difficult, but it is presumed that, within a short range of temperature variation, these variables might not produce an important perturbation to the qualitative conclusions derivable from such an experiment. Taking into account these limitations, the activation energy of the collision process resulting from the evaluation of the effect of temperature was evaluated. The experiments were all performed at a constant rate flow of emulsion, keeping the concentration of Mo in the precursor constant, and using a VGO whose gas fraction at the temperature range scanned is neglectable.

As a result two different set-ups for the continuous production of MoOx particles were used to evaluate the effect of temperature, one with a reaction zone diameter larger than the other. The values of Activation Energy derived for the process in the two set-ups rank between 30 and 35 Kjoules. An activation process of this low energetic requirement might be indicating that mass transfer (simple collisions) is at the bases of the agglomeration process.

Conclusions

The present study shows insight characterization of ultradispersed particles of Molybdenum for use as hydroprocessing catalysts. They were prepared via precursor emulsions or micro emulsions, using inexpensive aqueous Molybdenum salts in a continuous mode. These particles can have particle diameters in the nano size range, as low as 50 nm. An activation energy estimate performed by varying the temperature within a range of about 150 C, in two different set-ups and under continuous mode seems to reflect that simple collisions are at the bases of the agglomeration process.

Acknowledgments

The authors acknowledge the substantial financial support provided by several institutions for the creation of a research group on topics related to the content of this paper at the University of Calgary, Alberta, Canada. In particular this work is being supported by the Alberta Ingenuity Fund through the Ingenuity Scholar support provided to Professor Pedro R. Pereira-Almao. It is also acknowledged the financial support in the form of discovery and strategic research grants respectively, provided by the National Science and Engineering Research of Canada (NSERC) and the Alberta Energy Research institute (AERI) of the province of Alberta, Canada.

References

1. Lanier, D., Heavy Oil - A Major Energy Source for the 21st Century, UNITAR Centre for Heavy Crude and Tar Sands, No. 1998.039 (1998).
2. Tian, K.P., A.R. Mohamed, and S. Bhatia, *Fuel* 77(11), 1221 (1998).
3. Absi-Halabi, M., A. Stanislaus, T. Al-Mughni, S. Khan, and A. Qamra, *Fuel* 74(8), 1211 (1995).
4. Bianco, A.D. and N. Panariti, *Appl. Catal. A: General* 94, 1 (1993).

66d Synthesis and Thermal Reactions of Model Compounds for Bitumen Residue

Felaniaina Rakotondradany, Hicham Fenniri, Murray R Gray

The asphaltenic components of bitumen, which represent the highest molecular weight bitumen fraction, contain polyaromatic hydrocarbons rich in heteroatoms and polars groups resulting in strong self-association under extraction and upgrading conditions. Building on the techniques of supramolecular chemistry, we here report the synthesis and self-assembly of polyaromatic model compounds to provide new insight for the behaviour of bitumen residue fractions. The synthesis and characterisation of symmetric and non-symmetric alkylated hexabenzocoronenes, which can self-associate at temperatures up to 400 C, will be detailed along with a study of their alkyl-alkyl and pi-pi stacking interactions. The association behaviour of these model compounds will be related to bitumen behaviour and thermal chemistry under upgrading conditions.

A Case Study Using a New Titania Claus Catalyst

Elise M. Mophett, Almatis, Vidalia, Louisiana
Michael J. Pearson, Ph.D., Almatis, St. Augustine, Florida

ALMATIS

Abstract

After its installation over four years ago, Almatis DD-931 still shows outstanding performance in a gas plant setting where it is frequently exposed to excessive Benzene, Toluene and Xylene (BTX) carry over from the reaction furnace. High hydrocarbon levels in acid gas feed also produce elevated CS_2 levels entering the first converter. This specie is not readily converted by standard activated alumina, so titania catalysts are often employed. Unlike the traditional titanium dioxide extrudates, which rapidly deactivate from BTX, this catalyst maintains CS_2 conversion capability over time.

Introduction

Introduced in 2002, DD-931 is a new titania alumina composite Claus catalyst for the hydrolysis of carbonyl sulfide (COS) and carbon disulfide (CS_2).[1] Formulated to extract the catalytic advantages of both titania and alumina, DD-931 demonstrates a much higher capacity for conversion of these difficult carbon sulfur species than activated alumina alone. DD-931 is now installed in over 25 units world wide, representing over 8000 tons per day (TPD) of sulfur production.[2] This paper traces DD-931's performance in one unit in particular over the past 3 years.

Development work:

Driven by a demand from industry for a stronger, easier to handle titania catalyst, Almatis AC, Inc. (formally Alcoa) developed a composite titania-alumina sphere for use in Claus converters. Previously, only "pure" titania extrudates were available to treat systems that suffered from high concentrations of COS and CS_2. These species are a result of acid gas feeds with high carbon dioxide (CO_2) and hydrocarbons that react in the complex chemistry of the furnace to generate COS and CS_2 respectively. Conversion of these two species to hydrogen sulfide (H_2S) is required in certain cases, especially if overall recovery from the unit is to be optimized. DD-931 presents another, more cost effective, option for customers.

Initial testing of DD-931 was performed by Alberta Sulfur Research Ltd., ASRL, in Calgary, Alberta, Canada. Aged to resemble a catalyst after 3 years of service, DD-931 performance is compared to that of alumina and Almatis SRC 99ti, a "pure" titania extrudate. Test results are shown in Figures 1 and 2. This data reinforces that, with very reasonable adjustments in operating conditions, DD-931 is capable of achieving the same conversions of COS and CS_2 as a "pure" titania extrudate under standard Claus converter conditions. Titania catalyst service is common where there are hydrocarbons in the feed, which produce CS_2.

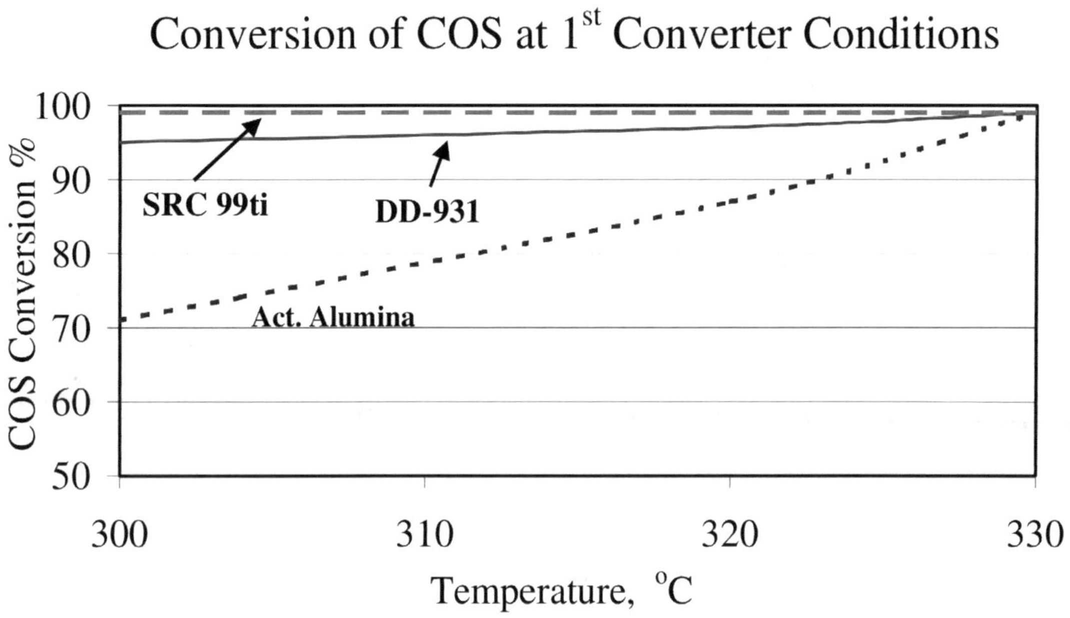

Figure 1: *Feed conditions: 8% H_2S, 4.5% SO_2, 0.5% COS, 0.5% CS_2, 25% H_2O, 309 ppm O_2 and balance N_2; GHSV = 1000*

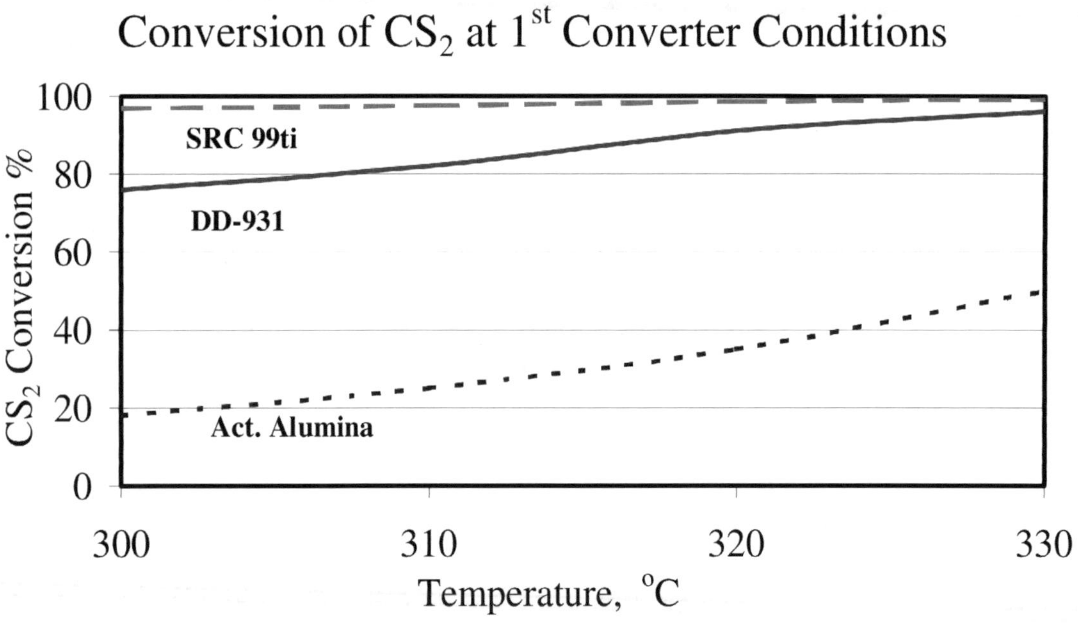

Figure 2: *Feed conditions: 8% H_2S, 4.5% SO_2, 0.5% COS, 0.5% CS_2, 25% H_2O, 309 ppm O_2 and balance N_2; GHSV = 1000*

Effect of BTX on catalyst performance

There is increased interest in sulfur processing circles on how BTX affects the performance of Claus catalysts. Two aspects demand attention: 1) are reactions within the combustion furnace and 2) fouling effects on the catalyst beds within the SRU. Especially common in the natural gas industry, the byproducts of BTX combustion generate most of the CS_2 found in the feed to the #1 Claus converter. If these hydrocarbons are not destroyed in the reaction furnace, they are carried over to the first converter where they crack and foul the catalyst. The result of this carryover is rapid catalyst deactivation.[3] In such cases, the rate of deactivation of the catalyst is an important factor for overall operations.

Past evaluations of split bed configurations show improved performance of DD-431/DD-931 over a split bed configuration of DD-431 and "pure" titania catalyst, under high BTX conditions[4]. Results are shown in Figure 3. More recent ASRL testing shows that catalyst deactivation from BTX occurs much more rapidly on pure titania then activated alumina.[5] Though originally conducted to evaluate titania catalyst performance in the presence of BTX, the tests also indicate conversion of CS_2 is enhanced when activated alumina is placed on the top of the titania layer. A possible reason for this may be that the alumina initiates the standard H_2S/SO_2 reaction, which decreases the SO_2 concentration in the gas stream. The CS_2/SO_2 ratio is then higher as it crosses the titania catalyst, generating higher conversion. It appears that this common split bed practice, normally done to protect the more expensive titania catalyst, may provide other positive aspects as well.

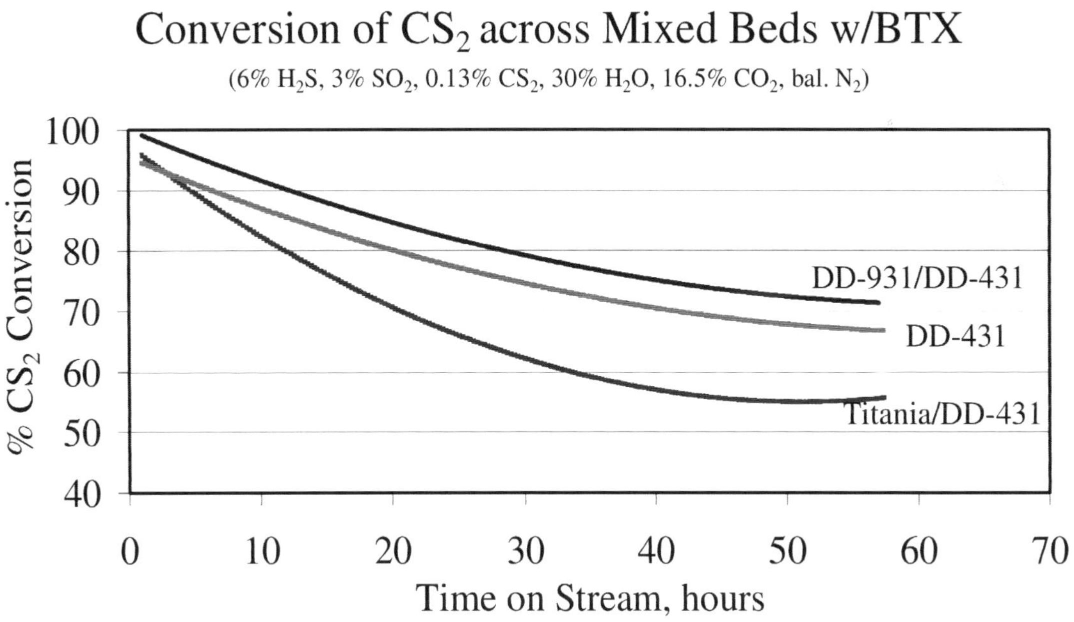

Figure 3: *Feed Compositions: 6% H2S, 3% SO2, 0.13% CS2, 30% H2O, 16.5% CO2, bal. N_2, Benzene 20,000 ppmv, Toluene 5,000 ppmv; GHSV = 1000*

Commercial Plant Layout

To evaluate the new catalyst, a plant with a history of BTX problems approached Almatis looking for an option to increase their recovery without major capital improvements. Located in the gas fields of Canada this facility has a design capacity of 1100 tons per day (TPD) of sulfur production. The sulfur recovery unit (SRU) complex is comprised of two independent trains; each with three converters and auxiliary acid gas fired reheaters, which feed a common, three bed Sulfreen Tail Gas Unit (TGU). The facility needed a way to convert the COS and CS_2 produced from burning BTX in their furnace and to prevent catalyst deactivation associated with BTX carryover onto the first catalyst bed. Typical acid gas feed and first converter feeds to the two trains are show in Table 1 and Table 2 respectively.

Table 1-
Typical Acid Gas Feed Composition

Component	Mol %
H_2S	50 - 56
CO_2	42 - 47
CH_4	.4 - .6
BTX	350 – 1250 ppmv

Table 2 –
Typical First Converter Feed Composition

Component	Mol %
H_2S	6
SO_2	2.8
CO2	26
COS	0.5
CS_2	0.6
BTX	20 – 70 ppmv

Unit improvements increased the combustion of BTX in the reaction furnace, but there is still BTX carryover onto the first converter bed. On average the total BTX concentration to the first converter is around 45 ppmv, but it sometimes reached over 160 ppmv during performance testing. Over time this poses a severe risk of catalyst deactivation. In addition, the BTX combustion produces high CS_2 and COS concentrations in the feed to the first Claus converter. Typical levels of 0.6% CS_2 and 0.5% COS are seen in the feed to the first converter.

This unit employs a TGU that is not capable of converting these carbon sulfur species, so to ensure high overall recovery it is necessary to decompose them in the first converter. A "pure" titania extrudate could be employed to address conversion of COS and CS_2 in the first converter. However, due to the incomplete combustion of BTX in the reaction furnace the customer chose to install a split bed of high activity activated alumina, DD-431, on the top of a layer of the titania-alumina composite, DD-931. In the secondary converters, #2 and #3, a multiple bed loading was also used. The acid gas fired reheats in the system were known to leak oxygen during operation. To limit the sulfation of the new activated alumina due to this excess oxygen, alumina impregnated with iron oxide, Almatis S-100 SR, was placed on the top of the bed. This catalyst reacts with any excess oxygen in the gas stream and reduces the sulfation of the activated alumina beneath. This type of impregnated catalyst is usually only used in the lower temperature secondary converters where sulfation of the alumina catalyst can be a real problem. There will be no further discussion of secondary converter performance in this paper.

After an initial stabilization period of about 3 months, the performance of the unit was tested over the temperature range 320 to 335°C. The CS_2 conversion for the first converter is plotted in Figure 4. For comparison, the past results over a bed of just activated alumina catalyst, also after 3 months of operation, are also show in Figure 4. In particular, 90% CS_2 conversion was obtained with the DD-931/DD-431 at 326°C compared to 70% for alumina. Figure 5 shows the COS conversion compared to CS_2. About 98% COS conversion was obtained over the whole temperature range for the DD-931/DD-431. The COS conversion represents an equilibrium condition for this relatively high CO_2 stream.

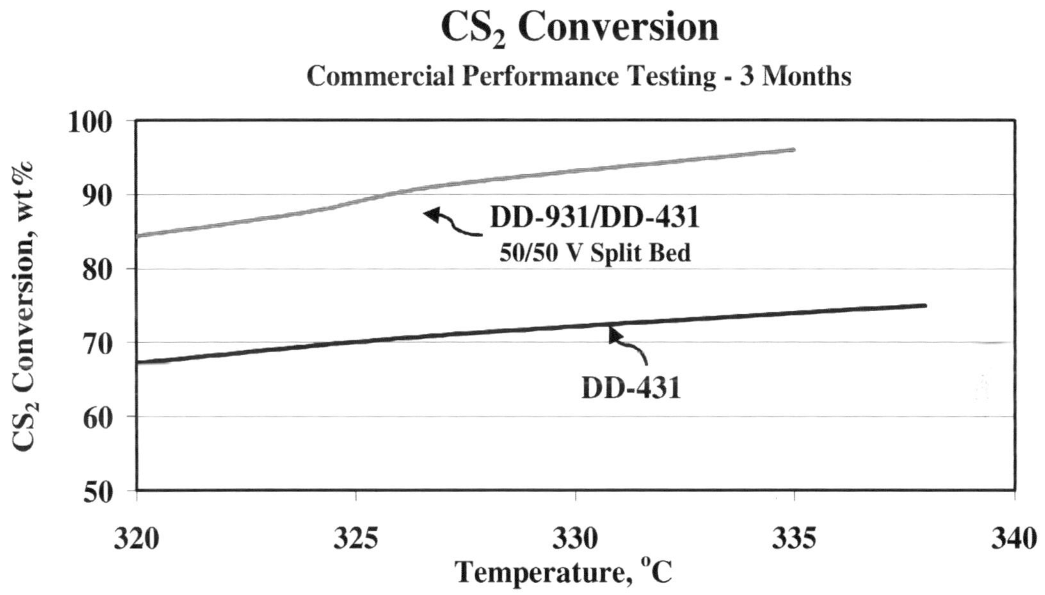

Figure 4

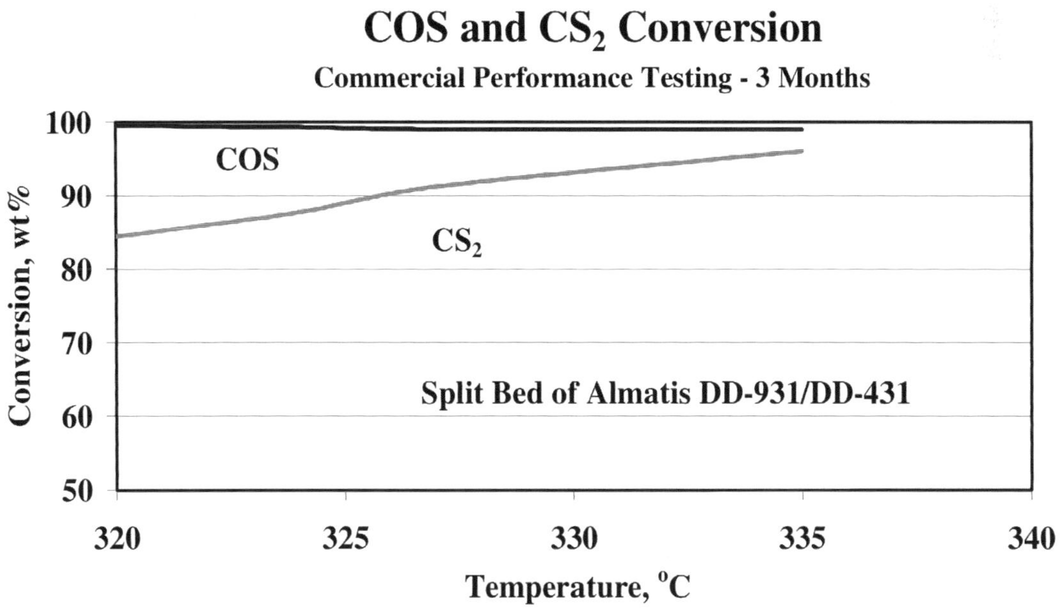

Figure 5

Annual Operation Evaluation

Year 1 –

Plant policy always allowed for annual testing of SRU overall performance. This gives the management and operations personnel excellent insight to the inner workings of their units. One year after the last performance test a similar evaluation was performed. The results are illustrated in Figure 6. Performance of the catalyst remained high with no measured decrease over the entire year's operation.

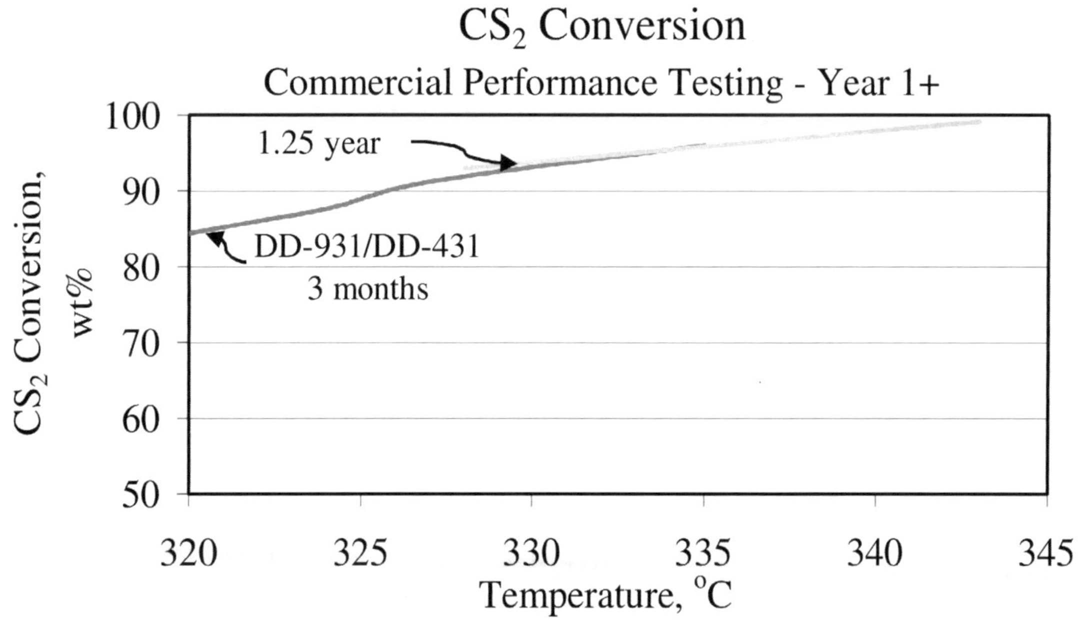

Figure 6

Another part of the first year study included an optimization of the auxiliary acid gas burners. The survey identified sources of oxygen infiltration into the second and third converter beds from the aux burners. This infiltration is thought to have no impact on the activated alumina catalyst loaded into the converters, because an iron oxide oxygen scavenging catalyst, Almatis S100-SR, is installed on top of the alumina. Optimizing the converters resulted in an overall recovery increase of approximately 0.2% for the SRU.

Year 2 –

Performance testing for the second year included the same temperature range as the previous year along with a detailed study of the BTX composition through to the first converters of each train. Conversion of CS_2 is shown in Figure 7. COS conversions during this test period ranged from 95 to 98%, similar to what was seen the previous year. The graph shows that a shift in temperature, approximately 10 °C higher, is required to achieve similar conversion to that under Start of Run (SOR) conditions. Levels of greater than 90% conversion were still capable, but the first converter outlet operating temperature had shifted from 326 °C to 333 °C.

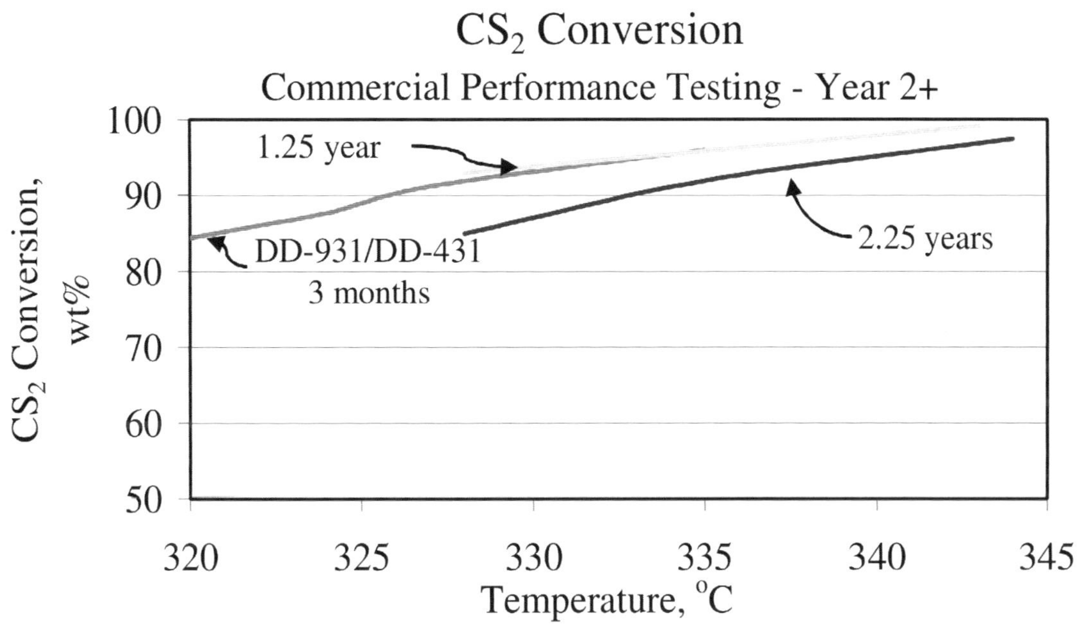

Figure 7

The BTX evaluation validated problems seen in an upstream separator. High levels of hydrocarbon were spilling over into the Claus unit. This variation in inlet feed concentration of BTX was common, though excursions into the higher range were usually short in duration. It appeared that the component known to play the most active role in catalyst deactivation, xylene, showed little decomposition across the first converter. This indicates that DD-931 has a low affinity to catalyze this reaction. This lowers carbon deposition on the bed and retains bed activity.

Conclusions

Through annual performance testing the plant is able to improve overall performance and counteract the negative effects of high BTX in their incoming feed. Optimization of existing reheat system decreased the infiltration of excess oxygen into their secondary catalyst beds and increased their overall recovery. Using a next generation catalyst, DD-931, allows them to achieve high conversion of CS_2 through their first converter, much greater than was ever-achieved using standard activated alumina catalyst. All these facets impact the overall operation of the plant, providing more stable operation and an increased ability to achieve their required compliance levels.

References

[1] "New Claus Catalyst Options for Higher Sulfur Recovery", M. J. Pearson, E. M. Mophett, P.D. Clark, N.I. Dowling and M. Huang, Sulphur 2000, October 2000

[2] "World Sulfur Production Survey", Oil and Gas Journal, June 28, 2004

[3] "Quantifying the Effect of Individual Aromatic Contaminants on Claus Catalyst", P.P. Crevier, N.I. Dowling and M. Huang, LRGCC, February 2001

4 "Introducing a New Titania Catalyst – DD-931", E. Mophett, M. Pearson, Brimstone Sulfur Symposium, September 2002

[5] "The Effect of Space Velocity and BTX on the Conversion of H_2S/SO_2 and CS_2 Over Titania Catalysts", P.D. Clark, N.I. Dowling and M. Huang, Brimstone Sulfur Symposium, September 2004

Flexibility of Liquid Redox Processing in Refinery Sulfur Management

Gary J. Nagl
Gas Technology Products LLC

Abstract

With over 25 years of experience, sulfur recovery via iron-based, liquid redox processing is a commercially proven approach to refinery sulfur management. The inherent characteristics of liquid redox processing, such as 100% turndown in respect to H_2S concentration, flowrate and sulfur loading and single stage removal efficiencies in excess of 99.9%, makes the process attractive as a stand alone device for applications of less than 15 LTPD of sulfur or in conjunction with a Claus unit at higher capacities resulting in an overall system having a removal efficiency of 99.9+% and 100% turndown capability.

Another attractive inherent feature of iron-based, liquid redox processing is its ability to process any type of gas stream such as fuel gas, amine acid gas, and sour water stripper gas. Consequently, for refineries with Claus plants, the liquid redox process can be employed as a tail gas treatment unit, with or without a hydrogenation/hydrolysis unit, while also directly processing the sour water stripper gas relieving the Claus unit of this burden. All of this can be accomplished without recycling gas back to the Claus unit, thus increasing the capacity of the Claus unit.

INTRODUCTION

With the installation of more and more desulfurization processes to meet ultra low sulfur specifications for gasoline and diesel in conjunction with more and more pressure from environmental agencies to reduce sulfur emissions, refiners are faced with the challenge of what to do with the additional H_2S being generated. In some cases, there is sufficient spare capacity in the existing amine and Claus units to process the additional load. And in other cases, the existing Claus unit(s) can be updated with oxygen enrichment technology to handle the additional load or in other cases, new sulfur recovery systems will be required to handle the incremental load. Another approach to this challenge, which is often overlooked, is the addition of liquid redox processing. Due to its excellent versatility, sulfur recovery via liquid redox processing can be employed to treat any type of sour gas streams. In a refinery this means that liquid redox technology can be use to treat sour, gaseous hydrocarbon-based streams directly or acid gas streams from amine units, or sour water stripper gas or even Claus tail gas streams. By integrating liquid redox processing into the overall refinery sulfur recovery program, refiners may be able to easily accommodate the additional sulfur recovery burden imposed by the production of ultra low sulfur fuels.

SULFUR RECOVERY VIA LIQUID REDOX PROCESSING

Liquid redox processes employ aqueous-based solutions containing metal ions, usually iron, which are capable of transferring electrons in reduction-oxidation (redox) reactions. Currently, the redox process of choice is the LO-CAT® process, which is licensed by Gas Technology Products LLC. In this process, a non-toxic, chelated iron catalyst is employed to accelerate the reaction between H_2S and oxygen to form elemental sulfur.

$$H_2S + \tfrac{1}{2}\, O_2 \xrightarrow[(Fe)]{} S^o + H_2O \qquad (1)$$

As implied by its generic name, liquid redox, all of the reactions in the LO-CAT® process occur in the liquid phase in spite of the fact that Equation (1) is a vapor phase reaction. In the process, the sour gas is contacted in an absorber with the aqueous, chelated iron solution where the H_2S is absorbed and ionizes into sulfide and hydrogen ions as follows.

$$H_2S + H_2O \Longrightarrow 2H^+ + S^= \qquad (2)$$

This reaction is mass transfer limited.

The dissolved sulfide ions then react with chelated, ferric ions to form elemental sulfur as follows.

$$S^= + 2Fe^{+++} \Longrightarrow S^O + 2Fe^{++} \qquad (3)$$

This reaction is very fast and is not equilibrium limited. In addition, since the reactions are occurring at ambient temperatures, the sulfur is formed as a solid.

The solution is then contacted with air in an oxidizer where oxygen is absorbed into the solution and the ferrous ions are converted back to the active ferric state as follows.

$$\tfrac{1}{2}\, O_2 + H_2O + 2\, Fe^{++} \Longrightarrow 2\, Fe^{+++} + 2\, OH^- \qquad (4)$$

Again, this reaction is mass transfer limited.

Adding Equations 2, 3, and 4 yields Equation 1.

TREATING ACID GAS STREAMS

As illustrated in Figure 1, Autocirculation type LO-CAT units are used when treating acid gas streams and streams, which can be mixed with air without creating a safety problem. In this type of unit, the absorber and oxidizer are contained in one vessel and separated by baffles. Due to the large differences in aerated densities between the liquids in the absorber and the oxidizer large circulation rates are achieved between the various compartments of the vessel without having to employ pumps. The acid gas enters the absorber section of the vessel (centerwell) where it is contacted with oxidized LO-CAT solution and where the H_2S is converted to elemental sulfur. The partially reduced solution then circulates to the oxidizer section where it is contacted with air, which reoxidizes the iron. The exhaust air from the oxidizer and the sweet acid gas from the absorber are combined and exhausted to the atmosphere.

In the conical portion of the vessel, the sulfur will settle into a slurry of approximately 15 wt%. A small stream is withdrawn from the cone and sent to a vacuum belt filter where the sulfur is further concentrated to approximately 65 wt% sulfur. Some units stop at this stage and sell the sulfur cake as a fertilizer. Drier sulfur cake can be formed by employing pressure filters. If molten sulfur is required, the cake is reslurried and melted as shown in Figure 6.

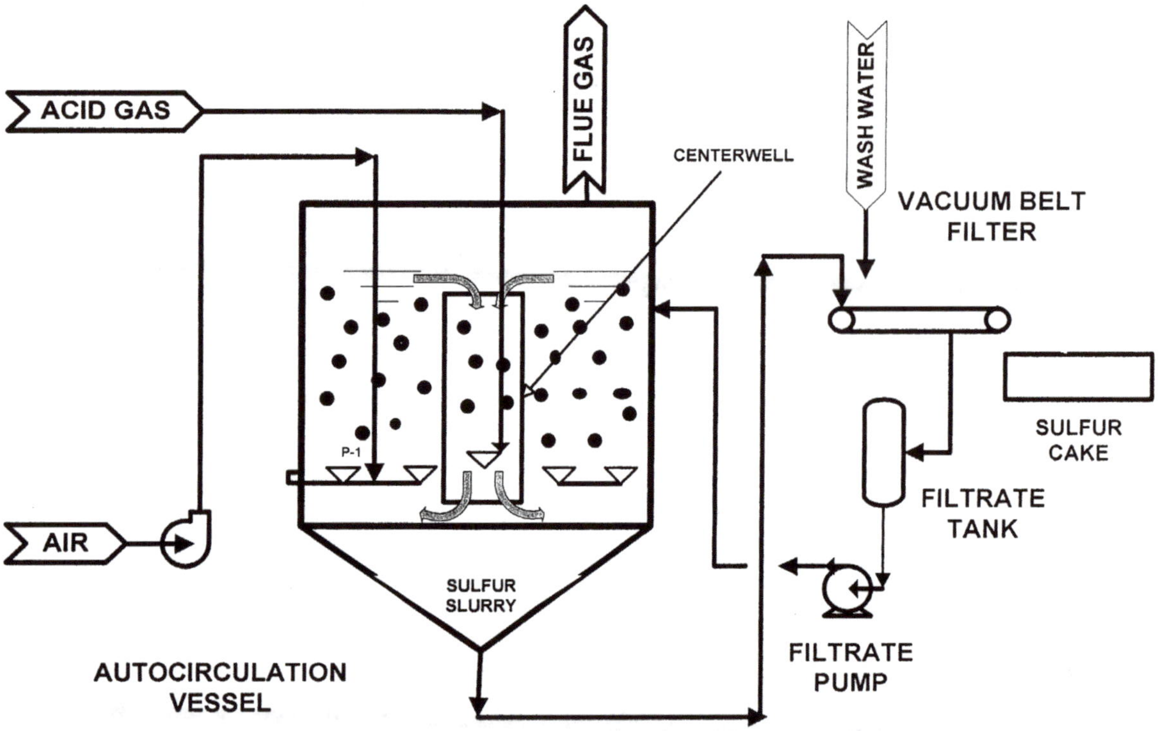

FIGURE 1
AUTOCIRCULATION LIQUID REDOX UNIT

Removal efficiencies of greater than 99.99 % and turndowns of 100% can easily be achieved with an Autocirculation unit.

TREATING SOUR WATER STRIPPPER GAS STREAMS

Contrary to Claus units in which sour water stripper gas (SWS) can create operational problems, processing SWS gas in a LO-CAT unit is actually beneficial. As previous described, all reactions in a liquid redox process occur in the liquid phase and are very rapid; however, the mass transfer of the H_2S from the gas phase to the liquid phase is relatively slow, and is thus the rate limiting step in the process. One method of increasing the mass transfer rate is to increase the alkalinity or pH of the circulating liquid solution. This is usually accomplished by adding caustic in the form of NaOH or KOH to the system, which creates an operating expense; however, due to the ammonia, which is usually contained in the SWS gas, no caustic addition is required when processing SWS gas, thus the operating cost is reduced.

Gas from sour water strippers, which contain reflux overhead systems, can be processed in a LO-CAT unit as illustrated in Figure 2; however, for non-reflux sour water strippers, the SWS gas will require cooling to reduce the temperature to approximately 140 - 150°F (60 -65°C), otherwise a water balance problem will be created in the LO-CAT system. The condensed water from this cooling step would be recycled back to the sour water stripper.

As shown in Figure 2, the SWS gas is contacted with LO-CAT solution in a venturi absorber, which serves two purposes. First, the venturi is an excellent, non-plugging absorber for this service and second, the venturi will produce a draft, which reduces the pressure requirement of the sour water stripper. The oxidizer and sulfur removal is the same as previously described.

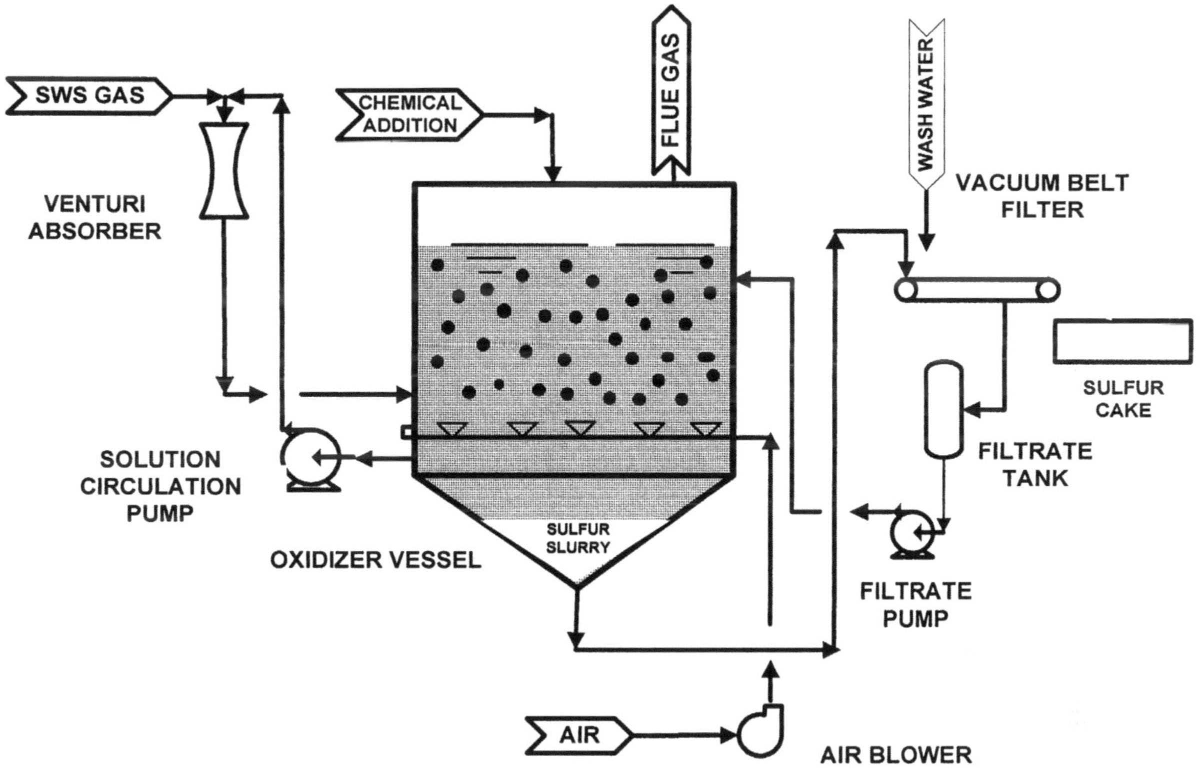

FIGURE 2
SOUR WATER STRIPPER GAS TREATMENT

The effluent gas from the Oxidizer will contain ammonia and water vapor and a good deal of dilution air. Depending on the amount of ammonia present and local regulations, this stream may need to be process through a catalytic reduction unit to remove the ammonia or combusted in a Claus incinerator if one is available. Again removal efficiencies of 99.99% and turndowns of 100% can easily be achieved with this processing scheme.

Removing the SWS gas from the Claus unit and processing it in a liquid redox system will not only free up capacity in the Claus unit but it will also eliminate the operational problems associated with processing this gas stream.

TREATING REFINERY FUEL GAS

When treating combustible gas streams such as fuel gas streams or recycle gas streams in which the gas stream cannot be mixed with air, a conventional processing scheme as illustrated in Figure 3 is employed. This processing scheme is similar to that of the sour water stripper gas processing scheme previously described with the exception that the venturi absorber is replaced with a non-plugging, liquid full absorber.

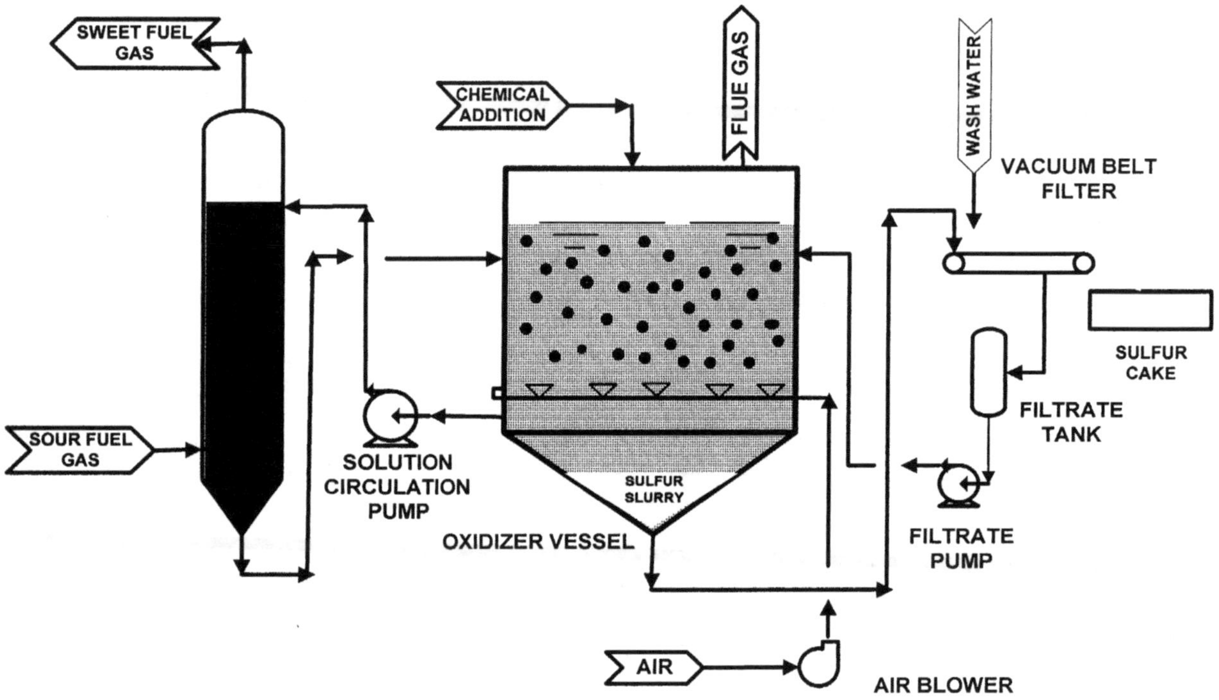

FIGURE 3
REFINERY FUEL GAS TREATMENT

Fuel gas, generally at 60 – 70 psig (4 – 4.8 barg) is passed through the liquid full absorber. If so desired, the H_2S concentration of the fuel gas can be decreased to less than 4 ppm (V/V) in this type of absorber. The spent solution from the absorber is directed to the oxidizer, which is simply a liquid full, atmospheric tank with a cone bottom through which air is sparged. The flue gas from the oxidizer is moist air that is slightly depleted of oxygen and contains no H_2S. The flue gas is normally discharged to the atmosphere; however, if the fuel gas contains BTX components, it will need to be incinerated. Sulfur is separated and removed from the system in the same manner as with the previously described Autocirculation processing scheme.

High pressure gas streams up to several hundred pounds of pressure can be treated in this type of processing scheme; however, a flash drum will be required between the absorber and the oxidizer. For very high pressure gas streams, it is generally more economical to treat the gas in an appropriate amine unit and then treat the acid gas in an Autocirculation LO-CAT unit.

TREATING CLAUS TAIL GAS

There are two approaches to coupling a liquid redox process to a Claus unit[1]. The first method involves processing the Claus tail gas through a cooler and then directly into a liquid redox unit. The second method involves processing the Claus tail gas through a hydrogenation/hydrolysis reactor, which will convert all of the SO_2, CS_2 and COS to H_2S followed by a cooler and a liquid redox unit. Both methods will result in removal efficiencies exceeding 99.9%. The only difference will be the operating cost of the first approach will be greater than that of the second. Both approaches will also result in turndown capabilities of 100%. This is accomplished by by-passing the Claus unit when it has reached its turndown capability and directing the acid gas to the redox unit.

When considering liquid redox to treat Claus tail gas without the inclusion of a Hydrogenation/Hydrolysis reactor, the amount of SO_2 in the tail gas is an important operating parameter. Since liquid redox units operate at alkaline pH's in the range of 8 to 9, any SO_2 in the tail gas will be easily absorbed, and form sulfates and sulfides in accordance with Reaction 1 and 2.

$$SO_2 + 2NaOH + \frac{1}{2}\,O_2 \Longleftrightarrow Na_2SO_4 + H_2O \qquad (1)$$

$$NaOH + H_2S \Longleftrightarrow NaHS + H_2O \qquad (2)$$

It is important to note that SO_2 does not interfere with the liquid redox chemistry and consequently does not affect the H_2S removal efficiency of the process. However, Reactions 1 and 2 do affect the operating cost of the process in two ways. First, caustic is consumed for each mole of SO_2 absorbed, which increases the operating cost of the unit. Secondly, the resultant sulfate/sulfide product will accumulate in the liquid redox solution, and eventually a continuous blowdown will be required resulting in loss of valuable catalyst solution, which must be replaced, again, increasing operating cost even further. Consequently, if this process configuration is to be employed, it is advantageous to minimize the formation of SO_2 in the Claus unit by operating the Claus unit with sub-stoichiometric quantities of oxygen, thus increasing the H_2S:SO2 ratio in the unit.

A flow diagram of a typical LO-CAT liquid redox unit for treating Claus tail gas directly is shown in Figure 4. Since the liquid redox system is aqueous-based, elevated temperatures will cause water balance problems; consequently, the tail gas is first passed through a quench tower where the gas temperature is reduced from approximately 135°C to 50°C.

For direct treatment of Claus tail gas, the LO-CAT process would employ a venturi absorber followed by a proprietary Mobile Bed Absorber (MBA). The venturi not only supplies much needed draft to the system, but it also provides a fair amount of H_2S removal. The MBA employs hollow, ping-pong-like spheres as contacting media which, when fluidized, are self-cleaning

In the event that the Claus unit is unable to operate due to turndown requirements beyond its capabilities, the system can be designed to bypass the Claus unit entirely and route the acid gas directly into the LO-CAT unit. In this mode of operation, the MBA will be bypassed, and the LO-CAT unit will operate as an Autocirculation unit as illustrated in Figure 1. To accommodate this operating scheme, the operating capacity of the Autocirculation vessel will need to be equivalent to that of the turndown capacity of the Claus unit.

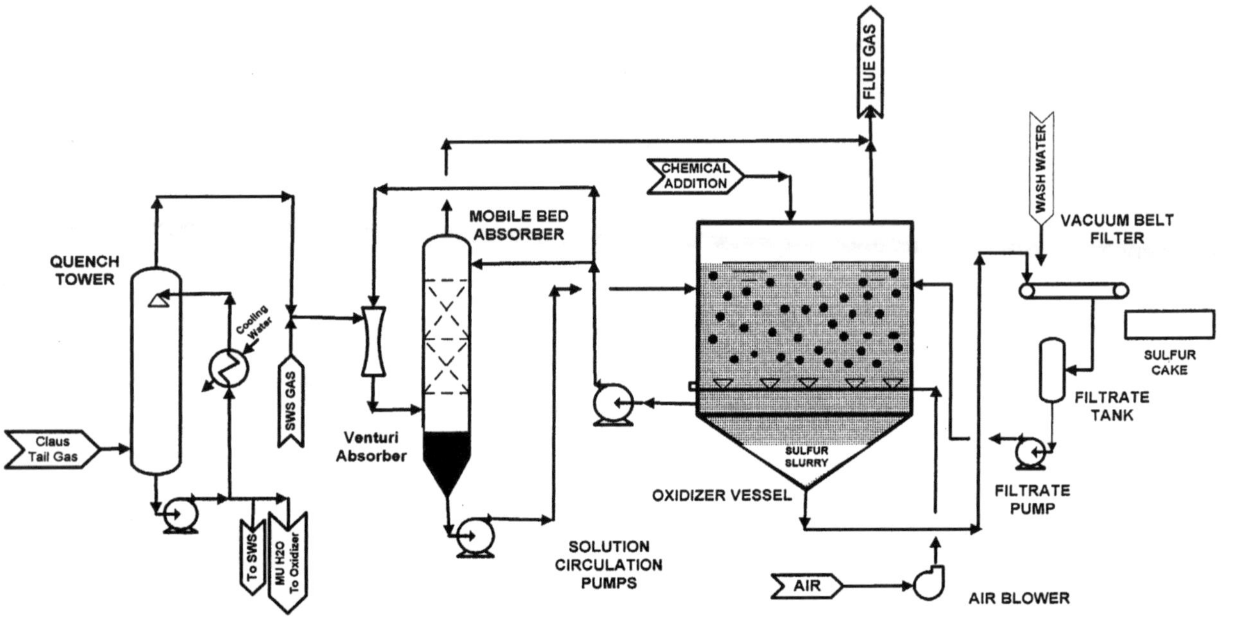

FIGURE 4
DIRECT TREAT OF CLAUS TAIL GAS

For direct treatment of Claus tail gas, the LO-CAT process would employ a venturi absorber followed by a proprietary Mobile Bed Absorber (MBA). The venturi not only supplies much needed draft to the system, but it also provides a fair amount of H_2S removal. The MBA employs hollow, ping-pong-like spheres as contacting media which, when fluidized, are self-cleaning

In the event that the Claus unit is unable to operate due to turndown requirements beyond its capabilities, the system can be designed to bypass the Claus unit entirely and route the acid gas directly into the LO-CAT unit. In this mode of operation, the MBA will be bypassed, and the LO-CAT unit will operate as an Autocirculation unit as illustrated in Figure 1. To accommodate this operating scheme, the operating capacity of the Autocirculation vessel will need to be equivalent to that of the turndown capacity of the Claus unit.

This mode of operation (Figure 4) will still yield H_2S removal efficiencies of 99.99^+%. In addition, the effluent gases from the liquid redox unit will not require incineration since the tail gas will contain only a very small amount of H_2S and no SO_2, and it will be diluted with the effluent air from the Oxidizer. It is also important to note that there is no recycle stream back to the Claus unit; hence the capacity of the Claus unit will not be reduced by the addition of a liquid redox system to treat the tail gas. In addition, by coupling the treatment of the tail gas and the SWS gas into a liquid redox system, addition Claus capacity is created.

In indirect processing scheme all sulfur compounds in a Claus tail gas are converted to H_2S by passing the tail gas through a hydrogenation/hydrolysis, catalytic reactor at elevated temperatures. Reactions 3 and 4 (hydrogenation) and Reactions 5 and 6 (hydrolysis) represent the major reactions, which occur in the reactor.

$$SO_2 + 3H_2 \Longleftrightarrow H_2S + 2H_2O \qquad (3)$$

$$S_2 + 2H_2 \Longleftrightarrow 2H_2S \qquad (4)$$

$$CS_2 + 2H_2O \Longleftrightarrow CO_2 + 2H_2S \qquad (5)$$

$$COS + H_2O \Longleftrightarrow CO_2 + H_2S \qquad (6)$$

In this processing scheme (Figure 5), a fuel gas is subjected to partial oxidation, which not only generates sufficient heat to raise the tail gas to reaction temperatures but also generates sufficient hydrogen to satisfy the requirement of Reactions 3 and 4.

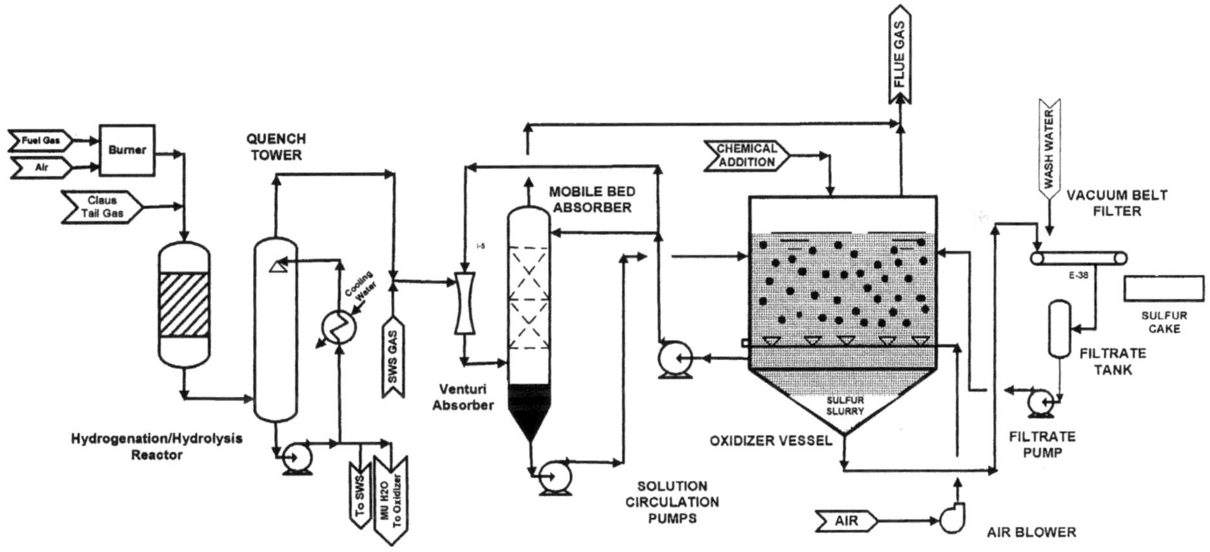

FIGURE 5
INDIRECT TREAT OF CLAUS TAIL GAS

After passing through the reactor, the effluent gas must be cooled to approximately 50°C, which will generate sour condensate. Again, a portion of the sour condensate may be used as makeup water for the liquid redox unit; however, some of it will need to be sent to a sour water stripper with the vapor being routed back to the liquid redox unit. The operating scheme of the LO-CAT unit will be identical to those described for the indirect treat treating case.

In summary, employing a liquid redox process as a tail gas cleanup yields the following benefits:

- 99.99+ % overall efficiency without having to recycle gas back to the Claus unit.
- Insensitivity to SO_2 break through.
- By incorporation of sufficient design capacity, 100% turndown for the sulfur recovery unit can be achieved.

SULFUR DISPOSAL

Because of its poor quality, sulfur produced from liquid redox processes has a bad reputation, which in some cases is well earned. However, due to the relatively small quantities of sulfur produced in liquid redox installations, most liquid redox sulfur has been either landfilled or disposed of as solid, agricultural sulfur; hence, not a lot of effort has been exerted to improve its quality. However, great progress has been made in improving the quality of sulfur produced in LO-CAT units[(2)(3)].

Sulfur is produced as a solid in a liquid redox unit. Since the reactions are not gas phase, there is no dissolved H_2S in liquid redox sulfur, thus sulfur degassing is never required. The sulfur is normally filtered and washed to produce a filter cake, which is 65% to 85% sulfur depending on the type of filter used, with the remainder being water and dissolved salts, albeit the salts are also good fertilizers and soil nutrients. It is not possible to simply dump this cake into the Claus sulfur pit, since there is insufficient heat in the pit to evaporate the water and to melt the sulfur. And even if the moisture is removed prior to dumping the sulfur in the pit, the solid particles have a tendency to float on top of the molten sulfur thus making heat transfer and consequently melting very difficult. Due to these problems, the sulfur from the liquid redox system must be disposed of as a solid or melted prior to being introduced into the sulfur pit. A typical melter system for a LO-CAT system is shown in Figure 6. It is worth noting that solid LO-CAT sulfur is in demand for use as a fertilizer and consequently, is demanding prices of around $35/LT.

CHEMICAL OPERATING COSTS

Chemical operating costs for a LO-CAT unit are generally between $100 and $200/LT of sulfur processed in the LO-CAT unit. If a Claus unit is operating at 96% conversion with the remaining 4% being processed in the LO-CAT unit, the chemical operating cost of the LO-CAT unit will add approximately $6/LT of total sulfur processed in the overall sulfur recovery unit.

SUMMARY

Liquid redox systems are extremely versatile and capable of achieving very high removal efficiency (99.9+ %) while yielding 100% turndown. These capabilities result in a proven technology that is capable of meeting today's sulfur management requirements in refineries. By incorporating liquid redox processing in refineries, the capacities of existing

Claus units can be increased by the removing sour water stripper gas and the tail gas treating unit recycle gas from the Claus unit, thus creating surplus Claus capacity.

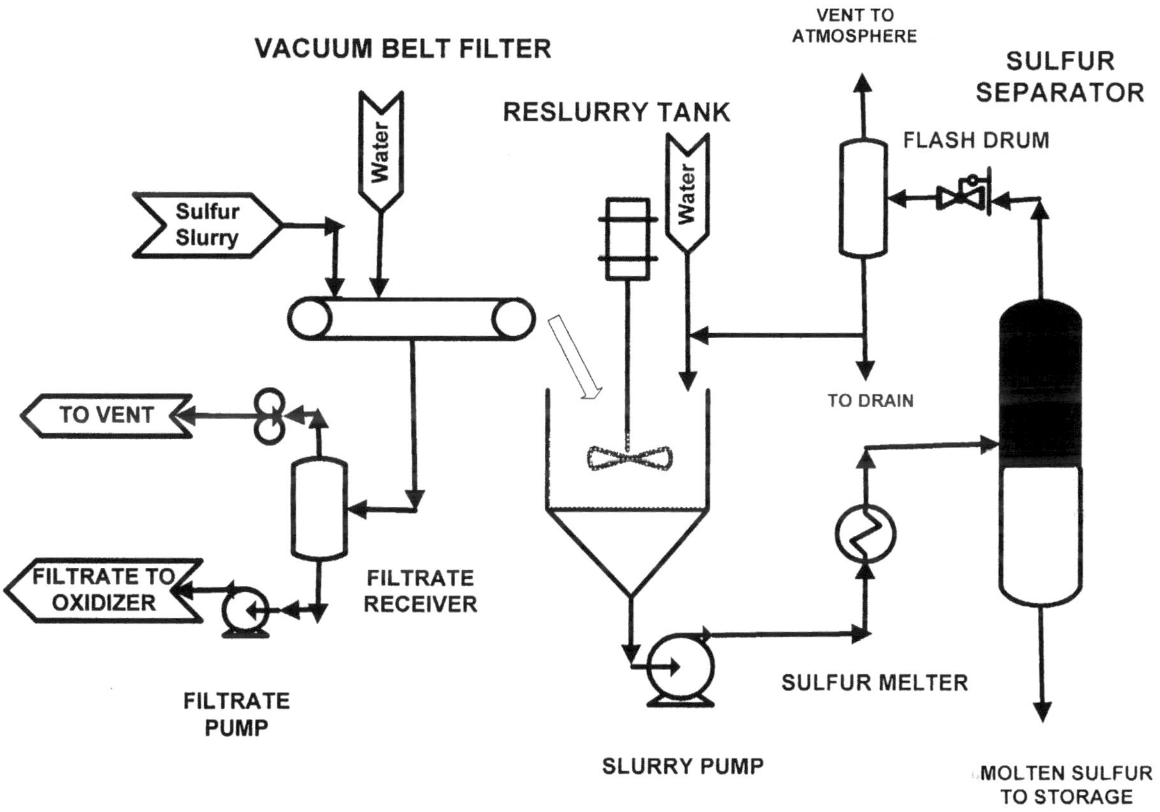

FIGURE 6
SULFUR MELTER

REFERENCES

(1) Nagl, G. "Employing Liquid Redox as a Tail Gas Cleanup Unit", Sulphur 2001 Conference, Marrakech Morocco, October 2001.

(2) Nagl, G. "Emerging Markets for Liquid Redox Sulfur", Sulphur 1997 Conference, Vienna Austria, November 1997.

(3) Connock, Lisa "Sulphur as a Fertilizer and Fungicide", Sulphur, January – February 2005, PROPRIETARY PROCESSES 13-18.

Steady State Simulation of DEA Plant using HYSYS to Reduce H_2S Emission

M.S. JASSIM

Department of Chemical Engineering, University of Bahrain, Manama, Bahrain
majeed3000@yahoo.com

Abstract

This manuscript shows the development of a steady-state simulation of an existing industrial Diethanolamine (DEA) plant in the Refinery of Bahrain Petroleum Company (BAPCO) using HYSYS simulation environment. The motivation for the work is the expected new environmental regulations that will reduce the maximum allowable concentration of Hydrogen Sulfide in the Refinery fuel gas. This will result in a need to reduce the Hydrogen Sulfide in the DEA plant sweet gas stream (which goes to fuel gas) from 500 ppm to 150 ppm. In order to meet this new specification, different scenarios were investigated by manipulating process variables to minimize the hydrogen sulfide leak in the base-case simulation.

Introduction

HYSYS is a simulation package used for educational purposes in the chemical engineering department at Bahrain University. It is also one of the simulation environments used in the Technical Services Department at Bahrain Petroleum Company (Bapco).

The expected new regulation of reducing the Refinery fuel gas Hydrogen Sulfide content to 150 ppm, would have an impact on the existing DEA plant. Hence alternative process solutions were investigated. One of the useful and cheap tools for performing sensitivity analysis is the simulation environment. Therefore, HYSYS package has been chosen to model the Refinery's amine plant.

A 30wt% aqueous diethanolamine (DEA) is employed to treat the refinery gases. This chemical solvent has the following advantages in comparison to Monoethanolamine (MEA): resistant to degradation, lower vapour pressure, lower cost, and reduced corrosive nature. However, the limitations of using DEA are: lower chemical reactivity, non-selectivity in acid gas removal, higher circulation rate, and the degradation products of carbon dioxide and DEA at high temperature are non-reclaimable by conventional reclaiming techniques.

Gas Sweetening Plant

Bapco's existing amine plant has two absorbers: Low-pressure (LP) and High pressure (HP) absorber and a single regenerator. The reason for such design is different operating pressures of the sour gas streams that are routed to the amine plant. A block flow diagram (BFD) of the simulated plant is shown in Figure 1. Table 1 shows typical sour gas characteristics for LP and HP absorbers. The concentration of hydrogen sulfide in the sour gas routed to LP absorber is almost tenfold more than the HP stream.

Table 1: Sour-gas characteristics

Process Variable	**HP**	**LP**
Composition (mol%)		
Methane	13.0	17.7
Ethane	1.9	12.8
Propane	0.8	12.2
Butane	0.3	7.9
Pentane+	0.2	1.9
Hydrogen	73.6	19.6
Water Vapor	0.0	0.07
Hydrogen Sulfide	2.5	24.5
Nitrogen	7.7	2.9
Temperature (F)	128	110
Pressure (psig)	900.7	85.7
Flowrate (kgmol/hr)	7592.0	323.9

The sour-gas from the refinery is introduced at the bottom of the absorbers whereas the chemical solvent (DEA) is introduced from the top of the absorbers hence a counter-current mass transfer accompanied by chemical reactions takes place throughout the absorber. The rich amine solution is then routed to a regenerator before being recycled back to the absorbers as lean amine. The reaction between the hydrogen sulfide and amine is instantaneous as it involves only proton transfer (reaction d) thus sieve tray is sufficient medium for gas-liquid contact. The set of reactions that take place in the absorber are as follows:

a) Water hydrolysis: $H_2O = H_3O^+ + OH^-$
b) Bisulfide formation: $H_2O + H_2S = H_3O^+ + HS^\bullet$
c) Sulfide formation: $H_2O + HS^- = H_3O^+ + S^{-\bullet}$
d) Proton transfer: $R_2NH + H_2S = R_2NH^+ + HS^-$

Where R is C_2H_4OH.

The characteristics of absorbers and regenerator are shown in Table 2.

Table 2: Characteristics of existing columns

	LP absorber	HP absorber	Regenerator
Diameter (ft)	4	6	7.5
Number of trays	20	2 x 11ft beds (Assumed 15 trays in model)	17
Tray Type	Sieve	Packed Bed	Sieve

Base Case

Table 3 summarizes a comparison for important process variables between actual field data and HYSYS simulation for LP absorber.

Table 3: Comparison between real process data and simulation for LP absorber.

	Real	Simulation	% Error
Sour Gas flowrate (Mlb/hr)	19.800	19.743	0.29
Sour Gas Temperature (°F)	110.0	109.63	0.37
Sour Gas mol% H_2S	24.560	24.557	0.01
Rich solvent Temperature (°F)	253.0	251.39	0.63
Amine circulation rate (Mlb/hr)	125.90	127.95	-1.63
Amine strength (wt%)	30	30	0
Lean amine loading (mol H_2S/mol DEA)	0.0504	0.0504	0
Rich amine loading (mol H_2S/mol DEA)	0.5796	0.5703	1.6
Sweet Gas (mol% H_2S)	0.0500	0.0506	-1.2

The above Table shows a low discrepancy between real field data and simulation. Thus, the simulation could be used as a reliable base case for further investigation.

Optimization Scenarios

Three different approaches were investigated to achieve the new governmental regulation for H_2S specification.

1) Effect of varying lean amine temperature

Figure 2 shows a diagram that illustrates the effect of varying the lean amine temperature on the concentration of hydrogen sulfide at fixed circulation rate. Lowering the lean amine stream temperature from 55°C to 30°C reduces the hydrogen sulfide concentration to 240 ppm due to the increase in the solubility of the gaseous species in the reactive solution. The consequence is increasing the duty of lean/rich heat-exchanger (or addition of air or water cooling) and hence the need for capital investment.

2) Effect of varying regenerator reboiler duty

Increasing the duty of the reboiler leads to lowering of the lean amine loading i.e. improves the performance of the amine solution; and hence reduces the hydrogen sulfide concentration in the sweet gas stream. Figure 3 illustrates the decrease of H_2S concentration with the increase of reboiler duty at constant lean amine temperature of 132°F. The increase of reboiler duty by 23% leads to reduction of hydrogen sulfide concentration by 83.3%.

3) Effect of lean amine temperature and regenerator reboiler duty

It is clear from the previous two scenarios that the target reduction was not achieved with the variation of a single process variable. Figure 4 shows the variation of hydrogen sulfide concentration with both lean DEA temperature and reboiler duty. It was possible to achieve the 150 ppm new specification with the existing columns but with at least 90°F lean DEA temperature and a reboiler duty higher than 31 MMBtu/hr.

Conclusions

HYSYS was successfully implemented in simulating an existing DEA amine plant. Sensitivity analysis was then applied to the steady state base-case in order to find out the best scenario for reducing the new H_2S regulation. The combined effect of low lean amine temperature and higher regenerator reboiler duty enabled achieving the new expected H_2S specification.

Acknowledgement

The author would like to thank Hafedh Al-Qassab and Savaş Soydaner from the Technical Services Department in Bahrain Petroleum Company (Bapco) for providing process data and helpful technical discussions. Also, the author would like to extend his gratitude to his students: Ali Abdulla, Mohammed Radhi, and Ahmed Najaf for their contribution to set-up the simulation model.

References

Bolhar-Nordenkampf, M., A. Friedl, et al. (2004). "Modelling selective H2S absorption and desorption in an aqueous MDEA-solution using a rate-based non-equilibrium approach." Chemical Engineering and Processing **43**(6): 701-715.

Lunsford, K. M. and J. A. Bullin (1996). Optimization of Amine Sweetening Units. AIChE Spring National Meeting, New York.

Weiland, R. H. and J. C. Dingman (2001). Column Design Using Mass Transfer Rate Simulation. 80th Annual Gas Processors Association Convention, San Antonio, TX.

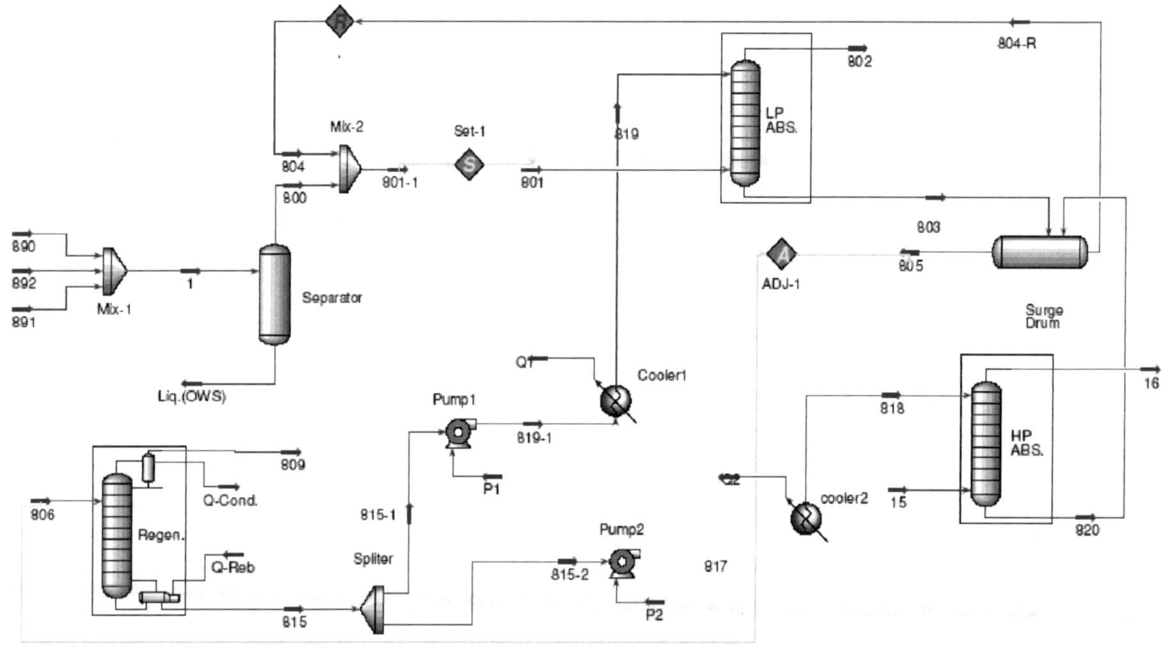

Figure 1: BFD of the amine plant.

Note: The BFD is not per actual design.

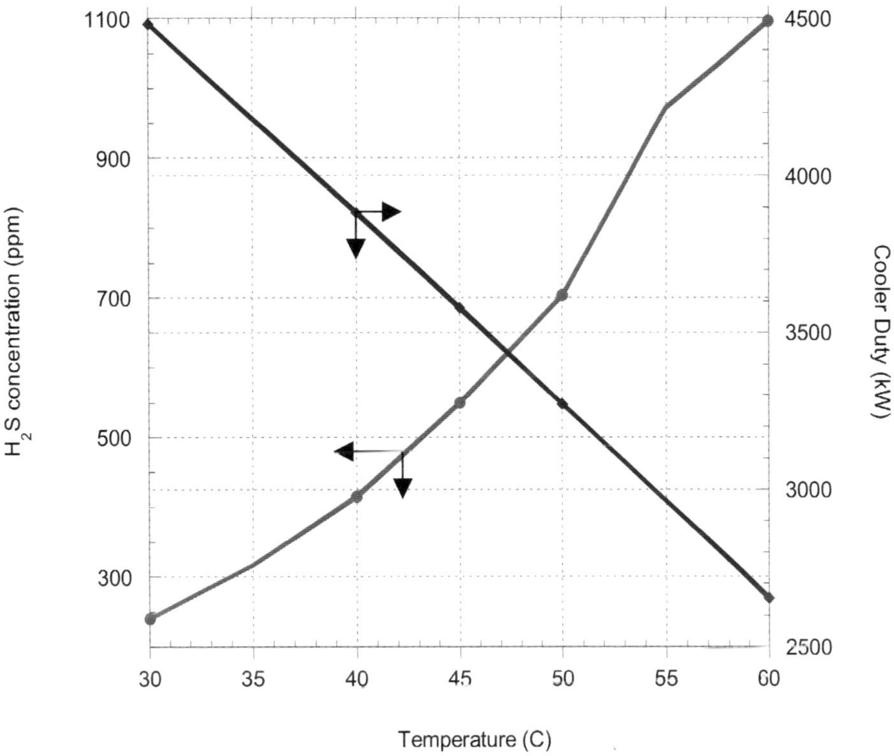

Figure 2: Effect of varying the lean amine temperature on H_2S concentration and cooler duty.

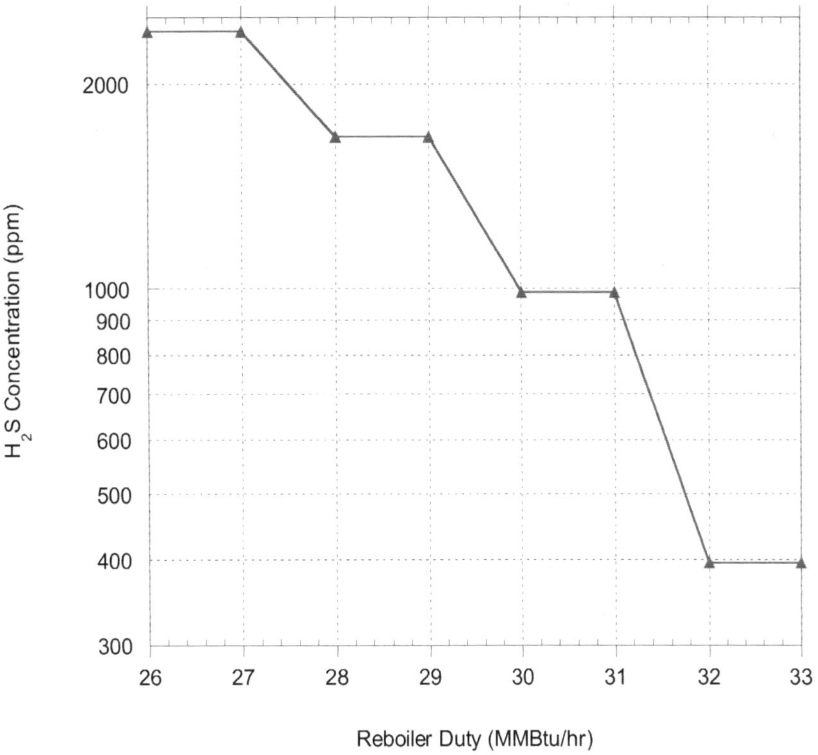

Figure 3: Effect of varying the regenerator reboiler duty on hydrogen sulfide concentration.

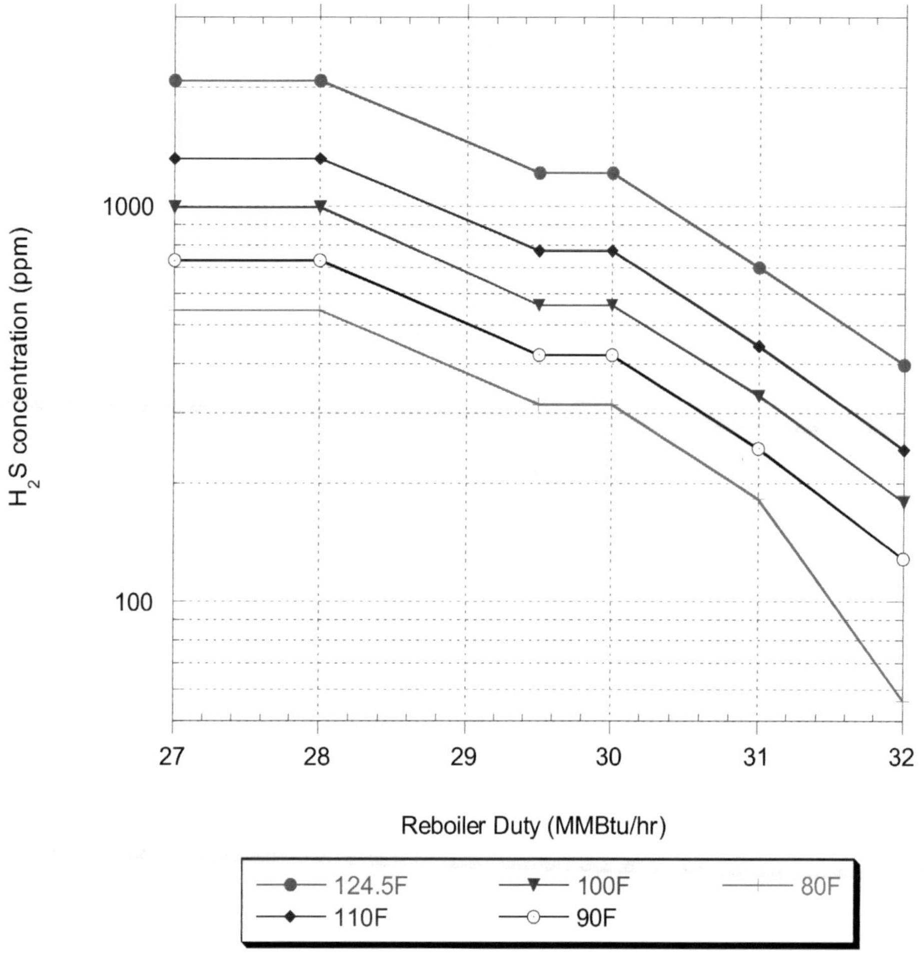

Figure 4: Effect of varying the regenerator reboiler duty on hydrogen sulfide concentration at different lean amine temperatures.

SRU OVERPRESSURE IN A WASTE HEAT BOILER FAILURE

Justin Lamar, Black & Veatch, Overland Park, Kansas USA

Protection of a sulfur recovery plant from overpressure involves special considerations. Sulfur vapors, which can solidify, and corrosive hydrogen sulfide gas make it difficult to assure reliable operation of pressure safety relief devices. The low operating pressure of sulfur recovery plants precludes venting relief devices to closed flare headers to contain hazardous process gases. These difficulties force the engineer to carefully consider the overpressure scenarios, resulting relief flows, and potential overpressure to ensure a safely-designed plant.

Usually the largest relief load results from a tube leak in the high-pressure waste heat recovery boiler installed immediately downstream of the thermal reactor. This boiler cools hot (above 1,800° F) process gases leaving the thermal reactor by raising high-pressure (usually above 400 psia) steam on the shell side of a set of tubes conducting the gas flow. Although this waste heat boiler is typically ruggedly designed, loss of water level can quickly overheat tubes and cause them to fail. Other events such as corrosion can also cause tubes to fail and leak. High-pressure steam and water then leak into the process side, which typically is designed for a pressure of 50 psig or less. The high-pressure steam and water must have somewhere to leave the system, or overpressure is possible. One available relief path is to simply allow the tube leakage flow to proceed down the normal process path to the incinerator and stack. This paper discusses the size and nature of the tube leakage flow, and how to evaluate the use of the conventional process flow path to safely discharge the leakage flow.

An interesting debate can address the questions of just what is the tube leakage flow, steam, water, or both phases, and how large can the failure be. Rather than answer these questions, this paper will instead focus on how to arrive at a realistic and accurate overpressure determination given a plant's configuration and an assumed failure size.

Differential pressure between a boiler shell and the process tube side is high enough to produce a sonic-velocity, choked flow through a leak. When the leak occurs, boiler water level, if it has not run dry before the leak, and boiler pressure both begin to drop. Dropping boiler pressure causes the leakage flow to fall as time proceeds, if the leak size does not grow. As boiler pressure drops, hot saturated water within the boiler will flash off steam. The steam/water equilibrium is quickly established and for practical purposes is instantaneous. Flashing boiler water is the chief source of steam flow supplying the tube leakage flow.

The choked flow rate through a tube failure can be large for a big leak. However, the boiler's capacity for sustaining the high flow is limited. The principal effect occurring in a tube failure is a depressuring of the boiler. Falling boiler pressure reduces the differential pressure producing flow through the leak and consequently the leakage flow decreases. Also, there is a fixed amount of steam and water within the boiler supplying the leakage flow. The boiler discharge stop-check valve, if it operates, prevents steam header backflow to the boiler from maintaining boiler pressure. The boiler feedwater control valve may open in response to dropping water level; however, the feedwater flow is typically much less than the tube leakage flow for a large failure. Low boiler water level or elevated SRU front end pressure typically will quickly shut down the thermal reactor feeds, and the absorbed duty from hot process gas flow through the tubes quickly diminishes. Although all these things limit the boiler's capacity to sustain leakage flow, the flashing water in the depressuring boiler is still a significant source of flow for a short period, as shown in an example calculation below.

The steam/water phase mixture supplying the leakage flow depends on what is in the boiler shell and where the failure occurs. A slowly dropping water level caused by instrument failure could expose, overheat, and fail top row tubes, leaking primarily steam. Or, corrosion from boiler solids accumulation could fail bottom row tubes, leaking primarily flashing water. Since we cannot anticipate what the leaking phase mixture will be, we must consider all possible mixtures and design for the worst effects.

Leakage flow rate is directly proportional to the total tube failure perforation size. Therefore the perforation size assumption is most important. In a failure one tube could have a tiny hole, several tubes could have tiny holes, one tube could have a larger hole, etc. API 521 addresses this uncertainty, saying, "In practice, an internal failure can vary from a pinhole leak to a complete tube rupture." API 521 suggests a sharp tube break with flow through both ends for the rupture flow basis for heat exchangers. A sharp tube break with flow through both ends is unrealistic for an SRU waste heat boiler failure, for several reasons. Sulfur plant boiler tubes are typically made from heavy-walled seamless pipe, which is sturdier than drawn heat exchanger tubes. API 521 makes a distinction between gauge tubes and schedule pipe in heat exchangers; for instance, API 521 Section 3.18.6 says, regarding double-pipe heat exchangers with inner tubes constructed of pipe material, "units that use schedule pipe for the inner conduit or tube are no more likely to rupture the inner pipe than any other pipe in the system." Boiler tubes have higher pressure on the outside and are in compression, and will tend to collapse and obstruct the leakage flow path in a failure. Finally, an API 521-type break depressures the boiler very quickly, as shown below, and has likely never occurred in a sulfur plant waste heat boiler. Assumed failure size for sulfur plant waste heat boilers should be based on failure analysis and examination of actual failure events. Below I give an example study of an actual failure event.

Given boiler pressure and an assumed perforation size we can calculate leakage flow for various steam/water phase mixtures as follows. Equations used for choked nozzles, as in for example relief valve sizing, are applicable to calculating rupture flow rates. For an all-vapor flow, equations for choked discharge through an orifice are adequate. For a saturated liquid or a steam/water mixture, the omega method described in API 520 Part I Appendix D is adequate. An interesting feature of the omega method is that it reasonably matches results obtained with choked vapor flow equations through an orifice.

Consider first an all-vapor leakage flow. Say we have a 2-square-inch rupture size in a 600 psig boiler. Using equation 5-1 from the GPSA Data Book 10th ed., for choked vapor flow through a nozzle,

$$W = \frac{A(C_1)(K)(P_1)(K_b)\sqrt{MW}}{\sqrt{(T_1)(Z)}}$$

where $C_1 = 343.16$ for $k=1.26$, $K = K_b = 1$, $A = 2\ in^2$, $P_1 = 614.7$ psia, $MW = 18$,
$T_1 = 489°\ F = 949°\ R$, $Z = 0.82$,

gives $W = 64{,}163$ lb/hr.

Next consider an all-saturated-liquid leakage flow. Again say we have a 2-square-inch rupture size in a 600 psig boiler. Using the omega method, section D.2.2 of API 520 Part I, with a downstream pressure of 8 psig, we calculate a flow of 191,227 lb/hr of saturated liquid, of which 54,359 lb/hr will flash into steam leaving 136,868 lb/hr of water. The volume flow rate of the flashed vapor, 11,345 ft3/min, is much greater than the volume flow rate of liquid, 43 ft3/min. Generally this is true for all rupture flows; the vapor flow rate is most significant. The water, if present, will settle out in the bottom of the SRU equipment, and to some degree will become entrained in the escaping steam flow through the unit. Given the equipment and piping configuration, it is infeasible to set up an entrained-flow hydraulic model. Since the liquid volume flow is much less than the vapor volume flow for all leakage mixtures, we can reasonably use a vapor-flow hydraulic model. Figure 1 shows leakage flow results obtained with the omega method for leaking mixtures of 0-100% vapor for our example 600 psig boiler. The worst case in terms of producing backpressure in the sulfur plant is likely the 100% vapor leakage case, since it has the highest vapor flow rate.

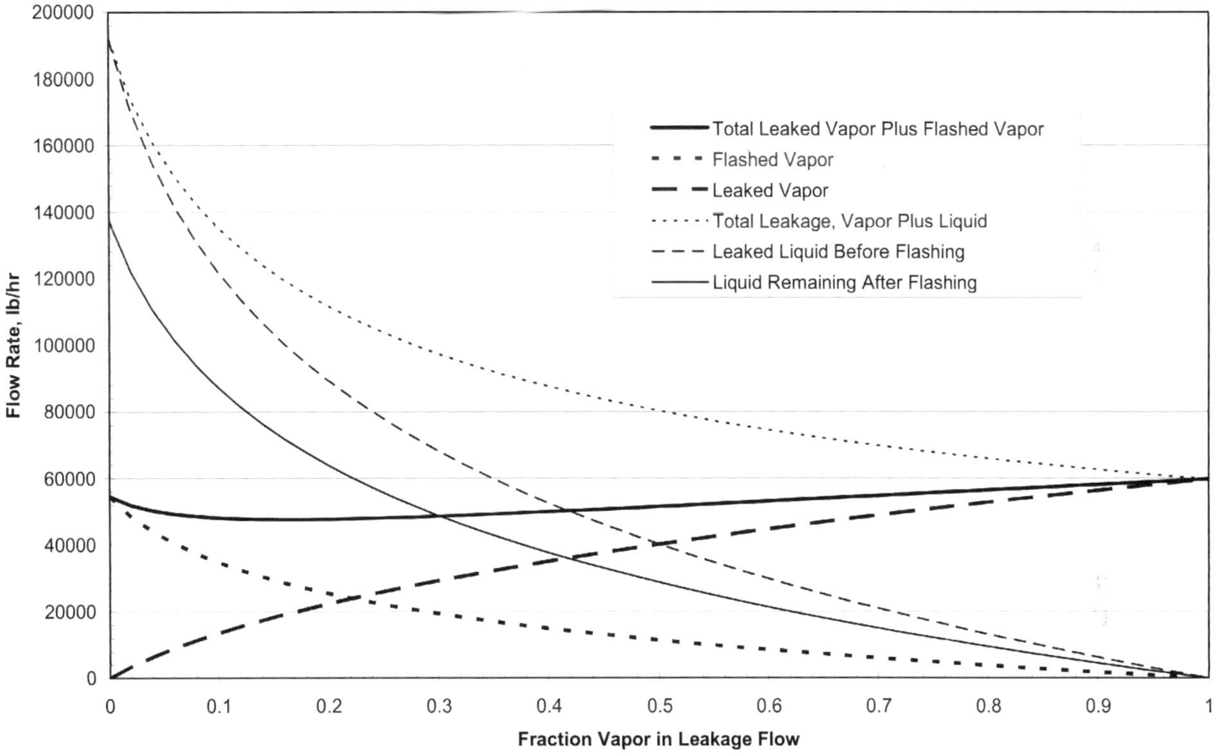

Figure 1: 600 psig Boiler Leakage Through 2 sq. in. Rupture with 22.7 psia Downstream Pressure.

We can combine the method above for leakage flow calculation together with an unsteady-state calculation to obtain the pressure and leakage flow decay profiles for the ruptured boiler. First assume a set of conditions existing at the instant of boiler failure: mass of water in boiler, boiler pressure, boiler feedwater flow rate, boiler blowdown flow rate, absorbed process heat duty, and perforation size. Next calculate the rupture leakage flow rate. Then, over a small time increment, calculate the mass of steam released through the rupture, the mass of blowdown removed, the mass of boiler feedwater added, and the absorbed process energy. Perform an energy balance as follows to calculate the enthalpy of the boiler water after the small time increment:

$$New\ h_{bw} = \frac{M_{bw}\ h_{bw} - M_{stm}\ h_{stm} - M_{blowdown}\ h_{blowdown} + M_{bfw}\ h_{bfw} + Q}{M_{bw} - M_{stm} - M_{blowdown} + M_{bfw}}$$

where M_{bw}, h_{bw} = mass and enthalpy of water in boiler, lbs. and Btu/lb,
M_{stm}, h_{stm} = mass and enthalpy of steam released through the rupture, lbs. and Btu/lb.
$M_{blowdown}$, $h_{blowdown}$ = mass and enthalpy of blowdown, lbs. and Btu/lb,
M_{bfw}, h_{bfw} = mass and enthalpy of boiler feedwater, lbs. and Btu/lb,
Q = absorbed process energy, Btu.

The new boiler water enthalpy value is slightly lower than the starting value because some water flashes to produce the steam supplying the rupture flow. With the new boiler water enthalpy value after the small time increment, find the new saturation pressure of the boiler water. This is the new boiler pressure after the small time increment. Now repeat the calculation for succeeding time increments to find how the boiler pressure, rupture flow, and other variables change as time proceeds. The procedure is made easier with a computer.

To illustrate the calculation, and the effects of various perforation sizes, consider the following example, taken from a paper presented by Joy Hansen and Ed Zamejc at the Brimstone Conference in 2001. In that paper the authors present a case study of a failure incident in 1999, and provide a great graph of measured boiler pressure data during the event, convenient to use for our example. Using the method outlined above, I applied a material and energy balance to that boiler, and calculated some pressure decay and rupture flow profiles with different perforation sizes. Figure 2 below shows some data points from the boiler pressure graph in the Hansen and Zamejc paper, with my calculated profiles for different perforation sizes. The 5.91 in^2 size represents a double-ended sharp tube break per API 521 in a 2" schedule 80 tube. Figure 3 shows my calculated rupture flow rates for the different perforation sizes. Figure 4 shows my calculated rupture flow rate for the 0.85 in^2 perforation size, together with my assumed boiler water mass, boiler feedwater and blowdown rates, and absorbed process duty for the calculations. The time shown on my graphs starts when the boiler pressure was at 530 psig in the 1999 data, when it began a smooth drop in pressure.

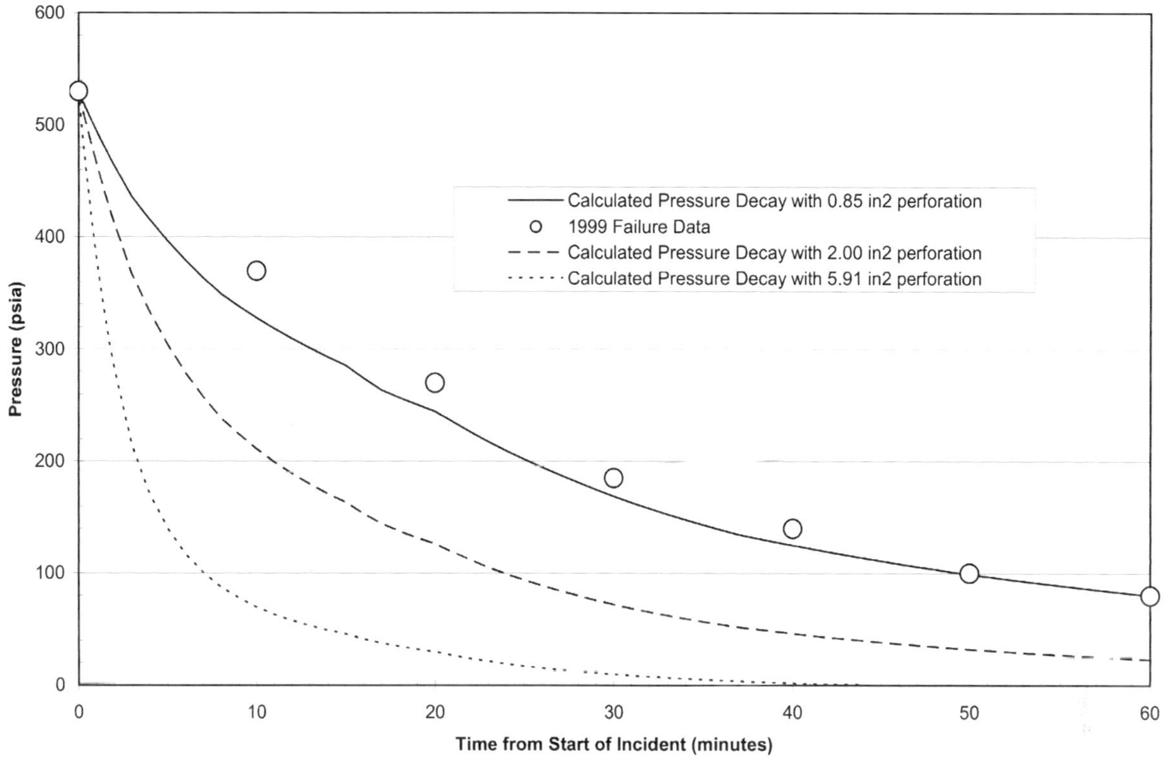

Figure 2: Calculated Pressure Decay Profile in Comparison with 1999 Failure Data.

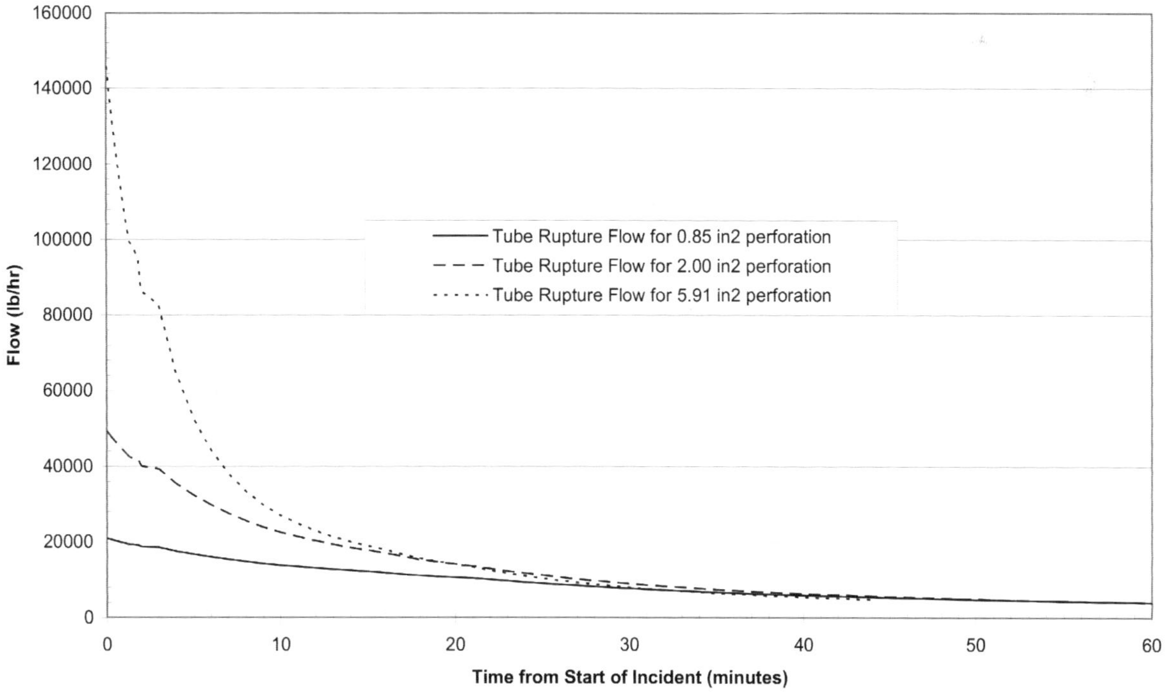

Figure 3: Estimated Leakage Flow for 1999 Example Event.

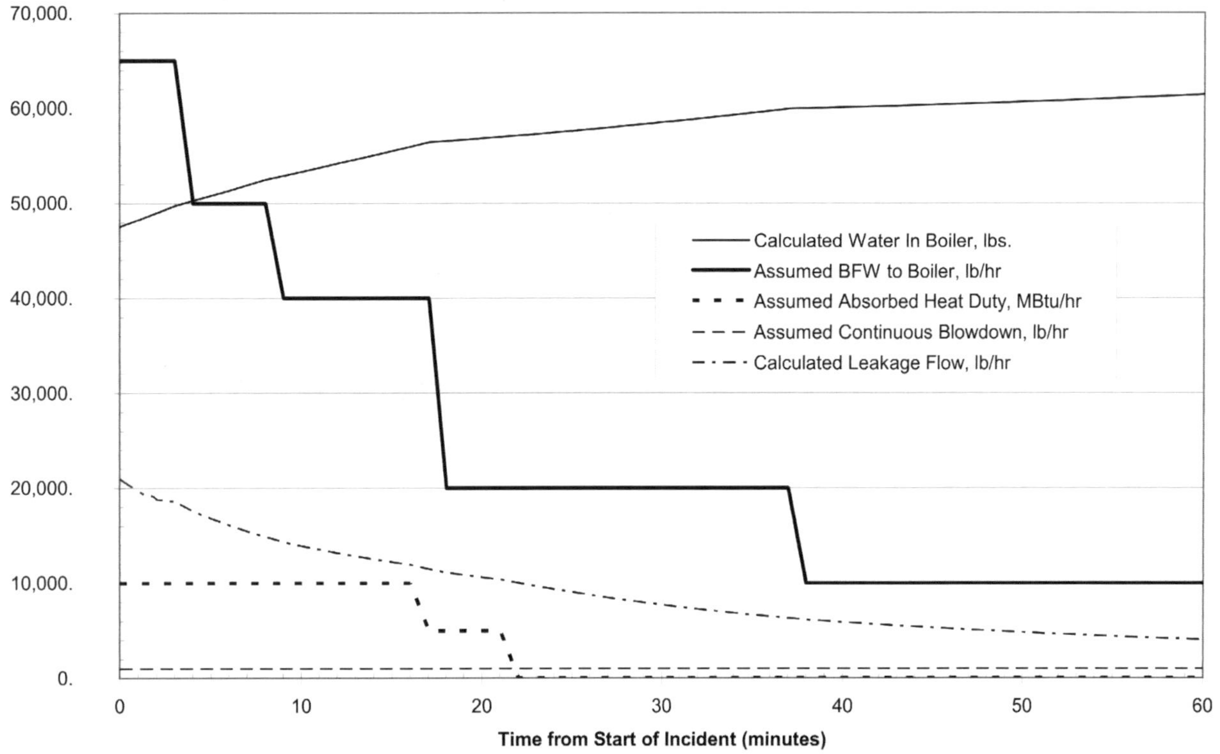

Figure 4: Input Data for 1999 Example Failure Calculation with 0.85 in2 Perforation Size.

My assumed values are based on the size of the boiler, and not on any actual data from the 1999 event. I estimate that the total perforation size in the 1999 event was 0.85 in^2. The resulting peak rupture flow rate, about 21,000 lb/hr, is comparable with the normal process flow through the SRU, which may explain why the SRU pressure during the 1999 event did not blow the sulfur seal legs.

With rupture flow estimates, we can calculate the backpressure developed in the SRU by the escaping steam flow. The easiest and most conservative backpressure calculation is to assume steady-state flow conditions develop through the sulfur plant and out the stack. We can take the initial rupture flow, calculated with the boiler at full operating pressure before its pressure drops from the rupture, and calculate the backpressure developed in the sulfur plant from this flow at steady-state. A reasonable calculation method ratios the pressure drops developed by the escaping steam flow to the normal design process flow pressure drops. The equation for the pressure drop ratio is as follows:

$$\frac{\Delta P_2}{\Delta P_1} = \frac{m_2^2}{m_1^2} \frac{P_1}{P_2} \frac{T_2}{T_1} \frac{MW_1}{MW_2} \frac{f_2}{f_1}$$

where ΔP_1, ΔP_2 = pressure drop for flow condition 1 and 2
m_1, m_2 = mass flow rate for flow condition 1 and 2
P_1, P_2 = absolute pressure for flow condition 1 and 2
MW_1, MW_2 = gas molecular weight for flow condition 1 and 2
f_1, f_2 = friction factor for flow condition 1 and 2.

The temperature, molecular weight, and pressure factors may be combined into a density factor $\rho_1/\rho_2 = P_1T_2MW_1/P_2T_1MW_2$. Table 1 below summarizes an example calculation.

Table 1: Example Steady-State Backpressure Calculation.

Item	Process Flow	Process Density	Process Inlet Pressure	Process DP	Steam Flow	Steam Removed	Steam Temp	Steam Inlet Pressure	Steam Density	Steam DP
	(lb/hr)	(lb/ft3)	(psig)	(psi)	(lb/hr)	(lb/hr)	(° F)	(psig)	(lb/ft3)	(psi)
Thermal Reactor	11,452	0.0219	10.23	2.38	31,808		309	21.91		2.48
Waste Heat Boiler	11,457	0.0449	7.85	1.08	31,808	9,356	309	19.43	0.075	4.99
1st Condenser	11,457	0.0556	6.78	2.46	22,452	7,987	309	14.44	0.064	7.18
2nd condenser	9,773	0.0446	4.31	2.26	14,464	5,666	309	7.26	0.048	3.68
Final Condenser	9,277	0.0431	2.05	1.42	8,798	0	309	3.57	0.040	1.37
TGU Inlet	9,058	0.0437	0.64	1.89	8,798		309	2.20	0.037	1.93
Quench Column	9,058	0.0292	-1.25	1.02	5,798	6,000	309	0.27	0.033	2.77
TGU Booster Blower	6,793	0.0571	-2.27	-4.53	2,798		212	-2.50	0.030	-3.29
TGU Booster Blower Discharge	6,793	0.0682	2.26	2.26	2,798		212	0.79	0.039	0.79
Stack	6,466	0.0682	0.00		2,798		212	0.00	0.037	0.00

If backpressure developed in the SRU exceeds the static pressure of the liquid sulfur standing in the sulfur seal legs, the liquid sulfur will blow out, and steam from the rupture will escape through the seal leg to the atmosphere. Should this be found in the calculated rupture flow pressure profile, calculate the escaping steam flow through the seal leg and subtract it from the main flow downstream of the seal leg connection. If SRU backpressure rises high enough, escaping flow out the seal leg may be choked at sonic velocity, and the seal leg flow calculations should take that into account. Steam flow through the seal legs is very significant for a large rupture, particularly for the first one or two seal legs at the waste heat boiler and first sulfur condenser where developed backpressure is the highest. The escaping steam flow through the seal legs relieves backpressure in the SRU. This is a good reason to properly air-blow the sulfur plant and rundown piping at major turnarounds, especially when new catalyst is installed, to ensure they remain clear of obstruction. Rundown piping steam jacketing and insulation should also be kept in good condition. It is unlikely for solid sulfur to obstruct properly jacketed rundown piping, particularly since the rundown sulfur is usually well above its freezing point. Since waste heat boiler tube failure can cause a release through the seal legs, the plant instrumented safeguarding system must reliably block in the acid gas feeding the unit upon high SRU pressure to prevent a release of hazardous process gas.

The above steady-state backpressure calculation is conservative for two reasons. First, the boiler drops in pressure as shown above from the instant of failure, causing the rupture flow to decrease, lowering the developed backpressure. Second, the accumulation of pressure within the SRU absorbs some of the rupture flow, and it takes a period of time for backpressure to build up in the SRU. These two effects combine to produce the actual peak SRU overpressure. To gain a better understanding of their interplay I carried out a dynamic study of an example failure. To facilitate the exercise I implemented a model using commercial dynamic simulation software, HYSYS from Aspentech. Table 5 below shows results obtained from a simulation for an example small, 20 LTPD SRU with a 150 psig boiler and a large failure consisting of a sharp tube break with flow through both ends of a 3" Schedule 80 tube. A backpressure of 86 psig, calculated using the steady-state method with the initial tube rupture flow of 24 lb/sec (86,400 lb/hr), is shown to compare the conservative approach with the dynamic approach.

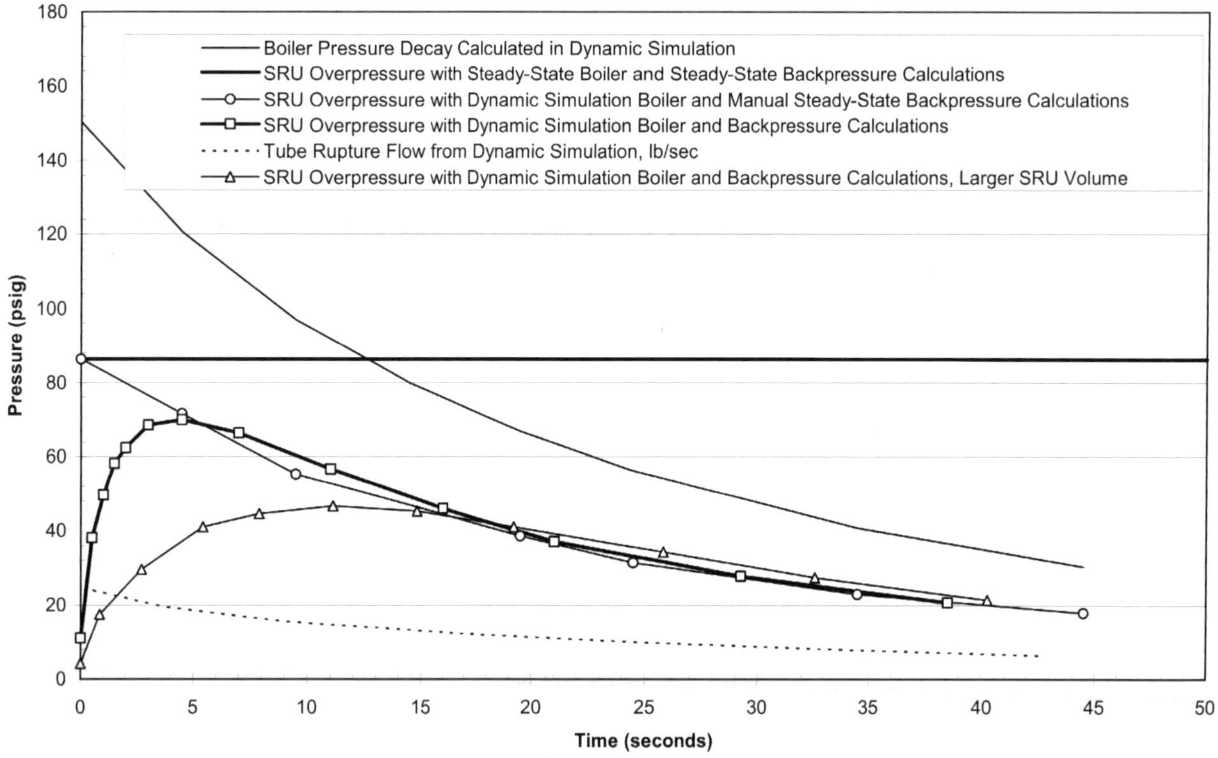

Figure 5: Comparison of Steady-State and Dynamic SRU Overpressure Calculation for a 150 psig Example Boiler, Sharp Double-Ended Tube Break in 3" Sch. 80 Tube.

The dynamic backpressure curve begins at the system normal operating pressure and rises after the tube rupture event at time = 0 seconds. It peaks about 5 seconds after the failure in this example. Thereafter the backpressure result from the dynamic simulation falls in step with the decreasing tube rupture flow rate. In fact, the dynamic simulation backpressure result at 5 seconds in this example matches what one would calculate using the steady-state methods described above per Table 1, for the rupture flow present at 5 seconds. It appears that once SRU backpressure develops and peaks, the backpressure decay profile can be calculated with a steady-state flow calculation given the tube rupture flow at a given instant. The curve marked by circles shows the results of such manual steady-state calculations for this dynamic example.

The curve marked by triangles in Figure 5 represents the dynamic simulation backpressure profile with everything identical in the model except one change: a much larger outlet channel on the waste heat boiler. Channel volume was increased from a 4 ft. ID x 3 ft. long size to an absurd 10 ft. ID x 10 ft. long size to illustrate the effect of plant volume in absorbing and smoothing the peak overpressure. With this size it takes longer for the backpressure to peak, about 12 seconds instead of 5 seconds, and the peak is at a lower pressure, 46 psig instead of 70 psig. Again the backpressure profile after the peak follows the decay profile which can be calculated with a steady-state flow calculation given the tube rupture flow at a given instant.

Dynamic backpressure calculations are more realistic than steady-state calculations because a waste heat boiler tube failure is not a steady-state occurrence. Dynamic calculation provides a lower backpressure result than conservative steady-state calculation because the dynamic approach takes into account the decreasing leakage flow and the build-up of pressure in the SRU, which act together to reduce backpressure. For SRU overpressure safety design, when steady-state calculations with reasonable assumptions provide unacceptable results, dynamic calculations provide a valid method to obtain a lower backpressure result which may be acceptable.

The author's work forming the basis for this paper was accomplished in discussion with Shell Global Solutions (US) Inc. The author expresses his appreciation for the support of SGSUS.

REFERENCES

Hansen, Joy and Ed Zamejc, "Seal Pit Elimination, Sulfur Recovery Unit, Texas City Refinery," The Brimstone Companies' Sulfur Recovery Symposium, Vail, CO, September 10-14, 2001.

Engineering Data Book, 10th ed., Gas Processors Suppliers Association, Tulsa, OK, 1994.

Guide for Pressure-Relieving and Depressuring Systems, American Petroleum Institute Recommended Practice 521, 4th Ed., Washington, D. C., March 1997.

Sizing, Selection, and Installation of Pressure-Relieving Devices in Refineries, Part I—Sizing and Selection, American Petroleum Institute Recommended Practice 520, 7th ed., Washington, D. C., January 2000.

THE ENGINEERED APPROACH TO SRU REFRACTORY LININGS

THREE KEYS TO RELIABILITY

S. J. Proctor, PE
J T Thorpe Company
Houston, Texas

Andy Piper
J T Thorpe Company
Houston, Texas

Prepared For Presentation at the 2005 AIChE Spring National Meeting
April 10-14, 2005
8th Topical Conference on Refinery Processing
Sulfur Management Session
Atlanta, Georgia

ABSTRACT

Sulfur Recovery Units and the refractories that line them are being pushed harder than ever before. Crude oil feed stocks contain increasingly higher levels of sulfur, while at the same time new regulations require lower sulfur levels in the end products. In order to process or convert this extra sulfur, throughput must be increased by means of oxygen enrichment (resulting in higher temperatures) or by adding new production units. At the same time throughputs are increasing, the Clean Air Act has restricted emissions during flaring operations. Penalties are severe enough to require major portions of a plant, if not the entire plant, to be shut down when the SRUs are not available. Consequently today's SRU linings must be robust enough to endure the high temperature upset conditions present in most refineries today, even in non oxygen-enriched units. This paper is intended to show 1) that the refractory linings for SRUs can be engineered, 2) the details that must be considered and 3) the level of craftsmanship required for their proper installation.

INTRODUCTION

For 30 years beginning in the mid 1950's, Thorpe was fully involved in the turnkey supply of reaction furnaces, re-heaters, incinerators and stacks. In the 1980s, we made a decision to focus on the refractory linings for these units and leave the ever-evolving combustion technology and controls to others. Thorpe has performed over 80 full reaction furnace lining installations involving the design, supply and construction of linings for both new and rebuilt units. Many of these include the steel vessel design and fabrication responsibility. Additionally, Thorpe has been involved in many localized refractory repairs and problem areas.

THREE KEYS TO RELIABILITY

1. Proper Material Selection of the Thermal Reactor Hot Face Lining. The three typical lining options are brick, plastic and castable. Many factors contribute to the success/failure of refractory linings so the various strengths and weaknesses of each system need to be evaluated. First, an explanation of the three hot face material options:

- Extra High Alumina Brick – prefired ceramic bonded shapes, mortared in place
- Extra High Alumina Plastic – phos-bonded moldable, rammed in place
- Extra High Alumina Castable/Gunmix - cement/chemical bond, cast or gunned in place

Tabular (high purity) alumina products are favored for the SRU application due to the extreme temperatures and reducing atmospheres to which the linings are exposed. As table 1 illustrates, brick is a superior choice for thermal reactors.

Table 1

	Aggregate	Bond	2700°F Hot Strength	Volume Stability
Brick	tabular Alumina	ceramic	excellent	excellent
Plastic	tabular Alumina	phosphate	fair	poor
Castable	tabular Alumina	cement/chemical	good/fair	fair/fair

The superior hot load strength of brick is derived in part from its purity. In contrast, the bonding systems in high alumina plastics and castables inherently contain impurities that reduce their performance in high temperature reducing atmospheres. Extra high alumina brick are virtually free of contaminates such as phosphoric acid, alkalies, iron oxide, etc. These impurities can be reduced in a reaction furnace environment. The new components that are formed typically have lower melting points thus reducing strengths when they are most needed. The newly formed compounds also cause volumetric changes that can be disruptive to the refractory lining. Volumetric changes that occur as a result of mineralogical changes can also be disruptive to the refractory lining. When considering volume stability, brick have a distinct advantage over monolithics because they are prefired to very high temperatures prior to use and have therefore already undergone the mineralogical transformations that can affect permanent growth or shrinkage. In considering high alumina brick products, some people make the erroneous assumption that the serviceability of a brick is directly proportional to its alumina content.

Table 2 compares the yield point (hot load bearing capability) of various high alumina brick products. As this table shows, the yield points are affected more by the level of impurities than by the alumina content. Note that there are some 85% alumina brick that could have disastrous implications for a reaction furnace application.

Table 2

Brick	Impurities (%)	Yield Point (°F) Reducing
90% Al_2O_3	0.5	2900
60% Al_2O_3	1.6 – 4.1	2700 - 2350
70% Al_2O_3	4.9	2200
85% Al_2O_3 (Phos Bonded)	8.9	1320

Typically, the products most widely used in today's furnaces are 90% alumina (extra high alumina, high purity) brick. Table 2 should not imply that all 90% brick are the same, as they are not. The ASTM C-16 100 hour x 3000°F x 25 psi test is used to compare hot load bearing capability of various brick products. Table 3 summarizes test results for various brick containing 90% to 98% alumina. The main point to be derived from this table is that not all extra high alumina brick are the same nor should all be considered acceptable for this specific application. A good understanding of these bricks and their properties is important to any design and the ultimate reliability of a lining.

Table 3

Brick Types	Deformation Under Hot Load*
90% mullite bonded	-0.2% to -2.0%
90% mullite bonded, phos added	-2.3% to -5.5%
99% corundum bonded	-1.5% **
90% alumina, 10% chrome, solid solution bond	-0.1% **

* ASTM C-16 - 100 hour test @ 3000°F

** Very limited data available

Other reasons for using brick instead of plastics or castables include:

- Lower porosity – helps to retard chemical attack
- Quality Assurance – not dependent on field mixing variations
- Minimize/eliminate metallic anchorage – Any "buried metal" in a reaction furnace lining should be viewed as detrimental, as metallics will not hold up long-term and can actually compromise the refractory lining. A self supported brick system does not rely on anchorage.
- Minimizes dryout concerns, time and costs.

The focus for this paper is on the hotface lining, however, mention of the backup lining is appropriate. The two most popular backup materials are insulating firebrick (IFB) and insulating castable. Both materials have successful track records but require different design considerations. Based upon our history, experience and successful track record, we prefer IFB for the following reasons:

- No metallic anchorage – IFB are self supporting by virtue of their geometry, eliminating the need for metallic anchors. If anchors burn up or corrode, castable can delaminate and break away, with the possibility of dislodging hot face brick during startups following maintenance outages.
- Less expensive – On an installed basis IFB are a more cost effective, less time consuming method.
- Reduced dryout time and de-watering concerns
- Proven track record – We have over 50 years experience with this type of construction

2. Engineered Lining Design – The first step is to work with the operator/engineer to gather the unit design parameters. This would include the steel vessel details, internal lining geometry and a review of the operating conditions and temperatures. For an existing unit with specific problems it is beneficial to perform a detailed inspection and failure analysis to address these issues.

The next step is to begin the refractory design process. The process includes establishing the working lining thickness and the back up lining type (IFB or castables) and thicknesses. A thermomechanical analysis should be performed to verify that the lining construction and refractory thicknesses are appropriate and will result in a stable system under actual operating conditions. It may be necessary to adjust lining thicknesses, change material qualities and/or to introduce expansion layers within the lining to alleviate circumferential stress and satisfy temperature limitations of the various components, including the steel shell.

A critical component of any successful design is the rain shroud (Diagram 1). The main purpose of the shroud is to moderate ambient conditions in the immediate vicinity of the reaction furnace shell (i.e., to mitigate the effects of climatic events such as drenching rains and temperature extremes). This is important for maintaining shell temperatures within a desired range; i.e. above the point where sulfuric acid will form against the steel shell but below the point where high temperature sulfidation can occur. A secondary purpose of the shroud is to prevent pinch spalling of the hotface brick. This could happen during a heavy rain that reduces the shell temperature, contracting the shell and overstressing the hotface brick. The design of the shroud must allow for the appropriate annulus between the vessel shell and sheetmetal and must also be designed to have sufficient venting at the top and bottom in order to prevent overheating.

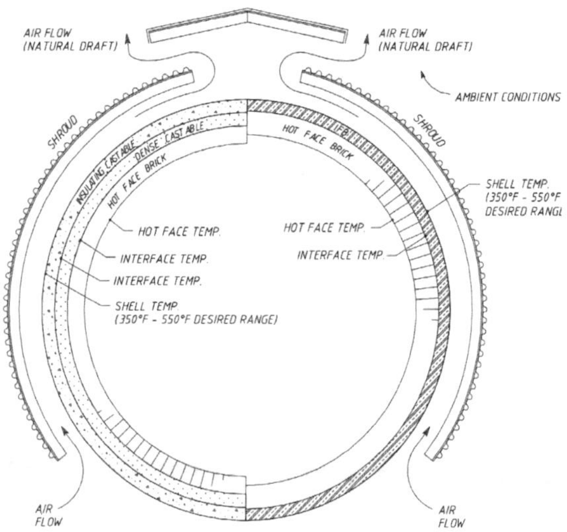

Diagram 1

Benefit of shroud

- Moderates shell temperature; minimizes low temperature & high temperature corrosion
- Eliminates pinch (rain) spalling of refractory hotface due to sudden shell contractions

Difficulties due to shroud

- Requires complex thermal calculations due to heat transfer mechanism within the annulus
- Maintenance of shroud; damaged areas need to be fixed and air flow paths kept open

After the material qualities and lining thicknesses have been decided it is time to begin the detailed brick design to maximize reliability. This is one of the most important yet least understood steps. In fact, many companies focus on monolithic linings as they have lost the knowledge/personnel familiar with complicated brick construction. Photo 1 shows an example of what can happen when bricks that are not properly tapered are used in the upper section of a reaction furnace.

Photo 1

It is very easy to pull tables out of a book that provide ring combinations to turn circles, but a full understanding of brick construction will provide a list of various options that can be evaluated and ranked in order of preference. Hotface ring combinations that utilize straights or bricks with insufficient taper in the upper 180° should always be avoided. If a brick is lost, the entire upper portion of the ring can fall out.

Illustration 1 demonstrates the simple difference between two common series of brick that turn the same circle.

Illustration 1

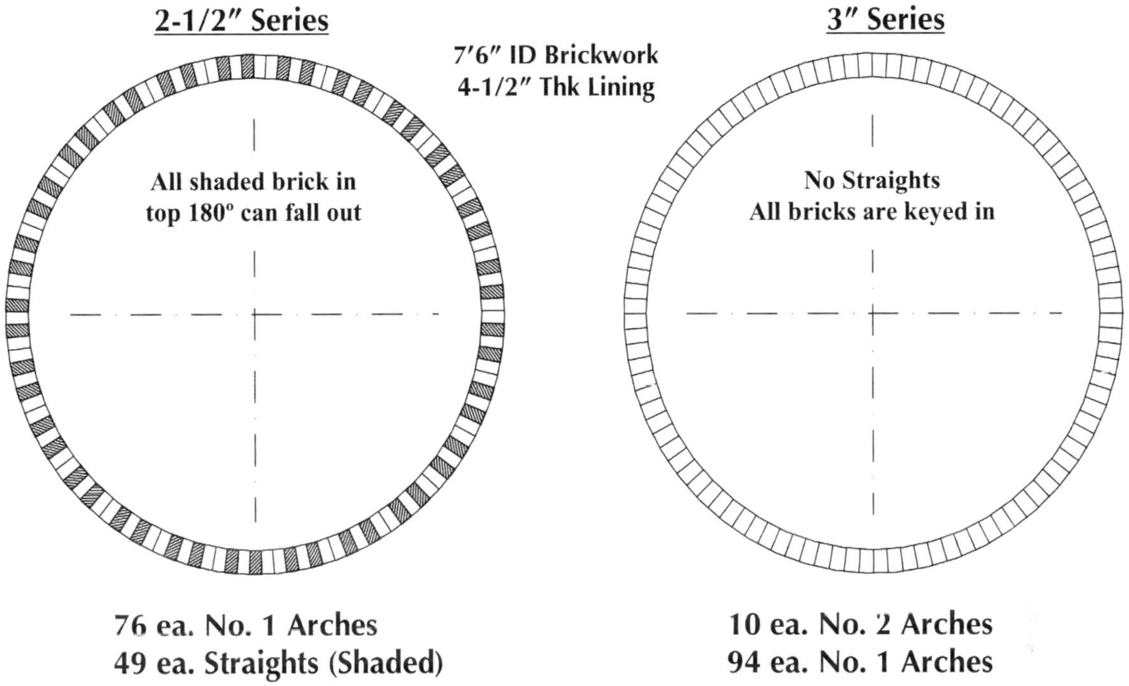

Reaction furnace linings are typically constructed using arch and wedge shapes, but design constraints and operating conditions may dictate the use of other refractory shapes that would have better keying action. The need for such special brick shapes should be evaluated on a case-by-case basis.

Common to these units are dished and elliptical heads. These are preferred over "flat" heads, since they have the potential of resulting in a much more stable brick lining that tightens against the steel shell during furnace operation. These head configurations also create unique design and installation challenges. The proper ring calculations cannot be determined simply by utilizing handbook tables. They must be calculated using special computer programs or in some cases, drawings in conjunction with hand calculations. Due to fabrication tolerances for the steel head and the brick shapes utilized, some custom cutting and fitting will always be required to achieve a reliable installation. Illustration 2 shows a typical non-engineered system common to these units.

Illustration 2

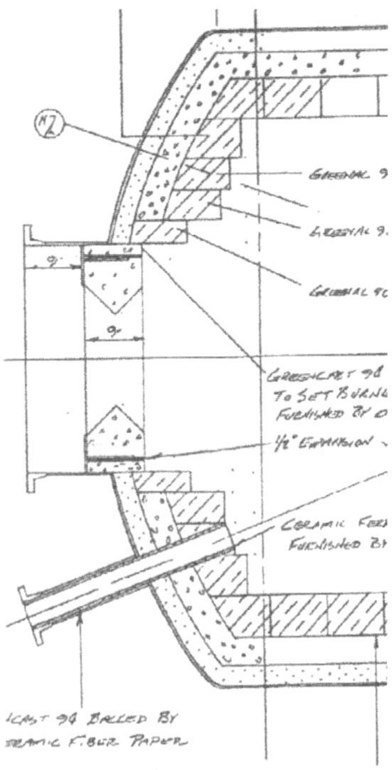

Problems with this non-engineered unit:

- No skewbacks – skewbacks are tapered brick affording a surface for the head brick to "push" against. The skewback is essential to make the head lining tighten against the steel head as lining temperatures increase and results in a stable lining configuration.
- Concentric head rings have not been constructed using double taper brick shapes. This simple ring construction will not result in the head rings pushing against the steel head during heatup. This can result in the formation of voids and looseness within the head lining.
- The drawing shows the back of the brick cut to fit the head contour. This is not the normal field result. Often voids not visible from the hotface are left open or filled with other materials such as ceramic fiber, plastic or castable. Even if brick are cut as shown, they will not tighten against the vessel head which is necessary to prevent the formation of voids.

This type of construction is problematic.

Photo 2

Photo 3

Photo's 2 and 3 show poorly designed elliptical head linings that indeed caused problems. Photo 3 is a particularly common failure. Many designers, OEM's & installers realize the importance of brick for these units but use monolithics (plastic or castables) as "fill-in" materials in the difficult areas because 1) they are not capable of designing (and/or installing) brick in these difficult areas or 2) to keep costs down. Whatever the reason, the result may be a lower installed cost but at the expense of reliability.

Proper design of such heads requires the use of double tapered brick shapes in conjunction with the skew brick to force the head lining against the steel shell during lining heatup. In a properly designed lining, an individual brick will have its' axis perpendicular to the surface of the steel shell. The brick shapes that are utilized in a properly designed application are shown in illustration 3. It is also important that the brick be properly placed, with at least one of the coldface chords of individual brick being larger than the corresponding hotface dimension. This self supported construction allows for the elimination of metallic anchors. This type of construction is also self-supporting around burners, boiler inlets and other connections. The complexity and number of shapes involved in this approach is apparent in illustration 4. Nine different shapes were required in this dished head alone.

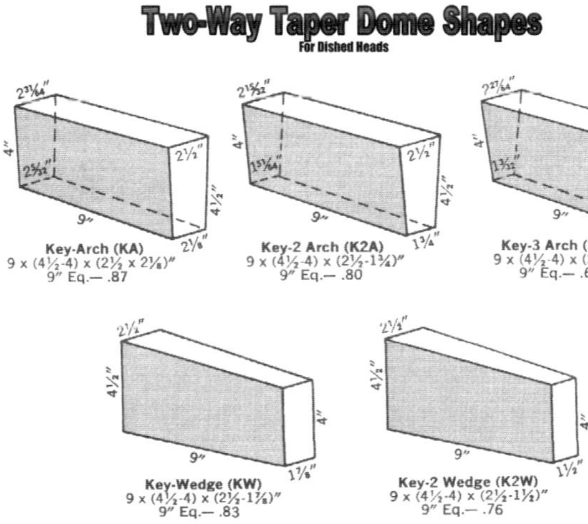

Illustration 3

Dished Head Bill of Materials

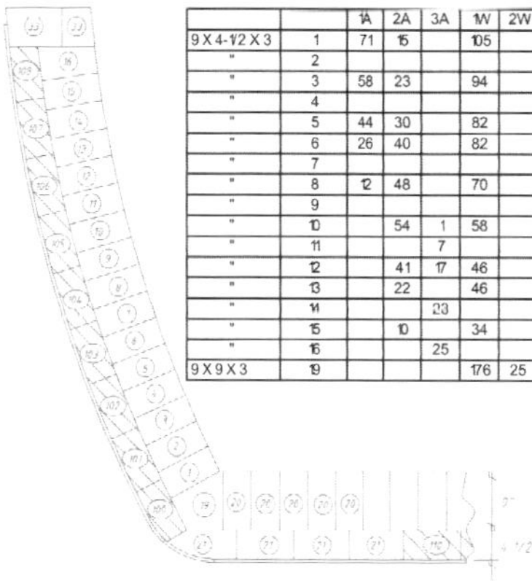

		1A	2A	3A	1W	2W	1K1A	1K2A	1K3A	1K1W	COMMENTS
9 X 4-1/2 X 3	1	71	15		105						DISCH. DISH
"	2						65	19		99	DISCH. DISH
"	3	58	23		94						DISCH. DISH
"	4						51	27		88	DISCH. DISH
"	5	44	30		82						DISCH. DISH
"	6	26	40		82						DISCH. DISH
"	7						19	44		76	DISCH. DISH
"	8	12	48		70						DISCH. DISH
"	9						6	51		64	DISCH. DISH
"	10		54	1	58						DISCH. DISH
"	11			7				47	4	52	DISCH. DISH
"	12		41	17	46						DISCH. DISH
"	13		22		46						DISCH. DISH
"	14			23				18	20	40	DISCH. DISH
"	15		10		34						DISCH. DISH
"	16			25				3	27	27	DISCH. DISH
9 X 9 X 3	19				176	25					SKEWBACK

Illustration 4

Choke rings and checker walls can also be problem areas. These areas are subjected to direct flame impingement and heating from all sides (no thermal gradient). The resulting thermal growth needs to be addressed in addition to the possibility of severe plastic deformation (melting). Realizing that these units can have upset conditions with a loss of combustion control, the choke ring/checker wall designs must be robust enough to absorb the resulting pressure wave to the extent feasible. This is accomplished by improving the structural design of the walls and by a keying or interlocking design. Photos 4 and 5 show a choke ring that was not engineered properly, resulting in 4-1/2" of movement at the top. We have seen choke rings move 8-9" and even some that have totally collapsed. Similar failures have been experienced in checker walls. Freestanding checker walls and choke rings need to be properly designed to allow for the vertical expansion of the brick while being mechanically keyed into place.

Choke Ring (back side)

Photo 4

Choke Ring (front side)

Photo 5

Other areas requiring special care include manways, peepholes and nozzle connections, which are characterized by the existence of a "bullseye" to support the surrounding brickwork. Depending upon the size and location of openings, "relieving arches" or "compression rings" may be required. Note in Photo 6 that the cylinder lining brick around the outside of the bullseye are designed to prevent "featheredge" cuts and the use of unacceptably small pieces of brick.

Photo 6

3. Experienced Installation Crews – The last component to assure a reliable lining is a knowledgeable, experienced and skilled installer. With the improvement in monolithic materials over the last couple of decades, compounded by the reduction of brick inventories, many contractors have not retained personnel skilled in proper refractory brick construction. Having lost these skills, these contractors and manufactures can only promote monolithics even though these have not proved to be as reliable as brick in reaction furnace service. The level of craftsmanship necessary for these complex brick linings is truly an art or craft. The best design cannot account for brick and steel variances. Only an experienced installer is able to understand what shape is required and properly cut and install the brick. The installer needs to be familiar with many different types of materials (brick, castables, plastics, steel anchorage, ceramic ferrules, ceramic fiber paper and blankets, mortars, etc.) and complex shapes and configurations.

Attention to detail is the difference between success and failure. Photos 7 and 8 show some of the complex construction shapes that are involved.

Photo 7

This photo shows the knuckle area of an elliptical head. This requires multiple cuts to change direction while keeping the long axis of the brick perpendicular to the shell.

Photo 8

This photo shows a congested area that includes a chase, two precast-prefired nozzle blocks, repositioned brick for a core drill and an expansion joint in the brickwork.

Realizing the importance of experienced and skilled installers, care should be taken to independently qualify and verify a potential installer. It is doubtful many installers will admit they cannot install a lining and miss a job opportunity, worse yet allowing a competitor into a plant. Most installers will also have a resume of SRU's that are quite long and impressive. Most of these are probably in older designed units that are not as complicated, were not run as hard and did not require the level of reliability demanded in most plants today. Detailed questioning and investigation will be required in this process.

One last area for discussion is the use of inspectors, even third party inspectors. An inspection that documents what, when and where materials were installed can keep a contractor honest and provides documentation in the event of a problem. Standards qualifying inspectors for monolithic installations do exist. These standards do not exist for brick construction. Brick inspectors should have hands-on experience in supervising and performing brick construction for horizontal cylindrical vessels such as thermal reactors, with all of the complexities introduced by dished heads, nozzles, cones, intersecting cylinders, etc. Many plants lacking refractory knowledge tend to promote the "inspector" to the position of engineer/designer. Many of their recommendations have been known to be incorrect, resulting in schedule delays and less reliable installations. Yet when a problem arises, inspectors have not been held accountable. Inspectors are typically not engineers and should not be expected to make design decisions unless so qualified.

Case History 1

Unit: Side-fired Claus Thermal Reactors (2)

Issue: Flat tied back walls

These two horizontal cylindrical units had flat end walls with brick laid in flat horizontal rows (see Photo 9). The flat walls failed due to the flat head design of the steel. Due to thermal gradients through a lining the hot face of the brick experiences higher temperatures than the back of the brick. This causes more expansion on the hotface of the lining resulting in a pulling effect toward the heat. In a flat wall design, there is a lack of the necessary brick geometry to make the expanding brickwork push against the steel. When the anchoring fails there is nothing to hold the lining to the shell and the brick pull "toward the heat", falling onto the furnace floor. To fix the problem a thermomechanical analysis was performed. The solution involved the selection of proper materials in conjunction with a new brick design. The new design entailed forming a shallow cone with the IFB back-up and then overlaying a 6" thick hotface layer of 90% alumina brick. The new wall was laid in concentric rings of various brick shapes (see photos 10 and 11). Due to the materials selected, the geometry of the shapes and the addition of the skewback, this lining is completely keyed in and will push toward the shell as it is heated facilitating the elimination of all metallic anchorage. Presently both walls are still in service and performing as designed. Barring major upsets, this customer should now safely make their 5-year runs.

Photo 9

Photo 10

Photo 11

Case History 2
Unit: TGU Heater (Reducing Gas Generator)
Issue: Burner Tile

Customer was experiencing 6-9 month runs on the burner tile. The 2 piece castable tiles supplied by the OEM were exposed to very high temperatures (2800°F - 3200°F), significant thermal shock and constant mechanical vibration. Photo 12 shows the original design and photo 13 shows the failure of the castable tiles.

Photo 12

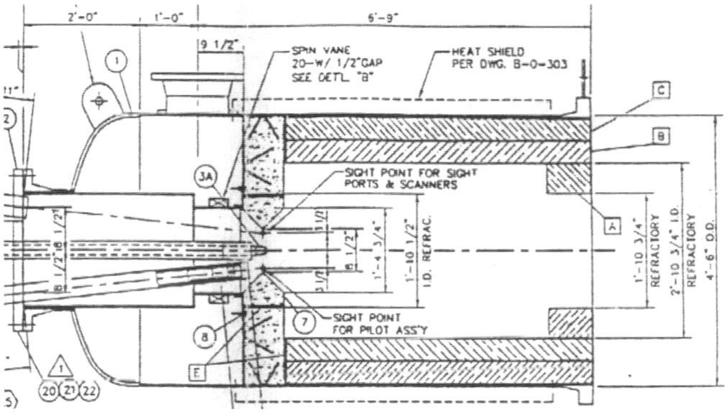

Photo 13

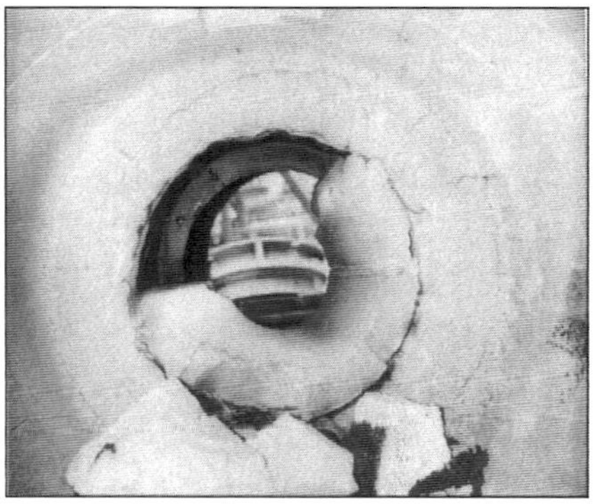

A special thermal shock resistant, high alumina brick composition was chosen to replace the castable in the new design. The new design included allowance for the brick expansion and the brick being keyed or locked in place. This lining lasted 38 months, a 400+% improvement. To further improve on this success, the design was modified to include special shaped high alumina, thermal shock resistant bricks formed in 4 pieces to improve the taper or keying action of each shape. Photo 14 shows the current improved design burner tile that has over three years of service.

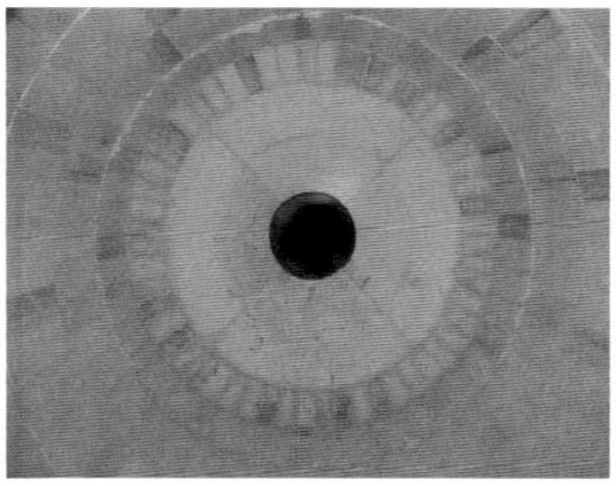

Photo 14

Case History 3
Unit: Thermal Reactor Elliptical Head
Issue: Elliptical head

A large thermal reactor head had an offset outlet to the boiler. The lining experienced numerous hot spots, failures, and unreliable service with only 1 to 2 years life. Hotspots were caused by loose brick due to a non-engineered lining installed by inexperienced installers. This was obviously a non-engineered lining installed by inexperienced installers. To solve the problem, a lining utilizing double tapered brick with skews was designed and installed, modified for the offset outlet to the boiler. This lining, installed in 1992, has performed with100% reliability, successfully meeting 5 year campaigns. Photo 15 shows the original problematic lining and Photo 16 shows the engineered solution.

Photo 15

Photo 16

CONCLUSION

Traditional linings designed and installed for ~2300°F service will not perform reliably at the higher temperatures and state-of-the-art conditions demanded by today's sulfur recovery units. To maximize lining life, operators should understand that the industry offers a wide range of refractory materials, engineering expertise and installation know-how. Many engineering companies and refractory installers have forgotten or lost the ability to design and install these complex linings. Fortunately, maximizing refractory reliability and life **can** be accomplished by following the three keys.

1. Proper material selection
2. Engineered lining design
3. Experienced installation crews

ACKNOWLEDGEMENTS

We would like to recognize the contributions and assistance of Philip Oberbroeckling, P.E., Operations Superintendent, Refinery Optimization and Logistics for Lyondell-CITGO Refining LP. in Houston, Texas.

Integrated Acid Gas Enrichment, Sulfur Recovery and Tail Gas Treating Technology for Processing Ultra Lean Acid Gases

By
Vincent W. Wong,
Senior Engineer, Process Engineering

Thomas K. Chow

Executive Director, Process Engineering

John Y. Mak

Technical Director, Process Engineering

Dick Nielsen

Executive Director, Process Engineering

Energy and Chemicals Group, Fluor Enterprises, Inc., Aliso Viejo, California

Abstract

This paper focuses on cost effective technology for the recovery of H_2S from lean acid gas streams by converting the H_2S to elemental sulfur. This new patent pending technology achieves and maintains a stable flame in the sulfur plant reaction furnace and reaches a reaction furnace flame temperature that is sufficient for complete destruction of contaminants such as BTX, ammonia and cyanides. This new process seamlessly integrates acid gas enrichment, Claus sulfur recovery and Claus tail gas treating processes to achieve an overall sulfur recovery for lean acid gas streams that exceeds 99.9 percent. The process can process acid gases with an H_2S concentration below 10%.

Several acid gas enrichment process configurations and various options for integrating sulfur recovery with tail gas treating are presented in this paper. Applications of these configurations will also be discussed.

Introduction

The modified Claus process is the best known and most widely used sulfur recovery process. However, it requires an acid gas stream relatively rich in H_2S and the maximum sulphur recovery is at best about 97 percent from a three stage design. There are many cases where acid gases are lean in H_2S; are contaminated with benzene, toluene and xylene (BTX); and contain ammonia and cyanide. Acid gases with these characteristics present unique operational issues and, in addition, need Claus tail gas treating technology downstream of the Claus unit to achieve sulfur recovery efficiencies greater than 97 percent.

For acid gas feeds with an H_2S concentration greater than 50 percent, a reaction furnace temperature in excess of 925°C (1700°F) can be achieved using the simple straight-through Claus process. However, if the acid gas H_2S concentration is significantly lower than 50 percent, the minimum required reaction furnace temperature of 925°C (1,700°F) may not be attainable without an upstream acid gas enrichment unit to increase the H_2S concentration of the Claus plant feed gas.

The required reaction furnace temperature is a function of the acid gas contaminants. The following figure shows the reaction furnace theoretical adiabatic flame temperature as a function of the H_2S content in the acid gas feed for a straight-through Claus plant configuration. The two horizontal dashed lines represent the temperature targets require to completely destroy BTX and straight chain hydrocarbons.

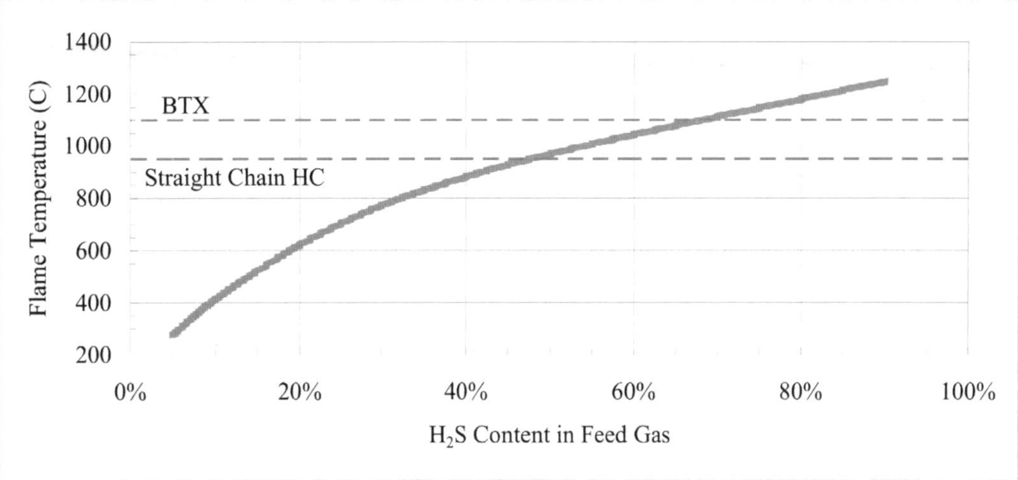

As the above figure demonstrates, the minimum flame temperature required to destroy BTX and heavy hydrocarbons cannot be maintained for the following conditions:

- Aromatics (BTX) cannot be destroyed if the acid gas H_2S concentration is less than 60 percent
- Paraffinic heavy hydrocarbons cannot be destroyed if the acid gas H_2S concentration is less than 50 percent.

Techniques such as supplemental fuel gas firing can raise the reaction furnace temperature. However, supplemental fuel gas firing introduces a number of negative factors. First, there is a decrease in overall sulfur recovery efficiency with supplemental fuel firing. This is mainly due to the formation of significant amounts of undesirable by-products such as COS and CS_2. In addition, the water and inert gases produced by combustion lead to less a favorable Claus equilibrium, which decreases the overall sulfur recovery efficiency. In addition, equipment sizes must be larger to handle the higher gas flow that results from co-firing with fuel gas. Finally, fuel gas supplemental firing usually requires more operator attention to ensure that coke formation does not occur.

Preheating the acid gas and/or the combustion air streams prior to being fed into the reaction furnace is another option for increasing the reaction furnace temperature. The maximum reaction furnace temperature increase that can be achieved by preheat depends on the type of heating medium used, but is limited to about 60 to 90°C (110 to 165°F). Preheating increases sulphur plant capital and operating costs and increases the total pressure drop across the unit.

Oxygen enrichment is another process option that can effectively increase the reaction furnace temperature. However, depending on the location of the facility and the quantity of oxygen needed, oxygen enrichment may be prohibitively expensive if an inexpensive source of oxygen is not readily available.

With the advent of selective amine treating systems, the H_2S concentration of a Claus plant acid gas feed stream can be increased by rejecting CO_2 and other contaminants. Selective amine treating systems can also be used to increase the feed gas H_2S concentration of existing Claus sulphur plants with the following advantages:

- No impact on the reaction furnace combustion air requirement
- No impact on sulfur plant capacity
- Stable operation in the reaction furnace
- Achieve the required temperature for destruction of contaminants such as BTX, ammonia and cyanides.
- Wide range of acid gas concentrations

New technology has recently been developed by Fluor that integrates an acid gas enrichment unit with the downstream tail gas treating unit. This technology utilizes a special "double absorption" design to enrich the acid gas. The process can be designed to process acid gas with a H_2S concentration below 10% and achieve an overall sulfur recovery that exceeds 99.9%.

Background of the New Technology

The process is designed to selectively absorb H_2S from lean acid gases that could contain less than 10% H_2S, such as sour natural gas, refinery gas, synthesis gas or other sour carbon dioxide streams to produce a high quality acid gas with an H_2S concentration up to 75%. This process can be used with various licensed selective amine treating technologies including formulated MDEA and proprietary selective treating solvents. The technology can also include a third stage absorber with an integrated Claus tail gas unit that uses a common regenerator.

Lean acid gas contains significant amounts of inerts such as CO_2. CO_2 lowers the net heating value of the acid gas and also reduces the concentration of SO_2 and H_2S in the reaction furnace, making sulfur conversion more difficult. Dilute acid gas feeds often contain contaminants such as ammonia and aromatics (benzene, toluene and xylene) that must be destroyed in the reaction furnace to protect the downstream catalyst beds from fouling. The low heating value of dilute acid gas streams makes the complete destruction of these objectionable components difficult. The minimum temperature for effective operation of the reaction furnace on "clean" acid gas should be above 925°C (1700°F). In extreme cases, when the H_2S content in the acid gas falls below 10%, the minimum reaction furnace temperature may become impossible to attain and additional processing steps must be employed to effectively convert the H_2S to elemental sulfur. In addition, the dilution effect of CO_2 in lean acid gases will increase the size of the sulfur recovery unit as the plant size is controlled by the total volumetric flow of acid gas. This, in turn, will significantly increase the cost of the sulfur recovery unit.

When the acid gas is too dilute in H_2S, a selective absorption technique may be used to enrich the acid gas prior to entering the Claus unit. By selectively absorbing H_2S from the acid gas and then stripping the rich solvent, two gas streams are produced. The gas passing through the absorber is primarily CO_2. This stream is sent to an incinerator for conversion of trace amounts of H_2S to SO_2 prior to discharge to the atmosphere. The gas stream leaving the regenerator is acid gas enriched in H_2S. This stream can be processed in a conventional Claus unit.

Another selective treating application is in the processing of tail gas from a Claus sulfur recovery unit. The sulfur recovery efficiency of a 3-stage conventional modified Claus unit is thermodynamically limited to about 97%. A common practice that is employed to comply with stringent sulphur emission environmental regulations is to convert the sulfur compounds in the Claus unit tail gas such as COS and SO_2 to H_2S using hydrogenation and hydrolysis. The H_2S is then removed from the converted tail gas and recycled back to Claus sulfur plant. This configuration can achieve sulfur recoveries of up to 99.9%.

Amine based solvents capable of selectively removing H_2S are used in these processes. These processes are based on MDEA or on sterically hindered amines that react rapidly with H_2S and slowly with CO_2. Typically, these processes concentrate H_2S by a factor of three to five.

Double Absorption Acid Gas Enrichment Process and Configurations

The following discussion describes the Double Absorption Acid Gas Enrichment and the Integrated Double Absorption sulfur recovery processes. These patent pending processes and configurations were developed by Fluor to increase the H_2S content of lean acid gas streams.

Double Absorption Acid Gas Enrichment Process

The Double Absorption process can be used as a standalone process to improve the quality of the acid gas from conventional acid gas removal units. This process recycles a portion of the acid gas to a second absorber to concentrate the H_2S.

Figure 1 shows the basic configuration of the Double Absorption process.

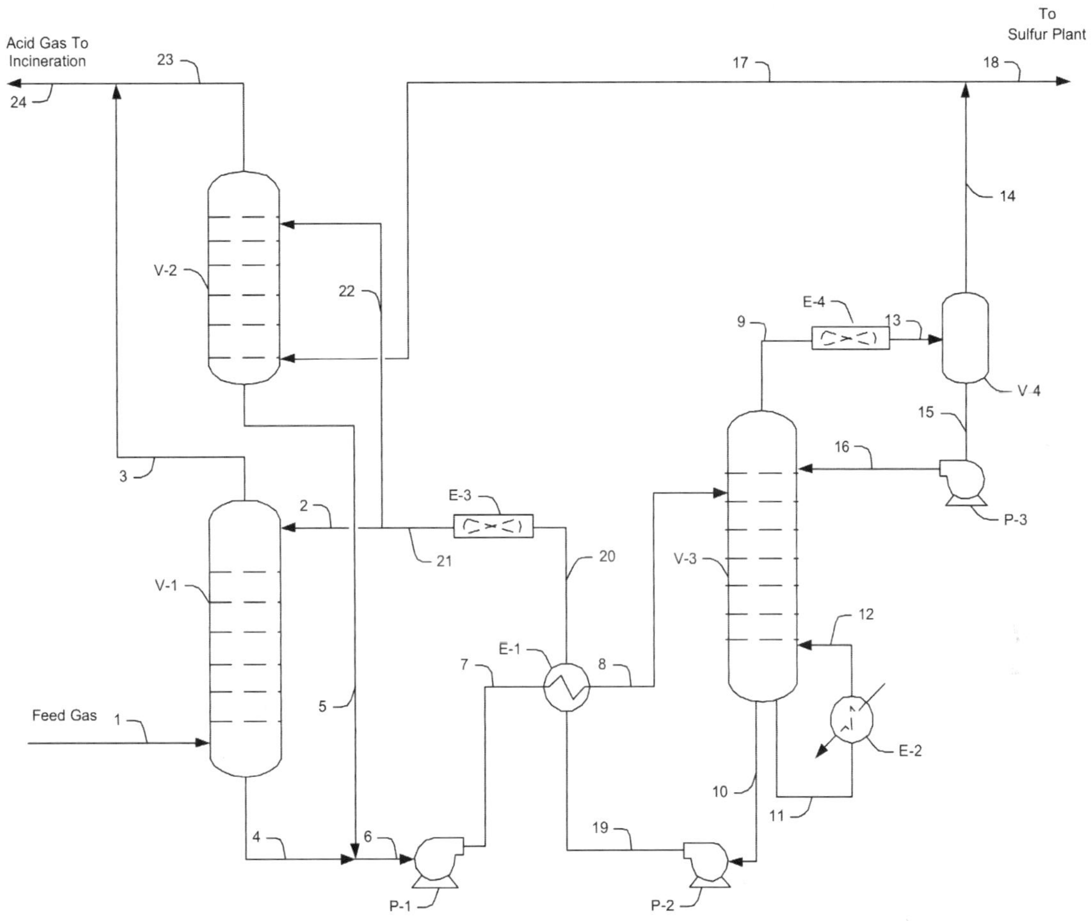

Figure 1 – Double Absorption Process

The acid gas feed stream enters the unit and is scrubbed in the first amine absorber, V-1, with a lean amine stream 2. The solvent typically consists of 40% to 50% MDEA, although other solvents such as a sterically hindered amine can be used. The amine absorber generally consists of 12 to 18 trays. About 85 to 90% of the feed gas CO_2 is rejected in stream 3.

The rich solvent stream 4 exits the bottom of the first absorber and combines with the rich solvent from the second absorber, V-2, forming stream 6. The combined stream is pumped and heated with the lean/rich exchanger, E-1, using the heat content of the lean solvent from the regenerator, V-3. The regenerator operates at a slightly higher pressure than the absorber. This allows recycling of a portion of the acid gas without using a compressor. The heated stream enters the top of the regenerator, which consists of 20 – 22 stripping trays and a wash section. Alternatively, other contacting devices such as packing can be used. The acid gas in the rich solvent is stripped with heat applied at the bottom reboiler, E-2, producing an overhead stream 9 and a lean solvent, stream 10. The lean solvent is pumped and cooled in the lean / rich exchanger, E-1. The lean solvent is further cooled in E-3. Air or cooling water can be used as the cooling medium. The lean amine should be cooled as much as possible as cooling favors the selective absorption of H_2S, thereby increasing the H_2S selectivity. The cooled lean amine is split into two portions, stream 2 and stream 22, which are fed into the first absorber, V-1, and the second absorber, V-2, respectively.

The overhead vapor from the regenerator, stream 9, is cooled in the overhead condenser, E-4. Liquid in the stream is separated in the reflux drum, V-4. The liquid stream, which is mostly water, is pumped and used to reflux the regenerator. The enriched acid gas is split into two portions, stream 17 and stream 18. Stream 17 is routed to the second absorber, V-2, for further enrichment and stream 18 is sent to the sulfur recovery unit.

The flow ratio of stream 17 to stream 14 ranges from 25% to 75%, depending on the H_2S concentration in the feed gas. For a low H_2S content feed gas, a higher flow ratio in the range of 75% may be necessary. The ratio can be reduced to less than 25% when the feed gas contains a higher H_2S concentration. For most applications, acid gas enrichment to about 75% H_2S can be achieved. In addition, over 90% of the hydrocarbons and BTX components can be rejected with the CO_2 stream. The H_2S enrichment and the absence of BTX and heavy hydrocarbons in the enriched acid gas are highly desirable for good performance of the Claus sulfur recovery unit.

Furthermore, depending on the feed gas composition and the acid gas loading of the semi-lean solvent, the overall circulation rate can be reduced by splitting the semi-loaded rich solvent stream 7 from the first absorber into two separate streams. One stream can be cooled and re-used for absorption in the second absorber V-2. The other stream, consisting of semi-lean rich solvent from the first absorber, which is still unloaded in terms of its H_2S content, can be fed to the lower section of the second absorber for bulk H_2S removal. The remaining semi-lean solvent can then be sent to the regenerator, V-3, for solvent regeneration. Figure 2 shows the configuration for this option.

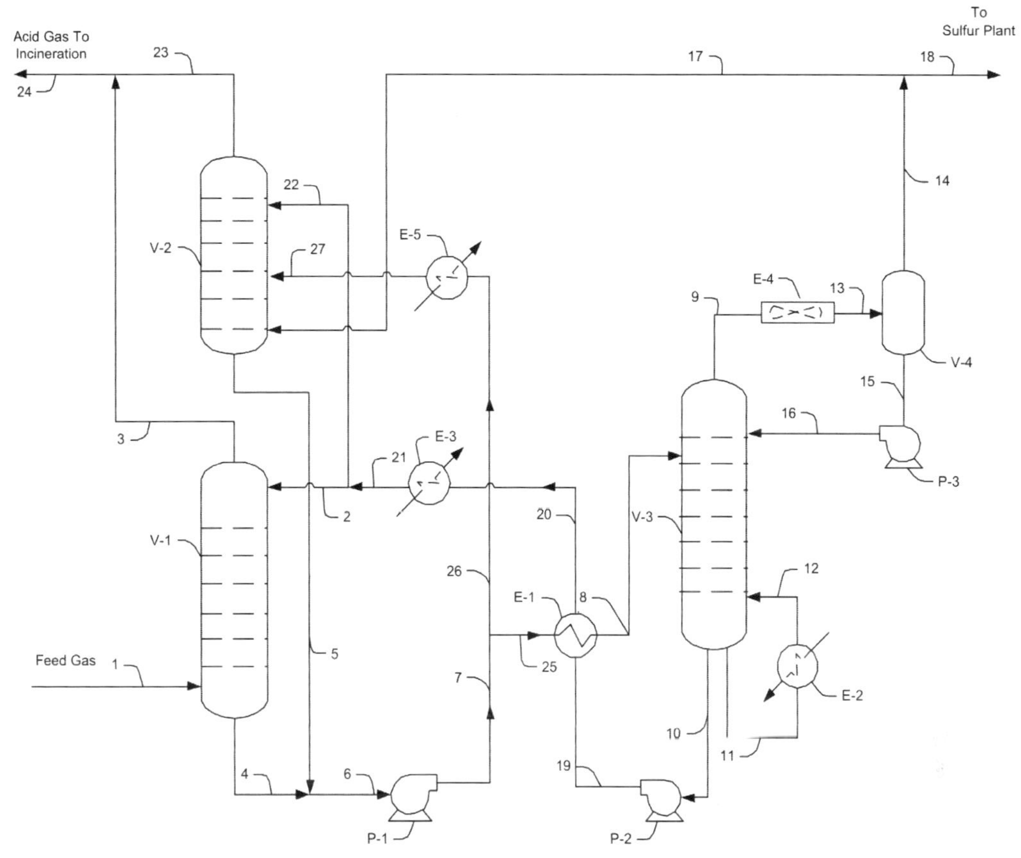

Figure 2 – Double Absorption Process Configuration Option

Integrated Double Absorption Acid Gas Enrichment / Sulfur Recovery Process

The Double Absorption acid gas enrichment configuration can be integrated with the tail gas unit to reduce the overall project cost. The semi-lean solvent from the tail gas unit, which is unloaded at the upstream absorber conditions, can be reused to reduce the overall solvent circulation while eliminating the requirement for a dedicated regenerator. With this option, a single regenerator can be used to regenerate the rich solvent streams from both the acid gas enrichment and the tail gas units.

Figure 3 shows a configuration where acid gas enrichment is integrated with the tail gas treating unit. The combination of the enrichment unit with a tail gas absorber processing the tail gas from a Claus unit can achieve over 99.9% overall sulfur recovery even when the feed gas H_2S concentration is low.

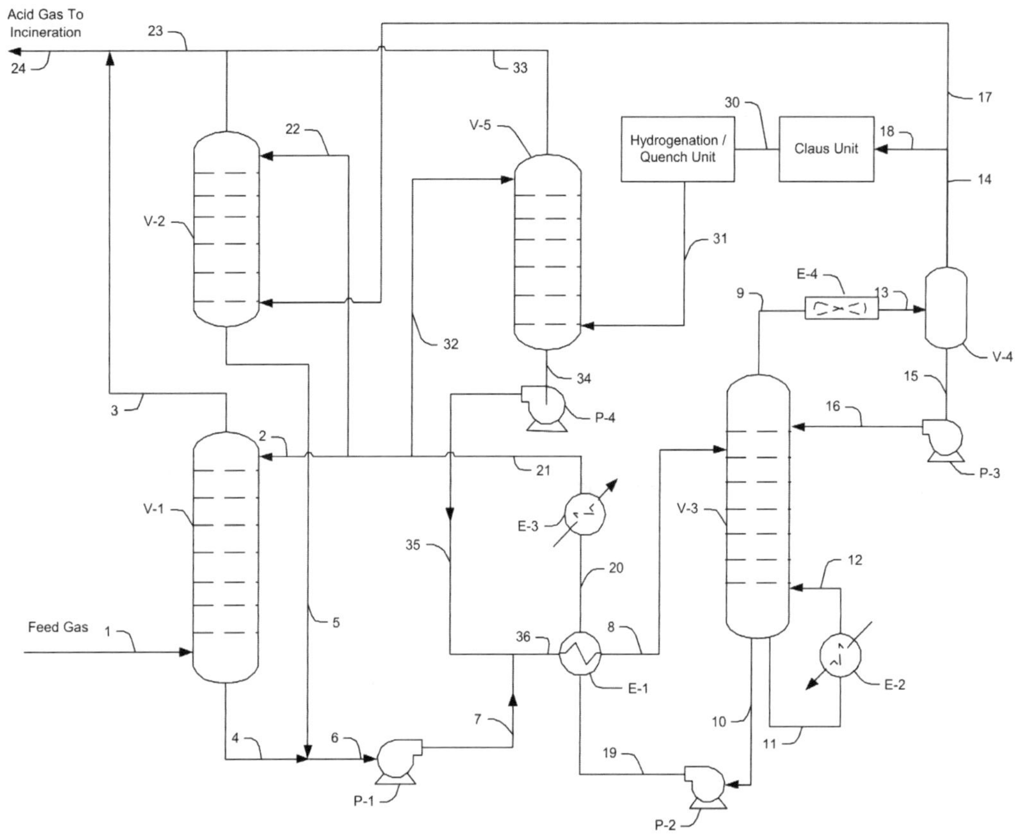

Figure 3 – Integrated Double Absorption / Sulfur Recovery Process

In this configuration, a two- or three-stage Claus sulfur recovery unit is used to process the enriched acid gas. A conventional modified Claus sulfur recovery unit would require more than 3 stages and other various additional processing steps in order to achieve at best 99% sulfur recovery. This integrated tail gas treating configuration significantly reduces the sulfur plant energy requirement and improves sulfur recovery efficiency while requiring less capital investment than a conventional design.

The effluent from the Claus unit, which contains trace quantities of H_2S, SO_2 and other sulfur compounds, is processed in the hydrogenation unit. The hydrogenated gas is quenched and extra water is condensed and removed prior to routing the gas to the tail gas absorber, V-5. Lean amine is supplied from the lean amine header. The effluent from the tail gas absorber, which contains environmentally acceptable levels of H_2S, can, depending on environmental regulations, either be vented directly to atmosphere or routed to an incinerator for disposal.

The rich amine from the tail gas absorber is pumped and combined with the rich amine streams from the first and second absorbers. The combined stream is heated in the lean / rich exchanger and fed to the common regenerator.

Depending on the actual feed gas conditions and sulfur recovery requirements, the semi-loaded solvent, stream 35, from the tail gas absorber, V-4, can be re-used in the second absorber, V-2. This configuration reduces solvent circulation and the solvent regeneration duty. With this configuration, the incremental amount of solvent used in the tail gas absorber can be reduced, improving process economics while maintaining high sulfur recovery efficiency. Refer to Figure 4 below for details.

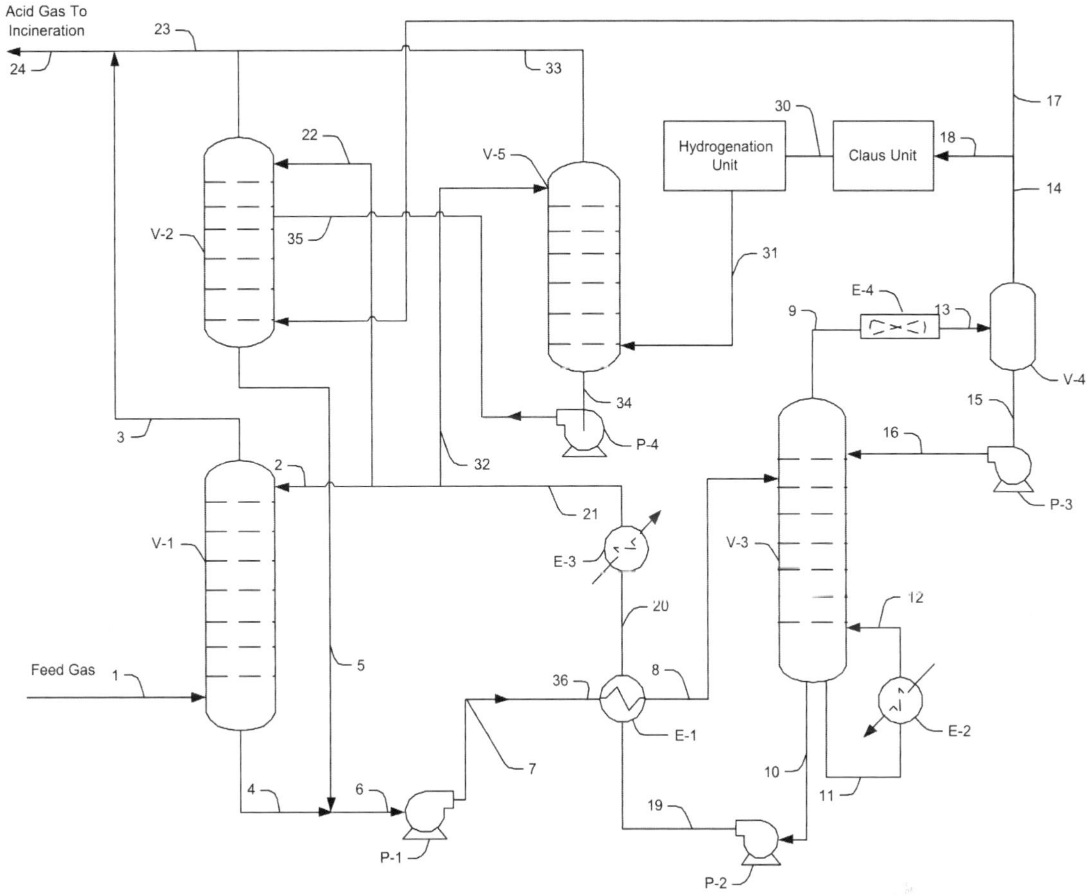

Figure 4 – Integrated Double Absorption / Sulfur Recovery Process Option 1

Depending on the acid gas composition and the semi-rich solvent loading, another option, shown in Figure 5, can be used to further reduce the total solvent circulation rate and the solvent regeneration duty.

This configuration re-routes a portion of the semi-loaded rich solvent stream 7 from the first absorber, V-1, to the second absorber, V-2. The stream is cooled prior to entering the lower section of the second absorber, providing a cost effective means for processing lean acid gas feeds.

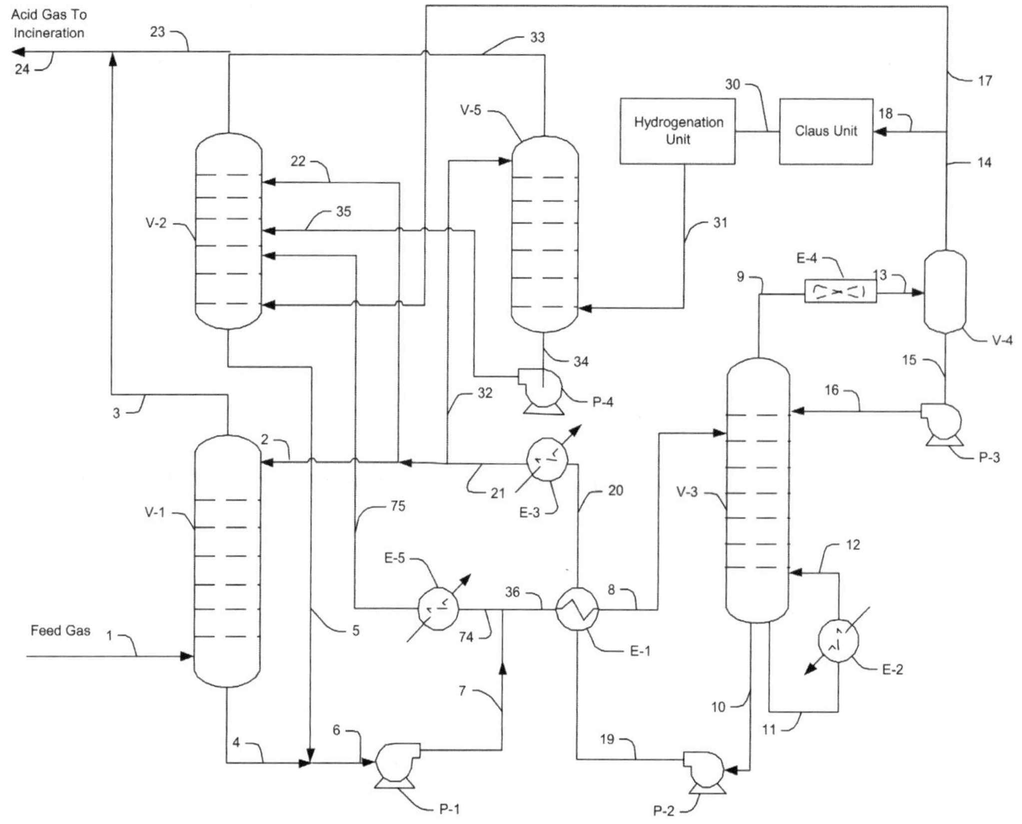

Figure 5 – Integrated Double Absorption / Sulfur Recovery Process Option 2

Conclusions

The Double Absorption process and the various configuration options outlined in this paper effectively produce acid gas enriched in H_2S from a lean acid gas feed stream. Acid gas can be enriched from less than 7% to over 75% H_2S. The various configurations also allow removal of hydrocarbons and BTX that are known to interfere with sulfur recovery unit operation. Furthermore, a carbon dioxide stream with environmentally acceptable levels of H_2S can be produced from the absorbers for disposal by incineration.

When integrated with Claus and tail gas treating units, the process is capable of reducing the number of Claus reaction stages and can achieve over 99.9% overall sulfur recovery. The Double Absorption process also solves the problems of low H_2S content and low acid heating value by providing a Claus plant feed with a high H_2S content. The process also eliminates the difficulties associated with fuel gas firing. The different Double Absorption process configurations are suitable for a wide range of applications such as gas plants, gas-to-liquids plants, gasification facilities and existing facilities that must meet more stringent emissions requirements.

References:

1. Mak, John; Nielsen, Richard; Chow, Thomas K.; Morgan, Oliver; Wong, Vincent W; "Methods and Configurations for Acid Gas Enrichment," U.S. provisional patent applications with serial numbers 60/538,105; 60/538,122; 60/538,608; and 60/565,053
2. Chow, Thomas K.; Lawrence, Clifford H.; Gebur, John A.; Wong, Vincent W.; "Cost Effective Sulfur Recovery Solutions for Canada's Greener Environment," Canadian International Petroleum Conference 55th Annual Technical Meeting, Calgary, Canada, 2004
3. Chow, Thomas K.; Gebur, John A.; Wong, Vincent W.; "Trends, Issues and Technology Advances in Sulfur Recovery for A Greener Environment," World Petroleum Congress Second Regional Meeting, Doha, Qatar, 2003
4. Klint, Bruce, "Hydrocarbon Destruction in the Claus SRU Reaction Furnace," 50th Annual Laurence Reid Gas Conditioning Conference, Oklahoma, 2000
5. Clarke, David; Iyengar, Jags; Al-Khaldy, Mohamed; Summers, Steve, "Qatargas Sulfur Recovery Improvement Project," 51st Annual Laurence Reid Gas Conditioning Conference, Oklahoma, 2001

68a Hydrogen Management for Clean Fuels and Profit

Alan Zagoria

As refiners modify their operations to produce clean fuels, hydrogen availability becomes a key issue. Demands generally far exceed the available today. Many refiners take for granted that investment in new on-purpose hydrogen production is required. The new hydrogen plant represents both a significant capital expense and continuing operating expense.

Effective hydrogen management is a systematic refinery-wide approach to getting the most out of each mole of hydrogen. New tools, such as hydrogen pinch and sophisticated hydrogen network models, enable significant improvement in hydrogen efficiency; minimizing the amount of on-purpose hydrogen required. Effective hydrogen management enables a refiner to capture and sustain this hydrogen efficiency improvement; however it can make an even greater contribution to the refiner's profitability. Increase in refinery gross margin above and beyond the reduced hydrogen production will be realized when a detailed understanding of the effect of hydrogen on the hydroprocessing units' performance is applied. When hydrogen usage is optimized, refinery profit is maximized.

Paper 68b

Managing Hydrogen Value Enhancement

by

Sanjiv Ratan

Technip USA Corporation
Calremont, CA

Prepared for AIChE Spring Meeting, April 2005, Atlanta GA

International Topical conference on Refinery Processing

Session 68 : Hydrogen Management

ABSTRACT

Hydrogen, as a critical utility for hydro-processing in refineries, carries substantial cost of its own. Moreover refiners also need to invest in order to comply with the clean fuels regulations and thus are on the look out for ways to stay competitive, while withstanding the recent high crude price environment. In such an endeavor, there is a strong focus placed on lowering the manufacturing cost of hydrogen, whether under the 'make' or 'buy' case, to enhance site-wide profitability.

Moreover, some of the refinery upgrade ventures, especially in the developing countries, can get the necessary commercial boost by extending the capabilities of the hydrogen facility which can render additional value and thus improve the economics of the overall project.

In responding to such needs, few proven options have been employed to improve capital efficiency of a hydrogen facility, both new and existing. The underlying aim is to create an 'added value' to reduce the net operating cost against cost-effective investment These include integration of potential Refinery-offgases for hydrogen generation, prospects for captive steam-power synergy and. finding possible credit for CO_2 as a by-product; in order to lower the net specific costs. Furthermore, most of these options carry inherent extra merits of reduced environmental emissions and increased reliability through custom modifications and exploitation of state-of-the-art technological advancements.

Some of such proven value-enhancement options are presented demonstrating the reduction in net unit cost of hydrogen, while also providing related benefits of improved reliability and better environmental impact.

1. CAPTURING POTENTIAL HYDROGEN FROM REFINERY OFF-GASES (ROG)

There has been an increasing trend of 'integrated' utilization of refinery off-gases (ROG) for on-purpose hydrogen generation. The bottom-line driver being that any hydrogen burnt as fuel is a loss of its 'asset' value above its heating value. Accordingly, by integrating ROG in a hydrogen generation plant, not only that most of its hydrogen content gets recovered, but also the price attached to the ROG is often lower than the premium hydrocarbon feedstocks for hydrogen production.

Typical reduction in net hydrogen cost can be between 2-10% depending upon the relative pricing of ROG versus the base feed, available quantity of ROG stream(s), hydrogen (or hydrocarbon) fraction, available pressure and level of impurities.

Thus it offers enough incentives to identify potential ROG streams in the refinery, which can be integrated cost-effectively in the hydrogen plant to enhance its economics. Utilization of ROG through integration with the hydrogen plant can be done in basically 3 routes:

1. Hydrogen recovery by mixing with the process gas upstream (common) PSA
2. Direct use as (part) feedstock for H_2 generation
3. Dedicated recovery PSA with integration of its purge gas as feed for H_2 generation

These routes are represented in Figure-1.

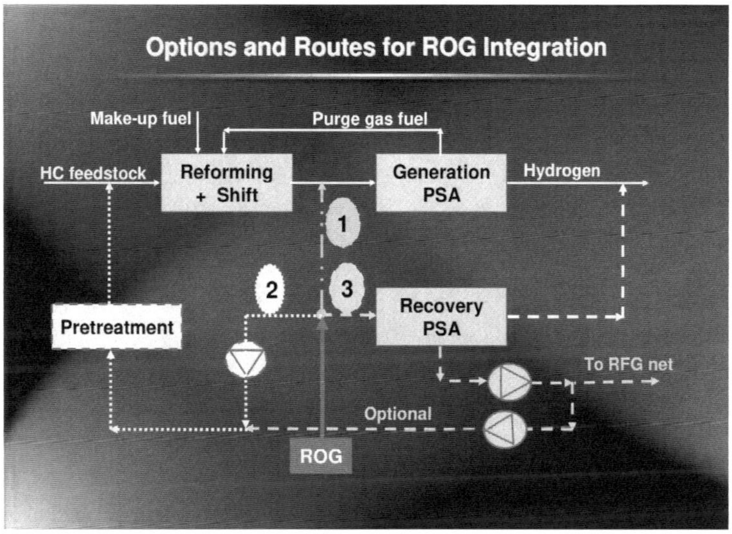

Figure-1

Several proven routes and potential options are in place, which can be qualified and selected based on their specific attributes. These are summarized in Table-1

Table-1 : Options for ROG Integration Guidelines

Application	Attributes #	H2 contribution
Direct hydrogen end-user	2, 3, 4	high
Common PSA recovery integration	3, 5	low
Direct feed for generation / reforming	3, 4, 6	Low-high
Dedicated recovery PSA	1, 3	Med-HIgh
Recovery PSA and integration of its purge gas as reformer	1, 3, 4, 5	Med
Membrane or PSA-membrane Hybrid	1, 3	Med-high
Cryogenic or Hybrid with PSA /Membrane	1, 7	Low
Retain for RFG net	1, 3, 4	low

Attributes #

1. Hydrogen fraction in ROG
2. Cost-effective end-use hydrogen purity
3. Available pressure (and required hydrogen pressure)
4. Level of (possible) contaminants
5. Ratio of SMR gas to ROG
6. Relative price of Hydrocarbon feedstock versus ROG
7. C_2+ credit and streams' composition

.A case history is described below for utilizing ROG (with average 40 vol % H_2) as primary feed for hydrogen generation based on medium level contribution. The available quantity (averaged over time) was too large for common PSA and not enough to justify dedicated recovery PSA. For full hydrogen generation, it needed to be supplemented by Natural gas under advanced front-end control for S/C ratio and capacity.

The comparative economics is presented in Table-2. Though the pricing of ROG was similar to NG in terms of LHV basis, the operating cost benefit due to potential contained hydrogen contribution was substantial (~190 $/h or ~7%).

The additional investment for ROG compression and pre-treatment (Chloride guard etc.) was easily paid-off within less than 3 years, without downsizing the reformer and downstream section, which were still sized for NG feed. The NG feed was supplied at the required B/L pressure in this case. However, if NG feed compression was needed; the payout for ROG integration would have become even more attractive. Other inherent benefits for operating with ROG feed mix included relaxation on recycle Hydrogen, easier start-up and longer reformer tube life.

Table-2 Case history : ROG integration as SMR feed
45 mmscfd H2

		ROG as Partial feed	NG feed only
Feed ROG	kg/h	6708	--
Feed NG	Kg/h	3763	14427
Fuel NG	kg/h	1011	1667
Export steam	t/h	29.5	36.8
Power	kWh/h	1470	690
Operating cost [1)]	US$/h	2455	2643
Additional TIC investment [2)]	MM $	4.1	Base
Pay-out	**Years**		**2.5** [3)]

1) Pricing data

ROG : 260 $/t (avg. 40 vol % H_2) Steam 8 $/t

NG 180 $/t Power 60 $/MW

2) For compression and pre-treatment

3) Shorter if NG feed needs compression

Various ROG integration schemes have been implemented in several of the hydrogen generation plants in America, Europe and Asia built or being built by Technip.

STEAM-POWER SYNERGY WITH HYDROGEN GENERATION

A hydrogen plant is typically a net exporter of steam based on the high temperature processing and related threshold heat recovery.. The amount of export steam can be adapted to a reasonably large extent by different design options to match the site-specific steam philosophy or the potential steam-power integration, especially in case of low-credit or excess refinery fuel gas. Cogeneration of power from this steam can represent a higher value end-use and such synergy becomes commercially more attractive on larger scale and thus for 'on-purpose' hydrogen plants say above 50 mmscfd capacity.

Synergy between the hydro-processing and potential power generation in refineries, especially for large hydrogen plants, can prove quite attractive under certain fuel-power scenario and for locations having insufficient or unreliable grid power supply. Such synergy not only offers site-wide power self-sufficiency and availability, but also offers to lower the specific cost of hydrogen under specific scenarios. Also it can provide few additional advantages like integrated project execution, better energy efficiency and emissions compliance.

The technical viability of co-generating power in hydrogen plants has been long established in several commercially operating plants. The export steam integration can be incorporated at different levels depending upon case-specific factors, demand and overall economics. However, having such integration, the operating hydrogen capacity dictates the 'steam dispatch', which tends to limit the ability to independently vary steam or power output.

Thus it needs to be studied how the export steam demand from the hydrogen plant affects its end-user operation. A limited amount of load variation can be done on condensing turbines; however, the power credit should only be accounted for the (minimum) guaranteed output. In some cases, especially for direct turbine drives on independent machines, partial capacity auxiliary boilers can be employed to handle swings in demand.

Such applications, especially topping turbines, typically call for high steam pressures in the range of 1500 psig, in order to achieve an efficient let-down ratio. The exhaust steam can be exported as either medium or low pressure steam, which is needed as heating medium in refinery processing. In case there is no demand for such steam, the turbine can be extended to a condensing turbine with the related vacuum system, which provides increased amount of power.

The steam turbine up can be implemented as a topping / back-pressure turbine or as an extraction / condensing turbine, with case-specific match for being a direct drive on large machines or coupled to an alternator for power generation. Further integration can be done with Gas turbines in a combined-cycle cogeneration scheme, with extended possibility of using gas turbine exhaust as combustion air in the hydrogen plant reformer. However, the best-suited arrangement for the steam-power synergy is governed by the following considerations:

- requirements for captive power
- (un)reliability of grid power
- possibility and credit for export power
- allowable flexibility in base-loads
- long-term fuel pricing scenarios

The different options and schemes for steam-power integration carry a distinct balance between their (annual) capital cost and the heat to power efficiency (heat rate). For the range of topping to condensing turbines, typical capital charges lie in the range of 1000 to 500 $/kWh, while withstanding the economy of scale, whereas the heat rates are in the range of 6000-12000 Btu/kWh respectively. Table-3 presents the summary of the steam-power integration economics in a large hydrogen plant of 80 mmscfd based on natural steam generation (i.e. without any additional features to either reduce or increase steam generation).

Several steam-power integration schemes have been implemented in quite a few large hydrogen plants, more in the US. Some of these involve auxiliary firing in the reformer convection section for increased steam generation, while in some even import steam in integrated in order to achieve the optimum synergy.

Table-3 : Economics of Steam-Power Integration
80 mmscfd H_2 plant; Natural steam make

Integration Option →		**Base Case (No synergy)**	**Case 1**		**Case 2**
			Topping Turbine		**Condensing Turbine**
Steam to Turbine [1] / Export	t/h	-- / 130 [2]	210 / 130 [3) 4)]		210 / 0 [4]
Power (-) Import / Export	MW	(-) 2.6	12		37
Operating cost credit [5]	$/h	910	1510		1850
Savings	$/h	base	600		940
	10^6 $/y [6]	base	5.1		8.0
Budgetary capital charges [7]	10^6 $/y	base	2.0 [8]	1.6 [9]	3.7
Net annual revenue	10^6 $/y	base	3.1	3.5	4.3
Reduction in H_2 price [10]	%	base	**~ 4**		**~ 5**

Notes:

1) At 1500 psig and 950 F
2) At 600 psig and 750 F
3) Back pressure of 220 psig for MP steam net
4) 80 t/h steam extraction at 600 psig for process use within hydrogen plant
5) Steam = $8/t (600 psig); $7/t (220 psig); Power = $ 50/MW
6) 8500 hours per year
7) Based on 20% per year
8) Based on STG + Alternator for power generation / export
9) Based on Direct drive(s) within refinery, includes cost adder for individual turbines
10) Based on $110/mmscfd H2, with added benefit of captive power reliability

CARBON DIOXIDE AS BY-PRODUCT

The use of hydrocarbon feedstocks for hydrogen generation leaves most of the incoming carbon content as carbon dioxide (CO_2) from the process steps of steam reforming and water-gas shift step. For the typical hydrogen plant based on PSA purification, the formed CO_2 ends up in the PSA purge gas, which is used as primary fuel for the reformer burners and thus eventually gets discharged in the flue gas to the atmosphere (along with the CO_2 formed from the make-up fuel firing).

For improving the economics of hydrogen, this CO_2 can be recovered as a by-product, in case its strategic demand can be explored around the plant / refinery location. Either a merchant gas supplying company can be engaged or it can be sold directly to an end-user. For liquefied and/or food grade CO_2 product, though it involves the extra investment for purification and liquefaction as well as its storage and logistics, it can provide a hike the returns by securing a high-premium credit.

The economics of CO_2 recovery is very site or case specific, governed by the commercial value it can fetch based on the opportunity for finding its outlet to a niche (long-term) market and the related product quality and economy of scale. Various potential outlets can be identified for realizing credit for CO_2, for instance:

- as industrial gas commodity for specific use
- as food grade quality for beverages (liquefied)
- additional (feed) stream for methanol, oxo syngas or carbon monoxide production, as well as in situations of leaner NG feed for ammonia having 'matching' Urea production etc.

CO_2 recovery is best suited from the process gas below the heat pinch in view of the available pressure and low-level heat. Its recovery from the purge gas is more tedious and costly due to its low pressure, and recovery from the flue gas is even more cumbersome due to presence of oxygen and other contaminants like SO_2, dust etc., apart from its very low pressure.

Hence CO_2 is normally recovered from (part of) the process gas upstream the cooing train, using typically a solvent based absorption/regeneration system like MDEA or Benfield process. However, care needs to be taken to limit the likely solvent carry-over within allowable limits for the downstream PSA unit. Very seldom dedicated or dual-service PSA units could also be employed, but are less preferred due to lower product CO_2 purity and need for vacuum system. The impact of (partial) CO_2 recovery on the hydrogen plant is reflected in:

- Smaller PSA and purge gas system
- Reduced reformer firing and export steam, consequently bit smaller steam system as well as flue gas path / load.
- Utilization of low-level heat below the Pinch (with supplemental external heat input in some cases)Reduced CO_2 emission
- Higher NOx level in flue gas due to (part) loss of the curtailing effect of CO2 in the fuel.

Figure-2 : CO_2 removal facility in a Hydrogen Plant

Table-4 provides an overview of the bulk CO_2 recovery in a 50 kNm^3/h hydrogen plant with NG feed (carbon number of around 105) at conditions ex-regenerator, based on 90% recovery (~2% CO_2-slip)

Table-4 : CO_2 as By-product
50 mmscfd Hydrogen plant

Hydrogen capacity	mmscfd	50
Available CO_2 [1)]	t/h	23
CO_2 credit [2)]	$/h	575
Additional operating costs [3)]	$/h	55
Net CO_2 credit	$/h Million $/y	520 4.5
Net additional investment charges [4)]	Million$/y	2.0
Net revenue	Million$/y	2.5
Net reduction in Hydrogen cost [1) 5)]	%	**~ 5**
Reduction in CO_2 emission [1)]	%	**~ 50**

Notes :

1) Based on full CO_2 recovery on Natural gas feed
2) Based on average base market value of $ 25/t
3) Includes solvent costs and low-level heat recovery within hydrogen plant;
4) Based on 20% depreciation per year; with adapted PSA size
5) Based on cost of hydrogen of $100/mmscfd (variable + fixed), at $ 5/mmbtu
6) For 50% recovery, reduction between 2-4% depending upon case-specific economics

As evident from Table-4, potential CO_2 recovery in a hydrogen plant as a by-product can lower the net hydrogen cost by 2-10% depending upon the strategic opportunity and striking the right deal. Such applications become quite prospective through involvement of a commercial gas companies, and much more sustainable in case they are also engaged for "over-the-fence"hydrogen supply. For instance, it has been implemented in the Air Products hydrogen supply facilities in UK and Spain, designed under the Air-Products -Technip alliance. (see Figure-2)

Another major merit of CO_2 recovery is the environmental benefit of reduced CO_2 emission per unit hydrogen, which can act as a promising candidate for receiving state subsidies for CO_2 reduction. Such grants are been strategically encouraged in several (western) countries which can boost the economics of such a venture.

REVAMP OF AN EXISTING PLANT FOR INCREMENTAL HYDROGEN

Achieving additional hydrogen through revamping of existing plants can prove to be attractive by providing reduction in the unit cost of hydrogen through cost-effective retrofitting. The actual economics however depends upon the degree of uprate, available design margins, condition of existing equipment and the level of modifications required. Normally a capacity revamp not only provides additional hydrogen at lower cost, but also the benefits of shorter time schedule based on the interfacing-facilities/ off-sites being already in place, and can also entail feedstock change or flexibility; improved efficiency and environmental performance.

There are various proven options for cost-effective augmentation of hydrogen capacity up to 30% and each case needs proper identification of the bottlenecks and assessment for selecting the most appropriate option.

In cases where the target capacity increase is substantial (>15%) with major limitations on the reforming section, one of the effective solutions without overloading the reformer is Regenerative reforming. The underlying concept is using the high-level heat of the reformed gas to reform additional feed through convective heat exchange, also known as post reforming. Furthermore, this option also provides 10-15% reduction in emission levels of CO_2, NOx and SOx per unit hydrogen. Such a retrofit for providing 25% additional hydrogen has been successfully implemented in a hydrogen plant in Spain.

In the typical 'incremental' hydrogen capacity range of 2.5 to 25 mmscfd, the option of post-reforming retrofit typically involves an investment of \$ 700-1400/ Nm^3/h depending upon the scale of additional hydrogen, available design margins in the existing equipment and level of modifications required in the plant. When compared with a new hydrogen plant for the same additional capacity, together with the necessary offsite / utilities and auxiliaries, such investment could be even less than half to three fourth depending upon the existing plant specifics

Table-5 : Typical overview of Hydrogen plant Capacity Revamp Options

Option	Typical incremental hydrogen (%)[1)]	Level of investment [2)]
LT Shift addition	3 – 5 %	Low-end
Pre-reformer retrofit	5 – 15 %	Medium
Post (regenerative) reformer integration	15 – 30 %	High-end

CONCLUSIONS

Attributing to the extensive (refinery) hydrogen requirements and the related costs, various proven options of 'added-vale' have successfully demonstrated improved economics of hydrogen production, without any compromise on operational safety, reliability and flexibility.

Such applications have the potential of reducing the unit cost of hydrogen by 2-10%, but need fair amount of effort to explore and establish a strategic niche demand and synergy around the hydrogen facility. Also needed are extensive expertise and experience in design, engineering and execution on the part of the technology and contracting company.

Most of these options hold good promise for future not only for improving site-wide profitability but also operational availability as well as environmental performance, especially for locations suffering from lack of power reliability.

EFFECTS OF HYDROCARBON FEED TYPE ON OPERATING COSTS AND ENVIRONMENTAL IMPACT ON A STEAM REFORMING BASED HYDROGEN PLANT

BY

P V Broadhurst, W J Cotton & M Hilton

Johnson Matthey Catalysts
PO Box 1, Belasis Avenue,
Billingham, Cleveland TS23 1LB,
United Kingdom

Prepared for Presentation at AIChE 2005 Spring Meeting
April 10 - 14, 2005
8th International Topical Conference on Refinery Processing
Session 68: Hydrogen Management

Unpublished

ABSTRACT

Increasingly, operators of steam reforming hydrogen plants wish to switch between hydrocarbon feedstocks to minimize operating costs, increase operational flexibility and utilize available streams. Many new refinery hydrogen plants are designed to use interchangeably three or even four available hydrocarbon streams but plants designed for a single feed can be modified with varying difficulty to switch to or add another feed type. Primary drivers for changing hydrocarbon feed are cost and availability, factors which are influenced geographically by local refinery stream availability and market conditions. Changing the feedstock has other implications and, in particular, impacts on the hydrogen plant's environmental performance in terms of energy usage and carbon dioxide emissions per unit of hydrogen product.

Building on earlier work, Johnson Matthey Catalysts has modelled the energy usage and carbon dioxide emission levels across the range of hydrocarbon feedstocks in the common steam reforming hydrogen plant flowsheets. The results provide plant operators with a guide to aid decisions about changes to hydrocarbon feed type.

Johnson Matthey Catalysts has participated in revamp studies examining changing feedstock in existing units. Using the models constructed to evaluate energy usage and carbon dioxide emission levels, Johnson Matthey Catalysts has assessed the environmental impact of this change in two cases. In the first case study, a refiner in the USA, driven by high natural gas prices in the USA, has successfully implemented a feedstock change from natural gas to a blend with off gases. In the second case study, a refiner in India is examining switching its hydrogen plants from naphtha feed to the increasing available piped natural gas derived from imported LNG. In both cases, the environmental impact of the feed change has been assessed.

1 Introduction

A steady stream of new steam reforming based hydrogen plants are being commissioned to meet the needs of the oil refining industry for more hydrogen to allow refineries to process sourer crude oils, to upgrade the heavier end of the barrel and to comply with tightening fuel specifications for sulphur and aromatics.[1,2]

The hydrocarbon feed for these hydrogen plants has varied around the globe due to a local availability or demand. In North America, natural gas has historically been abundant at economic price levels and so has been the principal feed. In areas such as Japan, Korea and India, however, natural gas has been unavailable and so naphtha has been a common feed for the hydrogen plants. In addition, the refinery may have a number of hydrocarbon streams available although this depends on the complexity of the refinery unit operations and the integration of these streams back into the refinery processes. Various off gases may be used and LPG is often retained as an optional feed for use when demand from other end uses is low, for example when blending into gasoline is suspended during summer to reduce the gasoline Reed vapour pressure (RVP).

In recent years, many new build refinery hydrogen plants have been designed so that a range of hydrocarbon feeds can be processed. On occasion, there may be three or even four feedstock options sometimes spanning the range from off gas through to naphtha and with a requirement for full flexibility between these feeds.[3,4] In addition, operators of plants designed with less flexibility are finding reasons to evaluate a change in feedstock type and the consequent impact on the hydrogen plant design. Recent high natural gas prices in North America are leading operators to consider alternative cheaper feeds[5] whereas operators of Indian plants are evaluating use of natural gas as imported LNG becomes more readily available.[6]

Making changes to the hydrogen plant feedstock has a wide range of actual or possible consequences. In particular where the intention is to accommodate a feedstock outside the original design intent, it is possible to encounter a range of operating limits which will necessitate modification of the plant operating conditions, hardware, equipment, catalyst selection and so forth[7]. There is also an impact on environmental performance as heavy hydrocarbons make more CO_2 per mole of hydrogen product, which in most hydrogen plants is emitted to atmosphere. This was considered in an earlier paper by Johnson Matthey Catalysts, in which the CO_2 emissions when using natural gas or naphtha feed were compared for the two prevalent hydrogen plant flowsheets. The first flowsheet uses high temperature shift (HTS) and a pressure swing absorption (PSA) unit to deliver the product hydrogen downstream of the steam reformer and the second flowsheet uses high and low temperature shift (HTS and LTS), wet CO_2 removal and methanation.[8]

In this paper, the work on environmental impact is extended to a wider range of feedstock types and examines both CO_2 emissions and net energy usage. Once in possession of this data, an objective view can be formed on the relative project and operating costs required to operate with different feed types.

2 Basis for and results from the modelling work

2.1 Basis for modelling

In order to provide wider flexibility to the modelling of the various feedstocks and flowsheet scenarios, a detailed hydrogen plant model has been established. This has been achieved using HYSYS for Hydrogen that Johnson Matthey co-developed with AspenTech[9]. HYSYS is a process simulation package in which a plant model is constructed by linking the units in the flowsheet and thereby provides a detailed heat and mass balance. HYSYS for Hydrogen adds in the Johnson Matthey reactor models and catalyst parameters, which allows for rigorous modelling of the catalyst unit operations within the syngas flowsheet. This allows for greater fidelity when determining the optimal flowsheet and operating conditions of syngas flowsheets based on steam reforming. Included as part of the HYSYS for Hydrogen model is the world-leading REFORM model that is a highly rigorous and detailed simulation tool for modelling both the process and fluegas side of steam reformers.[10]

The cases consider a target hydrogen production of 100 MMSCFD (110 kNm^3/h). The base case flowsheet includes a top-fired reformer with downstream HTS and a PSA unit to purify the product hydrogen. A range of feedstock types have been considered which represent a hydrogen-rich off gas (ROG), natural gas, butane, ROG/Butane mixture, light naphtha and naphtha. The compositions used in the modelling work are described in Table 1.

Table 1. Composition of feedstock cases

Feedstock composition (mol%)	ROG	Natural gas	ROG/ butane	Butane	Light naphtha	Naphtha
H_2	60	-	30	-	-	-
CH_4	20	89.8	10	-	-	-
C_2H_6	10	6.5	5	-	-	-
C_3H_8	5.6	2.23	2.8	-	-	-
C_4H_{10}	4.4	1	52.2	100	-	-
C_5H_{12}	-	-	-	-	18	-
C_5H_{10} cyclopentane	-	-	-	-	5	-
C_6H_{14}	-	-	-	-	50	-
C_6H_{12} (1-hexene)	-	-	-	-	5	-
C_6H_6	-	-	-	-	2	-
C_7H_{16}	-	-	-	-	20	-
C_9H_{20}	-	-	-	-	-	18
C_9H_{18} (nonene)	-	-	-	-	-	5
C_9H_{12} (cumene)	-	-	-	-	-	2
$C_{10}H_{22}$	-	-	-	-	-	55
$C_{11}H_{24}$	-	-	-	-	-	20
N_2	-	0.5	-	-	-	-

In order to provide a valid basis for comparison, it is necessary to define a number of set points and assumptions. These are described in Table 2. The natural gas case is defined as the base case in which the reformer exit temperature was fixed at 1616°F (880°C), inlet pressure at 20 Bara and steam to carbon ratio at 3.0. Natural gas and steam rate were adjusted to provide the target hydrogen output of 100 MMSCFD. For the other feedstock cases (unless otherwise stated) the same temperature, pressure and steam to carbon ratio were fixed. Methane slip and feed rates were allowed to vary to achieve the target hydrogen production. Where appropriate, hydrogen recycle was added at the level usually recommended for the feedstock type.

Table 2. Set points and assumptions used in modelling

Plant rate	100 MMSCFD (110 kNm^3/hr)
Flowsheet layout	steam reformer, HTS, PSA
Reformer details	top fired, 368 tubes, 5.1" ID (130 mm), **KATALCO**^{TM} Quadralobe catalysts
Reformer fuel	PSA purge gas plus hydrocarbon used as process gas
Steam to carbon ratio	3.0
Excess air	10%
Reforming section inlet pressure	20 bara
Reformer exit temperature	1616°F (880°C)
CO slip ex HTS	Optimized based on 664°F (351°C) inlet
PSA efficiency	88%
Product H_2 delivery pressure	40 bara

2.2 Modelling results

The modelling has been targeted at the relative differences in overall CO_2 emissions and overall energy usage in delivery of 100 MMSCFD H_2 at 40 bara. Other factors such as net export steam have been assessed also. It should be noted that, at this stage, no attempt has been made to optimize the operating conditions and flowsheet to minimize these values. Results are collated in Table 3 with additional relevant data and are discussed in the subsequent sections.

Table 3. Modelling results: CO_2 emissions, energy usage and net export steam

Feedstock type	ROG[#]	Natural gas	ROG/ butane	Butane	Light naphtha	Naphtha
Units: tonne/h unless stated						
Feed rate	24.1	30.0	30.5	31.3	31.9	32.0
Fuel rate	10.3	5.7	7.5	8.5	8.8	9.1
Feed + fuel	34.4	35.7	38.0	39.8	40.7	41.1
Process steam	154	95	111	116	120	122
Export steam	< 1	10.0	35.6	39.3	38.8	38.6
CO_2 emission	94	94	114	120	125	127
CO_2 /H_2 (tonne/tonne)	9.6	9.6	11.6	12.2	12.7	12.9
Import electrical energy (kW)	2675	3197	2645	2675	2710	2705
Efficiency (GJ/MMSCFD)	428.6	406.4	430.8	438.5	440.1	441.5

[[#] Modelled with butane fuel and a steam to carbon ratio of 6.]

The pure ROG case proved difficult to solve in terms of the heat balance. When run at the standard conditions (S:C 3:1, 10% XS air) there was insufficient heat flow along the waste heat recovery duct causing many temperature crossovers and a requirement for significant steam import. This results from the light reforming duty with the hydrogen-rich ROG feed leading to a very low enthalpy in the flue gas.

To overcome these problems, it is possible to increase excess air (to approx 30%) or to increase the steam to carbon ratio. Both would require increased firing in the reformer that would provide more heat along the duct. One case was run in which the steam to carbon ratio was increased to the point where there was sufficient heat available in the duct to eliminate the need for import steam such that the steam balance became zero (no steam export). The steam to carbon ratio required was about 6:1 but even so a large proportion of pre-heat along the waste heat recovery duct was removed amongst other issues.

This illustrates that even moving to an apparently easy feed for process side reforming leads to major potential issues if the plant has not been designed to accommodate this feed.

The flowsheet itself may have an influence on the performance. To investigate this, the two naphtha cases were modified to include a pre-reformer upstream of the steam reformer. In this case, the pre-reformed gas was taken through a re-heat coil in the flue gas duct to further raise the gas temperature following the exothermic naphtha pre-reforming reaction. Also, the steam reformer design was not altered whereas for a new plant designed for a set hydrogen output would generally require a smaller steam reformer where the design incorporates a pre-reformer. However, for an existing plant considering feedstock flexibility, there would be no such effect when adding a pre-reformer. The results are indicative of the flowsheet differences, however, and are presented in Table 4 before discussion in the following sections.

Table 4. Incorporation of a pre-reformer (naphtha cases only)

Feedstock type	Light naphtha	Light naphtha	Naphtha	Naphtha
Units: tonne/h unless stated		With pre-ref		With pre-ref
Feed rate	31.9	32.9	32.0	32.9
Fuel rate	8.8	6.2	9.1	6.6
Feed + fuel	40.7	39.1	41.1	39.5
Process steam	120	114	122	117
Export steam	38.8	19.6	38.6	20.2
CO_2 emission	125	120	127	122
CO_2 /H_2 (tonne/tonne)	12.7	12.2	12.9	12.4
Import electrical energy (kW)	2710	2720	2705	2695
Efficiency (GJ/MMSCFD)	440.1	423.1	441.5	424.4

3 Total CO_2 burden

One aspect of the environmental performance of a steam reformer is the amount of carbon dioxide emitted to the environment. Regardless of the design of the hydrogen plant, all of the carbon taken into the process whether as feedstock or fuel ends up as carbon dioxide. In older plants with an HTS/LTS/wet CO_2 removal/methanation flowsheet, CO_2 is ejected from the CO_2 removal system and methane in the product hydrogen usually ends up in steam reformer fuel gas purged from a downstream process. In flowsheets with a shift/PSA unit, PSA purge gas is fed to the steam reformer as fuel so that methane and CO are burned to CO_2.

The amount of CO_2 produced per unit of product H_2 is influenced by a number of factors. These include the feedstock and fuel composition, the process operating conditions and the flowsheet. In our earlier paper [8], we compared natural gas and naphtha feeds and were able to conclude that higher CO_2 levels resulted when

- switching to a heavier feedstock and fuel at a set hydrogen production rate and steam to carbon ratio;
- methane slip is driven down by increasing steam to carbon ratio or reformer exit temperature for a given feedstock and fuel;
- the flowsheet features HTS/LTS/wet CO_2 removal/methanation rather than HTS/PSA unit.

Product H_2 is stressed as some hydrogen produced by the exit of the final shift bed is consumed either in methanation of residual carbon oxides or is lost to the PSA purge gas as the product hydrogen recovery in the PSA unit is only 85-90% efficient. We also showed [8] that the majority of the CO_2 arises from the process side, more so in the HTS/PSA flowsheet where the process side methane, CO and CO_2 are returned to the fuel system in the PSA purge gas.

3.1 Theoretical considerations

Considering the reactions by which hydrocarbon is consumed in both steam reforming and combustion, the higher carbon dioxide potential of the heavier feed can be seen.

The general steam reforming and shift reactions are expressed as follows.

$CH_{2\alpha} + H_2O \rightarrow CO + (\alpha + 1)\ H_2$

$CO + H_2O \rightarrow CO_2 + H_2$

When combined the overall equation is thus.

$CH_{2\alpha} + 2H_2O \rightarrow CO_2 + (\alpha + 2)\ H_2$

Directionally, the amount of oxygen consumed in combustion follows the trend in the amount of heat released and so a rough measure of CO_2 burden for a given hydrocarbon fuel is the ratio of CO_2 produced to O_2 consumed. The combustion of the reformer fuel can be similarly expressed.

$CH_{2\alpha} + (0.5\ \alpha + 1)\ O_2 \rightarrow CO_2 + \alpha\ H_2O$

Considering the hydrocarbons used in the modelling, the increase in CO_2 burden as the feedstock and fuel becomes heavier can be illustrated from these equations (Table 5). To simplify the calculation, hexane and nonane have been used to represent the light naphtha and naphtha situations.

Table 5. Theoretical CO_2 per H_2 produced and per O_2 consumed

Feed type	ROG	Methane	ROG/ butane	Butane	Light Naphtha (C6)	Heavy Naphtha (C9)
α	2.34	2.00	1.42	1.25	1.17	1.11
CO_2: H_2	0.23	0.25	0.29	0.31	0.32	0.32
CO_2: O_2	0.46	0.50	0.58	0.62	0.63	0.64

3.2 Modelling results

As previously noted, the flowsheet used for the modelling has a top-fired reformer, HTS and a PSA unit with the PSA purge gas returned to the reformer as fuel to which make up fuel is added. Thus, all of the carbon from the process side passes out in the flue gas as CO_2. Referring to Table 3, as the feedstock and make up fuel become heavier, the total CO_2 emissions from the process increases such that the naphtha case emits 35% more CO_2 than the natural gas case. The CO_2 emissions are shown in Graph 1.

Graph 1. CO_2 emissions with varying feedstock

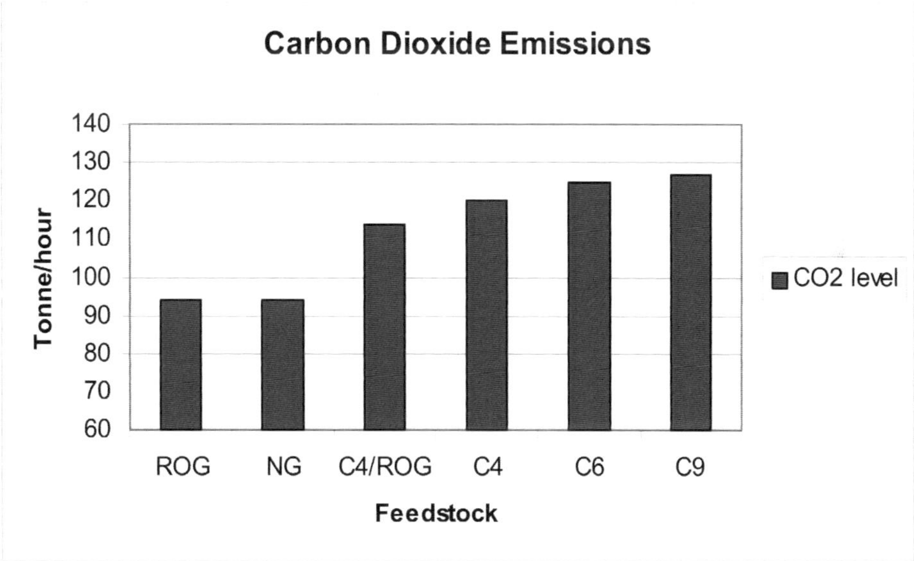

At equivalent conditions, the ROG case would be expected to give lower levels of but, as noted above, the model did not allow data to be generated without raising the steam to carbon ratio and thus the data for ROG is not directly comparable with the other cases.

Operating sites that are close to consent limits for CO_2 emissions must take particular care if proposing to switch feed types towards a heavier feed type. For example, butane is often used as feedstock during the summer months when it cannot be blended into gasoline but operation reverts to another feedstock in winter, for example natural gas. The modelling work indicates a difference of 26 tonnes of CO_2 per hour for the 100 MMSCFD plant case in moving from 100% butane to 100% natural gas as feedstock.

The general advice for sites that are seeking to minimize CO_2 emissions is therefore to consider moving to a lighter feedstock into the hydrogen plant if this is possible, however, there may be other issues. Examples may be

- hardware limitations because in general the steam reformer will run hotter with the lighter feed;
- reduction in net export steam with the lighter feed may necessitate additional steam raising by another means which itself may cause additional CO_2.
- a lighter feed may simply be unavailable.

Incorporation of a pre-reformer into the flowsheet was also considered for the naphtha cases (Table 4). This indicates that less CO_2 results. This is because the reformer firing is less. The pre-reforming of naphtha is an overall exothermic reaction that raises the temperature of the gas inlet the reformer. The gas temperature is increased further by pre-heating the pre-reformer exit gas to around 1112°F (600°C). These factors reduce the heat input needed to the reformer and so less fuel is burned which is reflected in the lower requirement for make up fuel. Thus, directionally, a pre-reformer flowsheet favours lower CO_2 production but again there will be other considerations such as

- reduction in net export steam which may necessitate additional steam raising by another means which itself may cause additional CO_2.
- relative Capex and Opex considerations.

4 Process efficiency

Another environmental measure for hydrogen production is the overall process efficiency measured in the total energy equivalent required to produce a set amount of hydrogen. This figure has been derived in the modelling and takes into account the hydrocarbon used as feedstock and fuel plus the electrical energy import required to run plant equipment.

With reference to Table 3, the clear trend is that the efficiency becomes worse as the hydrocarbon feed and fuel becomes heavier (Graph 2). A major factor in this trend is again the higher CO_2 generation as the feed becomes heavier. This means that the PSA purge gas going to fuel in increasingly CO_2 rich and this CO_2 must be heated. The ROG case is again atypical as it was modelled with a higher steam to carbon ratio and thus consumes more heat in the heating of the higher level of steam. The inclusion of a pre-reformer (Table 4) also leads to higher process efficiency.

Graph 2. Process efficiency with varying feedstock

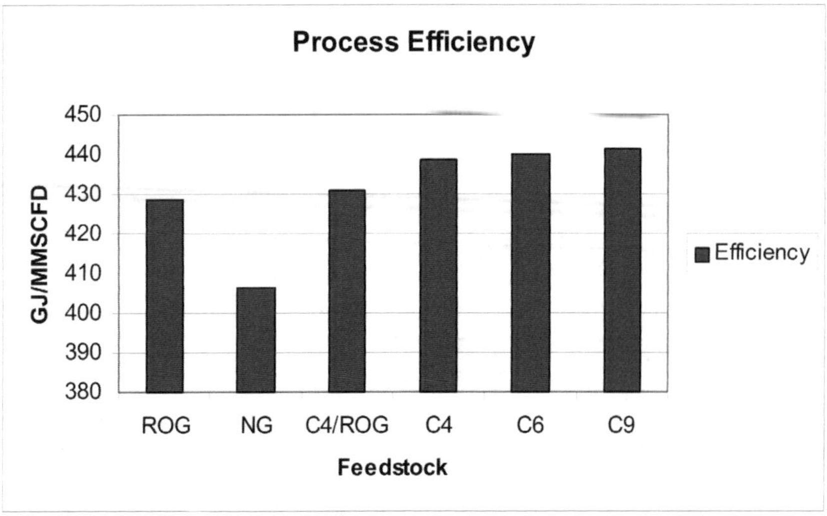

Again there are issues with switching feedstock type and including a pre-reformer in the flowsheet as outlined in the previous section.

5 Other factors

In addition to the influence of feedstock type on the environment using CO_2 emissions and plant efficiency as a measure, there are other factors to consider.

Export steam is another consideration that may have an impact. A steam reformer is a relatively efficient means of raising steam but the need to raise steam will depend on the steam balance of the site into which the hydrogen plant is integrated. If there is no end user for the export steam, then it makes sense to run the plant to make the minimum amount of export steam, that is to say no steam export. The modelling work indicates that this is favoured by using a light feedstock and incorporation of energy efficiency improvement technologies such as a pre-reformer in the plant design. Conversely, if the operator wants to maximize the steam available for export, then the opposite would be preferable. Any move towards a heavier feed leads to higher CO_2 emissions, however, it the additional steam would otherwise be generated in a less efficient package boiler, then the overall impact on CO_2 emissions should be assessed. For this reason and where a new hydrogen plant is being added to an existing site even if it has surplus steam, it may be preferable to design the new plant to generate steam if this allows less efficient boilers to be decommissioned.

6 Operating costs

In completion of this modelling, the tonnage requirement for feedstock and fuel was estimated to make 100 MMSCFD of hydrogen. Electrical power requirements were estimated based on known efficiencies and feedstock costs. In addition, there will be fixed costs, a value associated with net export steam and possibly a levy on CO_2 emissions.

Discussions with hydrogen producers have indicated that the fixed costs associated with operating a large hydrogen plant are in the order of US$ 500 per MMSCF. We will assume that the fixed costs are not affected as the feedstock type is changed. In checking the other costs, it becomes apparent that the variability plays a significant part in the relative attractiveness of various feedstocks. In North America, natural gas prices have generally increased in recent times and have oscillated between US$ 5 – 8 per MMBTU since the start of 2003. LPG and naphtha prices are variable also and have ranged between US$ 250 – 450 per tonne. Electrical costs have been estimated at US$ 0.055 per kWh. The value of a ROG feed depends on the options for its use but assuming it would be otherwise used for fuel, and then it may be valued against alternative fuels and for our purposes natural gas has been taken as the alternative. The relative value of net export steam was discussed also. If there is no requirement, then the value is nil but should the export be required or sold over-the-fence, then the value can be significant. In one case the price was approaching US$ 15 per tonne. Trading in emission allowances for CO_2 is in the order of US$ 10 – 20 per tonne and this is indicative of the cost in exceeding consent limits where such regulation is in place. On this basis, the costs associated with the feed types are estimated in Table 6. It must be remembered that this estimate is indicative only and that the modelling to date has not attempted to optimize costs.

Table 6. Estimated costs associated with hydrogen production

Feedstock type	ROG	Natural gas	ROG/ butane	Butane	Light naphtha	Naphtha
Units: US$/MMSCF						
Lowest prices						
Steam value	0.00	0.00	0.00	0.00	0.00	0.00
Hydrocarbon cost	0.00	1919.23	1140.00	2388.00	2442.00	2466.00
Energy cost	62.99	42.58	35.23	35.63	36.10	36.03
CO_2 charge	0.00	0.00	0.00	0.00	0.00	0.00
Overall	**62.99**	**1961.82**	**1175.23**	**2423.63**	**2478.10**	**2502.03**
Highest prices						
Steam value	3.70	37.04	131.88	145.58	143.73	142.99
Hydrocarbon cost	2955.65	3067.34	3684.48	4298.40	4395.60	4438.80
Energy cost	62.99	42.58	35.23	35.63	36.10	36.03
CO_2 charge	451.20	451.20	547.20	576.00	600.00	609.60
Overall	**3466.13**	**3524.08**	**4135.03**	**4764.45**	**4887.97**	**4941.44**

By far the largest elements of the costs are the hydrocarbon feed and fuel. The electrical energy costs are almost negligible by comparison. Concerning the ROG cases, in the "lowest prices scenario", ROG value has been set to nil, for example assuming it is being flared, however, when the ROG is valued on a par with natural gas it is much closer to the natural gas cost case. It generally appears that as the hydrocarbon becomes heavier the cost of production increases when considering the trends in either the "lowest" or "highest" tables but when these are compared it is clear that variation in hydrocarbon cost over time can bring huge differences. This illustrates the importance in a plant having a long-term supply contract in place for its hydrocarbon feed and fuel so that exposure to the swings in spot prices is minimized.

Where steam export is of value, the contribution can be significant and the additional export steam with heavier feeds may influence operating costs. In the "highest" case, the extreme example of US$ 15 per tonne has been used and it might be more realistic to use about half this value.

Another significant contribution to the cost equation may come from regulation on CO_2 emissions. In the "lowest" case, it is assumed that there is no financial penalty imposed on operation. In the "highest" case, however, a US$ 20 per tonne levy is assumed on the total calculated CO_2 emissions. Even with the lighter feedstock and fuel, the contribution this would make to overall costs would be significant. It is more likely that a plant will be allocated a certain consent limit for CO_2 emissions and a levy will be made for breach of the consent limit. Plants that operate within the allocated consent limit may be able to sell the unused allocation and this is already happening in certain parts of the world where appropriate legislation is in place. Thus, the ability to operate within emission consent limits offers an additional income stream at least in the short term until consent limits are tightened in line with plant performance.

7 Case studies

Johnson Matthey Catalysts has reviewed available data from operating plants to provide verification of the trends identified through the modelling work.

Case study 1

A US mid-western refiner operates a hydrogen plant in which available refinery off gases can be used in combination with natural gas feed. The flowsheet features a steam reformer, HTS and PSA unit. Johnson Matthey Catalysts has conducted reformer surveys at different times when the plant has been running predominantly on natural gas or with a blend of natural gas with two different off gases. The make up fuel is the same in both cases which is refinery fuel gas comprising approximately 30 mol% hydrogen, 50 mol% methane, 10 mol% ethane and diminishing amounts of $C_3 - C_6$ hydrocarbons. Relevant information about the operating conditions at the time of each survey is shown in Table 7.

Table 7. Data from reformer surveys

	Case 1	Case 2
Feed type	Natural Gas	Blended natural gas + off gases
Plant rate	12 MMSCFD (13.2 kNm^3/hr)	10.5 MMSCFD (11.6 kNm^3/hr)
Reformer fuel	PSA purge gas plus fuel gas (30 mol% H_2; 50 mol% CH_4; 10 mol% C_2H_6; $C_3 - C_6$ hydrocarbons	
Steam to carbon ratio	4.4	4.9
Excess O_2	2.4%	2.8%
Reforming section inlet pressure	22 bara	23 bara
Reformer exit temperature	1450°F (788°C)	1480°F (804°C)
CO slip ex HTS	2.1 mol% dry	2.0 mol% dry
PSA efficiency	86%	85%

Feed Composition	Case 1	Case 2
H_2	10.32	30.38
N_2	1.17	1.01
CH_4	84.58	56.65
CO	0.00	0.07
CO_2	0.80	0.29
C_2H_6	2.25	6.55
C_3H_8	0.39	2.75
C_4H_{10}	0.10	1.29
C_5H_{12}	0.00	0.33
C_6+	0.39	0.67
Total	100.00	100.00

In Case 1, the natural gas case, a small amount of a hydrogen rich off gas is added to provide the hydrogen necessary for desulphurisation in the purification section of the flowsheet. In this case, the amount of hydrogen at 10.32 mol% is rather higher than is usually recommended for a natural gas feed. In Case 2, the combined natural gas and off gases case, the off gases contribute a significantly larger amount of hydrogen but also increase the C_2+ content. Although the plant rate is not identical, the difference of 12% is deemed small enough to allow scrutiny of the CO_2 burden in each case. The operating conditions are more forcing in Case 2 with a higher exit temperature and steam to carbon ratio being reflected in a lower methane slip from the reformer, 3.5 mol% (dry) in Case 2 against 4.7 mol% (dry) in Case 1, despite the higher feed hydrogen level and slightly higher exit pressure in Case 2 which should tend to increase methane slip.

It was noted earlier that more forcing conditions that drive methane slip down tend to increase CO_2 emissions but that lighter feeds tend to decrease CO_2 emissions. In Case 2, the average carbon number for the feed is 0.89 whereas in Case 1 it is higher at 0.94. Re-analysis of the reformer survey data indicated that, despite the more forcing conditions in Case 2, the change to the lighter feed still proved beneficial in terms of CO_2. In Case 1, the CO_2 emissions are calculated as 9.78 tonne per tonne of H_2 and for Case 2 the value is 9.59 tonne per tonne.

Case study 2

Refineries in India are considering a switch from naphtha feed to natural gas. A typical plant flowsheet features a steam reformer, HTS and PSA unit. Johnson Matthey Catalysts has modelled the expected performance based upon a fixed exit temperature. The limiting factor in the switch to natural gas is likely to be the design temperatures for the original naphtha design case unless the reformer is modified to tolerate higher temperatures. Details of the study are given in Table 8.

Table 8. Data for Indian case study

	Case 1	Case 2
Feed type	Naphtha	Natural gas
Plant rate	53 MMSCFD (58 kNm^3/hr)	53 MMSCFD (58 kNm^3/hr)
Reformer fuel	PSA purge gas + naphtha	PSA purge gas + natural gas
Steam to carbon ratio	3.00	3.41
Excess O_2	2.0%	2.0%
Reforming section inlet pressure	24 bara	24 bara
Reformer exit temperature	1584°F (862°C)	1584°F (862°C)
CO slip ex HTS	2.0 mol% dry	2.0 mol% dry
PSA efficiency	85%	85%

Feed Composition	Case 1	Case 2
H_2	20.00	2.00
N_2	0.00	0.45
CH_4	0.00	87.97
C_2H_6	0.00	6.34
C_3H_8	0.00	2.20
C_4H_{10}	0.00	1.04
C_5H_{12}	0.00	0.00
C_6H_{14}	2.14	0.00
C_6H_{14} benzene	14.40	0.00
C_7H_{16}	63.46	0.00
Total	100.00	100.00

In addition to the constraint on exit temperature, the methane slip was also fixed at 3.5 mol% (dry) and the steam to carbon ratio and natural gas feed rate were varied to provide the required hydrogen production rate. Based on these constraints, the natural gas case resulted in only slightly higher peak tube wall temperature and heat flux (Table 9).

Table 9. Modelling results for Indian case study

Parameter	Units	Case 1	Case 2
Feed		Naphtha	Natural Gas
Naphtha Feed Rate	kmol/hr	217.5	n/a
Natural Gas Feed Rate	kmol/hr	n/a	832
Steam Rate	kmol/hr	3545	3156
Steam to Carbon	kmol/hr	3.0	3.4
Methane slip	dry mol %	3.50	3.50
ATE	°C	3.9	3.7
Maximum TWT	°C	889	894
Peak Heat Flux	KW/m²	109	112
TWT Margin	°C	92	87
Fluegas Temperature	°C	1020	1035
Reformer Duty	MW	112	114

At these conditions, the CO2 burden is much lower in the natural gas case. In the naphtha case, Case 1, it is calculated to be 12.37 tonne per tonne of hydrogen whereas in the natural gas case, Case 2, the figure is 9.94 tonne/tonne.

8 Conclusion

Modelling work on a hydrogen plant flowsheet demonstrates that, when changes are made to the hydrocarbon feedstock and fuel type, the environmental impact increases in terms of higher CO_2 emissions and lower process efficiency as the feedstock becomes heavier. The environmental impact is also influenced at the hydrogen plant itself by the particular design of the flowsheet. For example, when a pre-reformer is included the CO_2 emissions are lower and the process efficiency is improved. Other aspects of plant performance are affected also. With lighter feedstock or with a pre-reformer, the steam available for export is also diminished. In many plants, this export steam is required and its removal would require the steam to be raised elsewhere possibly in a relatively inefficient boiler and so that overall impact of any change must be considered rather than focusing on the hydrogen plant in isolation.

Having calculated the relative usage of feedstock, fuel and other energy usage, this allowed an estimate to be made of the relative costs as feedstock is changed. The principal conclusion is that feedstock/fuel costs dominate and that variability in the price of these commodities can have a significant impact on overall plant economics. The value of steam export and impact of forthcoming regulation and levies on CO_2 emissions should also be taken into account.

References

1. "Increasing Hydrogen Plant Throughput, Reliability and Cost Effectiveness by Optimisation of Operating Conditions and Catalyst Selection", P V Broadhurst and P E J Abbott, Proceedings of European Refining Technology Conference (ERTC) 6th Annual Meeting, Madrid, November 2001.
2. "Improving Hydrogen Plant Performance: Part 1", P V Broadhurst and P E J Abbott, Petroleum Technology Quarterly, Summer 2002, pp 130-135.
3. "Hydrogen Market Review", B J Cromarty; Synetix Technical Paper, 1999.
4. "Hydrogen Technology Overview", S Ratan, Proceedings of 6th International Conference on Refinery Processing, AIChE Spring National Meeting, New Orleans, April 2003.
5. "Alternative Lower Cost Feedstock for Hydrogen Production", K Chlapik and B Slemp, NPRA Annual Meeting, 2003.
6. "Feedstock Conversion for Indian Ammonia Plants. A Review of the Challenges", W J Cotton and V Singh, Proceedings of Fertilizer Association of India (FAI) Seminar 2003, New Delhi, December 2003.
7. "Implications of Feedstock Flexibility on the Design and Operation of Steam Reforming in Hydrogen Plant Flowsheets", P V Broadhurst and W J Cotton, submitted for publication.
8. "Feed for Thought", P V Broadhurst, W J Cotton and A Vasudeva, Hydrocarbon Engineering, March 2004, pp 37-40.
9. "Modelling Tools and Application for Synthesis Gas Plants", M Davies, B R Fisher and W J Cotton, Johnson Matthey Catalysts Technical Paper, 2004.
10. "REFORM – The Worlds Leading Reformer Simulation Tool", W J Cotton and C M Foster, Johnson Matthey Catalysts Technical Paper, 2004.

Impact of Gas Phase Impurities on Refinery Hydrogen Network Management

*Bhari Bhujan Singh, Nan Zhang**

Centre for Process Integration, School of Chemical Engineering & Analytical Sciences, The University of Manchester, P.O. BOX 88, Manchester, M60 1QD, UK

Hydrogen management in refineries has received significant attention in recent years, due to increasing hydrogen demand and decreasing supply. On one hand hydrogen demand is increasing as oil refiners need more hydroprocessing to face the challenges caused by the clean fuel regulations, increased processing of heavier sour crude and heavy-end upgrading. On the other hand, its availability is decreasing. For example, lower aromatic gasoline specification will decrease the operation severity in catalytic reformers thus reducing the by-product hydrogen. Under the new circumstances, the challenge is how the refiners should manage hydrogen in an efficient manner.

Previous research by Alves (1999) in the area of hydrogen network management proposed an approach for the analysis of hydrogen distribution systems. This method begins with identifying all the hydrogen sinks/sources in the network. The flowrate and purity for hydrogen sinks/sources in the network can be extracted from operating data and then hydrogen demand/supply profiles, hydrogen composite curve/hydrogen surplus cascade can be conveniently represented in a two-dimensional plot. These curves are used to check the feasibility of the network and to set the targets for hydrogen recovery and hydrogen plant production. To achieve the required target from the graphical approach, the network can then be designed by using Linear Programming. To overcome some of the limitations in Alves (1999) work, Liu (2002) proposed an automated design methodology for hydrogen network, which is based on the optimisation of a reducible superstructure. Another contribution of Liu (2002) is to integrate hydrogen generation and hydrogen purification processes into the automated hydrogen network design by using MINLP formulation of the problem. Liu (2002) proposed a simple solution procedure to overcome the difficulty of solving the complex MINLP problem.

* Correspondence concerning this paper should be addressed to Nan Zhang

However, in previous approaches, irrespective of the different impurity distributions (e.g. CH_4, C_2H_6, C_3H_8, C_4H_{10}, etc.) in the refinery gas streams, all the impurities are lumped as CH_4 and thus gases are treated as a binary mixture of H_2 and CH_4. In the previous optimisation approaches, hydroprocessor conditions are assumed constant, but different impurity distributions in the gas streams affect hydrogen partial pressure and hydrogen to oil ratio of hydrogen consumers. Hence neglecting the impurity compositions can lead to an infeasible network after optimisation.

In this paper examples are shown to prove the importance of considering light hydrocarbon impurities. Effect on the typical hydroprocessors' conditions e.g. hydrogen partial pressure, recycle conditions etc. with different impurity distribution is presented. A systematic methodology is presented to incorporate impact of impurities on hydrogen network management. A new simulation model of hydrogen consumer with impurities is proposed. Hydrogen network is optimised with the existing automated design approach (lumping impurities as CH_4) and then impact of impurities in the optimised network is exploited through iterations with simulation.

A case study is presented to demonstrate the effectiveness of the proposed methodology. Results presented in this case study show that impurities have considerable impact on hydrogen network management and the proposed methodology improves the accuracy of hydrogen savings in the network.

Literature cited:

Alves, J.J. (1999). Analysis and design of refinery hydrogen distribution systems. Thesis, University of Manchester Institute of Science and Technology, Manchester.

Liu, F. (2002). Hydrogen integration in oil refineries. Thesis, University of Manchester Institute of Science & Technology, Manchester.

Impact of Gas Phase Impurities on Refinery Hydrogen Network Management

Bhari Bhujan Singh, Dr. Nan Zhang

TC005: 68 - Hydrogen Management
Eighth International Conference on Refinery Processing

AIChE Spring National Meeting 2005
April 10-14, Hyatt Regency, Atlanta, GA, USA

Outline

1. Introduction
2. Review of previous work
3. Modelling hydrogen consumers with impurities
4. Hydrogen network optimisation with impurity considerations
5. Case study
6. Conclusions

1. Introduction

Many Refinery Operations Consume Hydrogen

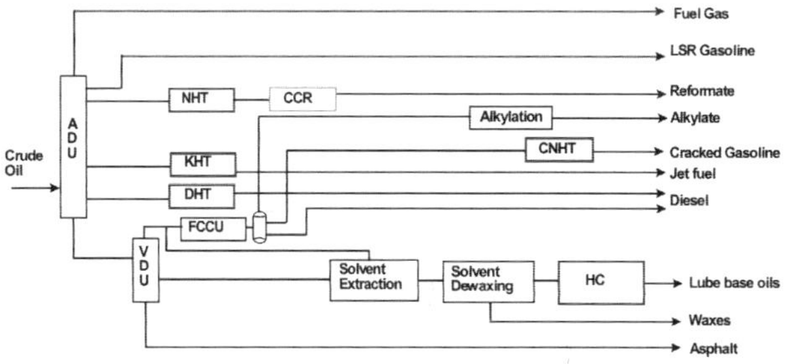

Consumers and Sources of Hydrogen

Consumers

- **Hydrotreating**
  Removal of impurities (including sulphur etc.) from streams
  Hydrogenation of aromatics and olefins
- **Hydrocracking**
  Break large molecules into smaller and higher value molecules
- **Other consumers**
  Isomerisation, lubricant plants, exports

Sources

- **Catalytic reformer**
  Generates hydrogen as a by-product
- **Recycle and recovery from off gas streams**
  Direct re-use or with some additional purification
- **Hydrogen production plant**
  e.g. steam reforming, partial oxidation
- **Imports**
  Buy hydrogen from external suppliers

Hydrogen Distribution System

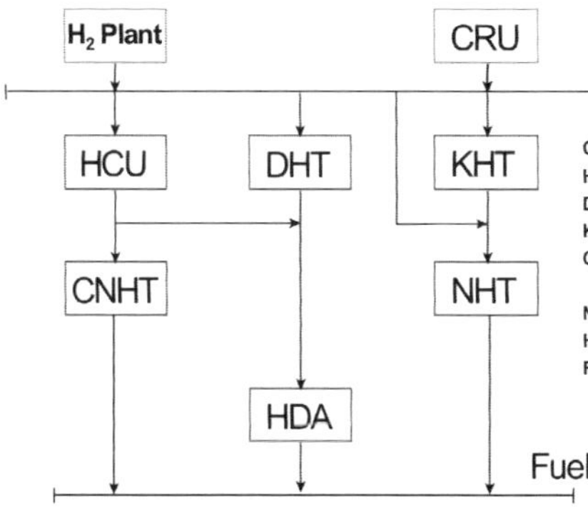

CRU Catalytic Reformer Unit
HCU Hydrocracker Unit
DHT Diesel Hydrotreater
KHT Kerosene Hydrotreater
CNHT Cracked Naphtha Hydrotreater
NHT Naphtha Hydrotreater
HDA Hydrodealkylation
Fuel Fuel Gas System

New Trends Increase the Hydrogen Demand

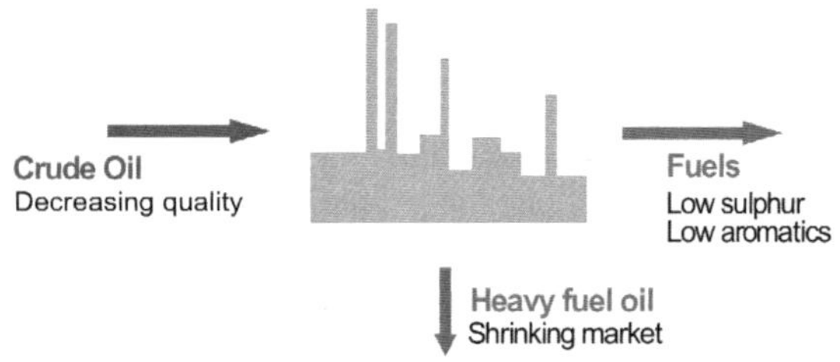

Hydrogen supply is also often a production bottleneck!

How can we make best use of hydrogen resources in order to meet new demands and improve profitability?

7

2. Review of Previous Work

8

Hydrogen Pinch Analysis

Pinch Approach for Targeting the Minimum Hydrogen Utility (Alves, 1999)

➤ Identify sources and sinks of hydrogen

A simplified diagram of the consumer

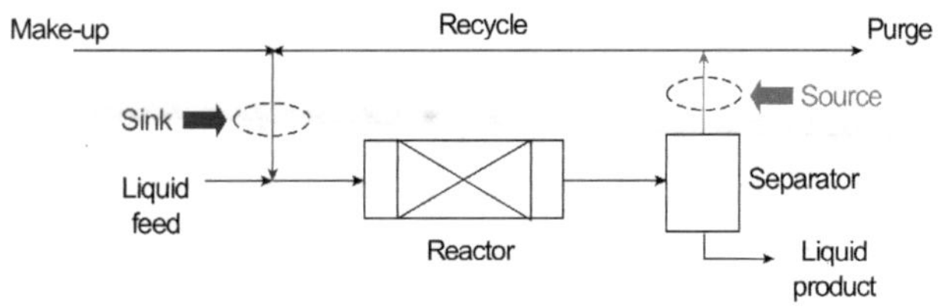

Alves, J., "Analysis and Design of Refinery Hydrogen Distribution Systems," PhD thesis, UMIST (1999)

9

From Hydrogen Composite Curve to Hydrogen Surplus Diagram

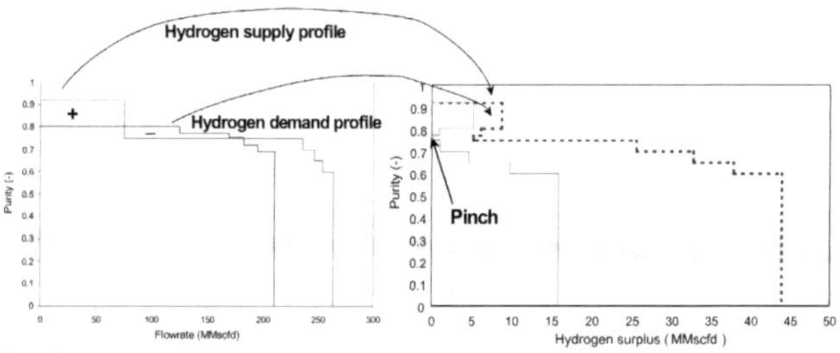

Remarks:

☺ Simple graphical method for targeting minimum hydrogen utility

☹ Targets can be unrealistic as pressure is not considered 

Alves, J., "Analysis and Design of Refinery Hydrogen Distribution Systems," PhD thesis, UMIST (1999)

10

Network Design using Linear Programming

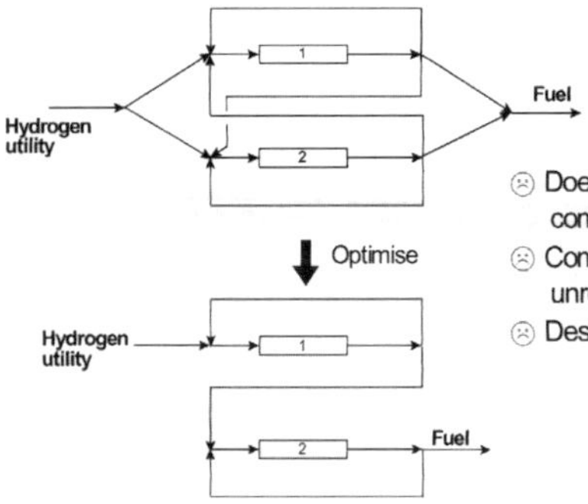

☹ Does not account fully for existing compressors

☹ Compression costs used may be unrealistic

☹ Designs often need to be evolved

Alves, J., "Analysis and Design of Refinery Hydrogen Distribution Systems", PhD thesis, UMIST (1999)

11

☺ **Automated hydrogen network design (Liu, 2002)**

- Pressure considerations included
- Systematic selection of retrofit options
- Can be used for cost savings and debottlenecking

☺ **Automated hydrogen purification network design with purifier selection and integration (Liu, 2002)**

- Superstructure-based
- MINLP formulation
- MILP/NLP iteration to overcome the computation difficulties

☺ **Integration of hydrogen plant with hydrogen network (Liu, 2002)**

- Superstructure-based
- MINLP formulation
- MILP/NLP iteration to overcome the computation difficulties

Liu, F., "Hydrogen Integration in Oil Refineries," PhD thesis, UMIST (2002)

12

Scope of Research

Improving the hydrogen network management by incorporating the effect of gas impurities in a hydrogen network

☞ Effect on hydrogen to oil ratio and hydrogen recycle/purge purity by considering refinery gases as multi-component mixture

13

3. Modelling Hydrogen Consumers with Impurities

14

Impact of Impurities on the Hydrogen Network

❑ **Sulphur**
- Effect on performance of hydroprocessors
- Effect on process recycle/purge purity and hydrogen consumption

❑ **Nitrogen**
- Effect on performance of hydroprocessors
- Effect on process recycle/purge purity and hydrogen consumption

❑ **Light hydrocarbons**
- Affect performance of the flash separators
- Effect on process recycle/purge purity

15

Dealing with Sulphur and Nitrogen

❑ Set maximum intake limit for each consumer

❑ Purify streams, e.g. Amine wash for sulphur removal

Must be included in optimisation

16

Dealing with Light Hydrocarbons

❑ Include high pressure flash calculations in hydrogen network management

- Simulation of the optimised network with commercial software e.g. HYSYS, Aspen Plus, etc.
- Model for high pressure flash in hydrogen pinch analysis software e.g. RefOpt

17

Modelling

What is required to model hydrogen consumer in simulation?

❑ Operating conditions for high pressure flash separator
❑ Hydrogen consumption in process
❑ Products and make-up flowrate
❑ Product properties
❑ Product composition
- Gas composition
- Oil assay data

❑ Make-up flow composition

Required information can be

☞ Taken from real plant data
☞ Derived from correlations*

* Petroleum Refining Process Correlations, HPI Consultants, INC., 1987

18

Simulation Model for a Hydrogen Consumer

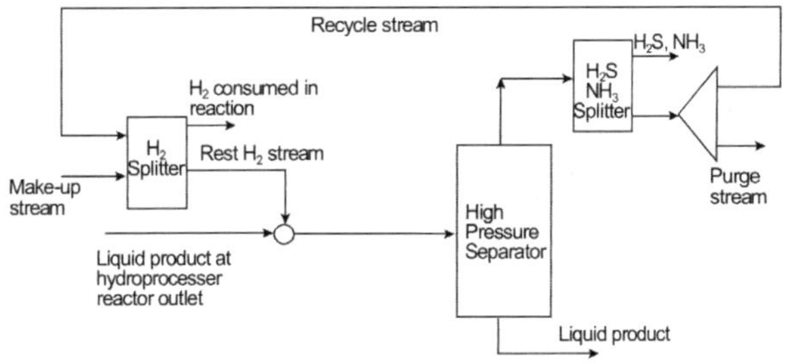

19

Example 1

Effect of light hydrocarbon impurities on Straight-run Naphtha Hydrotreater (SR NHT) purge purity with fixed Hydrogen/Oil ratio

20

Process Data

SR Naphtha Flowrate, bpd		36000
Properties		
Specific Gravity, 60°F/60°F		0.7347
°API		61.1
Sulphur, (wt%)		0.035
P (LV%)		66.8
O (LV%)		0
N (LV%)		19.9
A (LV%)		13.3
ASTM D86	LV%	°C
	IBP	87
	10	92
	50	119
	90	149
	FBP	165
Reactor Inlet		
H_2/Oil ratio, scfd/bpd		1150
High pressure flash conditions		
Pressure, bar (a)		18
Temperature, °C		50

21

Calculation Procedure

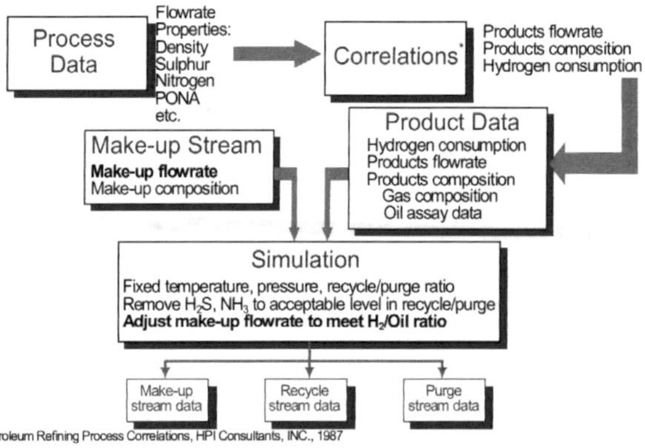

* Petroleum Refining Process Correlations, HPI Consultants, INC., 1987

22

Effect of Light Hydrocarbon Impurities on SR NHT Purge Purity with Fixed Hydrogen/Oil Ratio

☞ Two make-up streams with the same hydrogen purity but different impurity composition are compared

SR NHT with make-up flow 1

H_2/Oil ratio, scfd/bpd = 1150	H_2 Make-up	Recycle	Purge	H_2 Feed
	Flow rate, MMscfd			
	6.24	47.90	5.32	54.14
	Composition, vol%			
H_2	83.00	75.64	75.64	76.49
CH_4	17.00	21.41	21.41	20.90
C_2H_4	0.00	0.54	0.54	0.48
C_3H_8	0.00	0.50	0.50	0.44
$i\text{-}C_4H_{10}$	0.00	0.06	0.06	0.05
$n\text{-}C_4H_{10}$	0.00	0.06	0.06	0.06
H_2S	0.00	0.38	0.38	0.34
NH_3	0.00	0.00	0.00	0.00
C_5+	0.00	1.40	1.40	1.24

SR NHT with make-up flow 2

H_2/Oil ratio, scfd/bpd = 1150	H_2 Make-up	Recycle	Purge	H_2 Feed
	Flow rate, MMscfd			
	6.24	44.30	4.92	50.54
	Composition, vol%			
H_2	83.00	81.78	81.78	81.93
CH_4	7.00	10.52	10.52	10.09
C_2H_4	2.00	2.59	2.59	2.52
C_3H_8	6.00	2.96	2.96	3.33
$i\text{-}C_4H_{10}$	1.00	0.20	0.20	0.30
$n\text{-}C_4H_{10}$	1.00	0.16	0.16	0.26
H_2S	0.00	0.39	0.39	0.34
NH_3	0.00	0.00	0.00	0.00
C_5+	0.00	1.40	1.40	1.23

Remarks:

Recycle purity increases from 75.64 to 81.78 vol%
Recycle flow decreases from 47.9 MMscfd to 44.3 MMscfd (7.52%)

23

Example 2

Effect of light hydrocarbon impurities on SR Naphtha Hydrotreater (SR NHT) purge purity with fixed recycle flow

24

Calculation Procedure

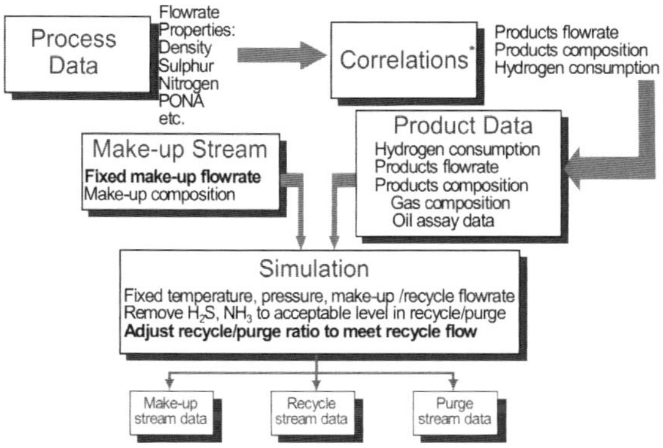

* Petroleum Refining Process Correlations, HPI Consultants, INC., 1987

25

Effect of Light Hydrocarbon Impurities on SR NHT Purge Purity with Fixed Recycle Flow

☞ Two make-up streams with the same hydrogen purity but different impurity composition are compared

SR NHT with make-up flow 1				
H_2/Oil ratio, scfd/bpd = 1150	H_2 Make-up	Recycle	Purge	H_2 Feed
	Flow rate, MMscfd			
	6.24	47.90	5.32	54.14
Composition, vol%				
H_2	83.00	75.64	75.64	76.49
CH_4	17.00	21.41	21.41	20.90
C_2H_4	0.00	0.54	0.54	0.48
C_3H_8	0.00	0.50	0.50	0.44
i-C_4H_{10}	0.00	0.06	0.06	0.05
n-C_4H_{10}	0.00	0.06	0.06	0.06
H_2S	0.00	0.38	0.38	0.34
NH_3	0.00	0.00	0.00	0.00
C_5+	0.00	1.40	1.40	1.24

SR NHT make-up flow 2				
H_2/Oil ratio, scfd/bpd = 1232	H_2 Make-up	Recycle	Purge	H_2 Feed
	Flow rate, MMscfd			
	6.24	47.89	4.92	54.13
Composition, vol%				
H_2	83.00	81.79	81.79	81.93
CH_4	7.00	10.52	10.52	10.11
C_2H_4	2.00	2.59	2.59	2.52
C_3H_8	6.00	2.96	2.96	3.31
i-C_4H_{10}	1.00	0.20	0.20	0.29
n-C_4H_{10}	1.00	0.16	0.16	0.26
H_2S	0.00	0.39	0.39	0.35
NH_3	0.00	0.00	0.00	0.00
C_5+	0.00	1.40	1.40	1.24

Remarks:

H_2/Oil ratio increases from 1150 to 1232 scfd/bpd (7.13%)
Recycle purity increases from 75.64 to 81.79 vol%
Purge flow decreases from 5.32 MMscfd to 4.92 MMscfd (7.52%)

26

Summary

☞ Real plant data as well as correlations (e.g. HPI correlations) can be used to model and simulate hydrogen consumers with commercial simulation software

☞ Sulphur and Nitrogen can be dealt with by putting constraints on its maximum possible level in recycle/purge streams

☞ Light hydrocarbon impurities have a strong impact on hydrogen to oil ratio and recycle/purge stream conditions in hydrogen consumers

This needs to be taken into account in hydrogen network management...

27

4. Hydrogen Network Optimisation with Impurity Considerations

28

Previous Approach

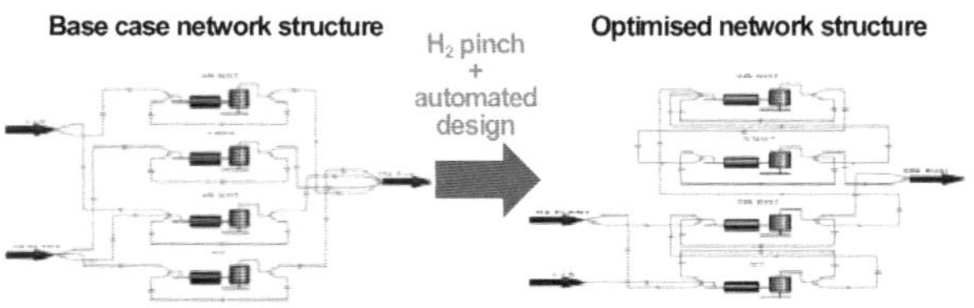

Optimised network structure with lumping impurities as CH_4...

How can we design hydrogen network incorporating impact of impurities?

29

Hydrogen Pinch Analysis for Base Case

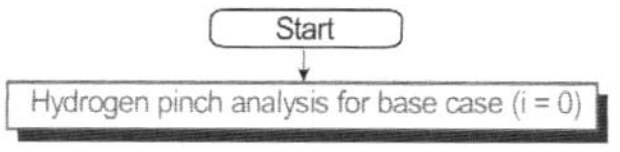

Perform hydrogen pinch analysis on base network to check scope for improvement...

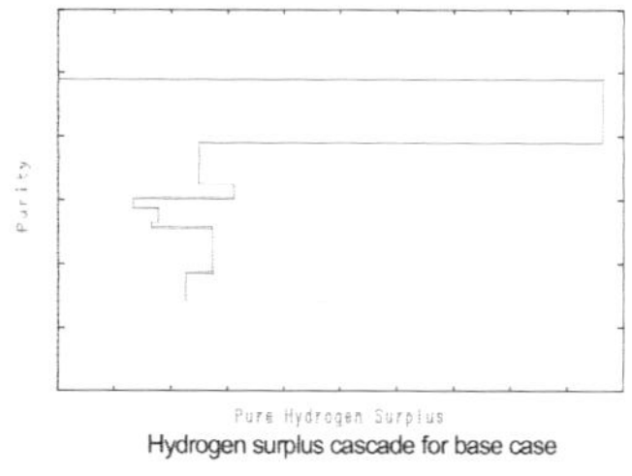

Hydrogen surplus cascade for base case

30

Check for the Scope of Improvement in Base Case

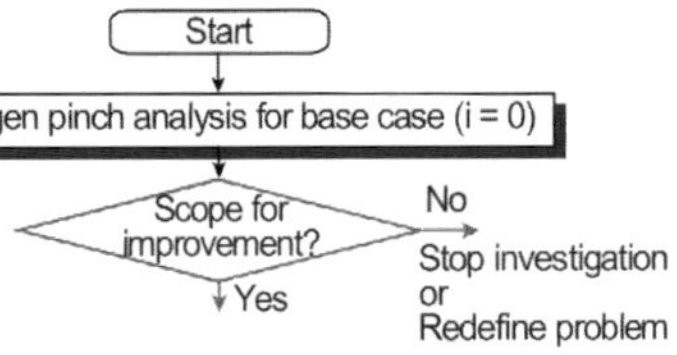

If there is scope for improvement...

Proceed to the next step

Else

Stop investigation or Redefine problem

(Network is already optimised but might try new purifiers or process changes)

31

Hydrogen Network Optimisation

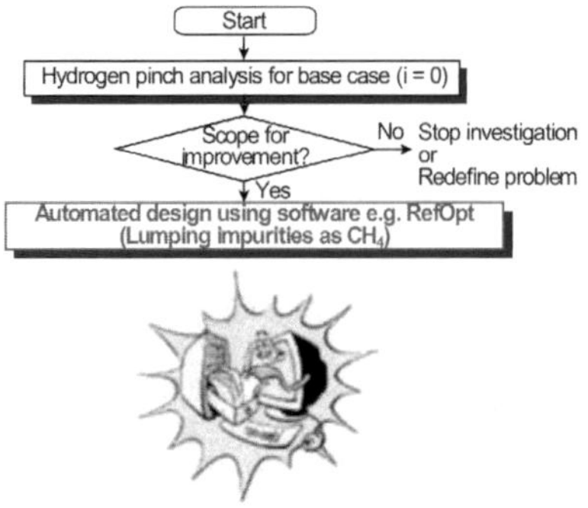

New network structure (Lumping impurities as CH_4)

Network includes realistic constraints like compression, piping, etc.

32

Hydrogen Pinch Analysis for Optimised Network with Lumping Impurities as CH_4

Scenario 1: Realistic Target

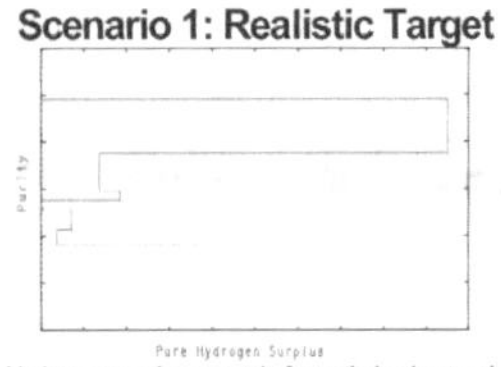

Hydrogen surplus cascade for optimised network considering realistic constraints

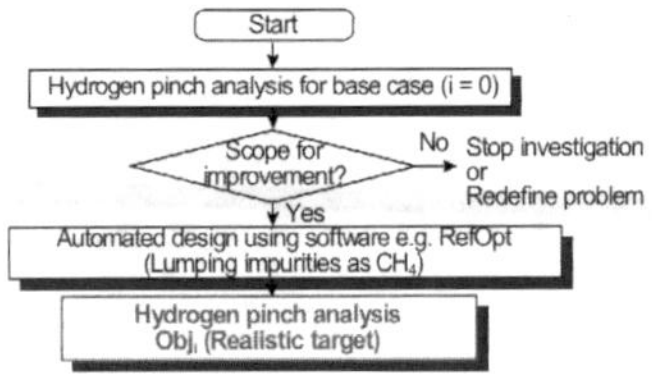

33

Remarks

☞ After optimisation, realistic constraints like compression, piping etc. may lead to a system not being pinched

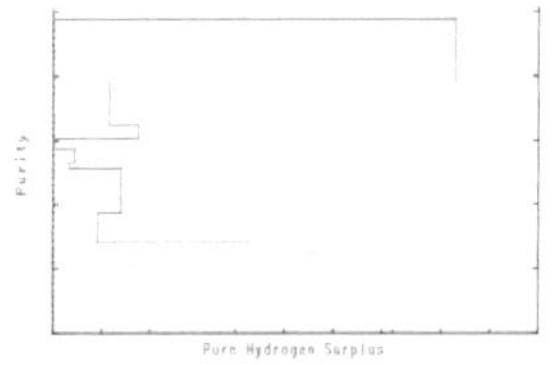

Hydrogen surplus cascade for optimised case without considering realistic constraints

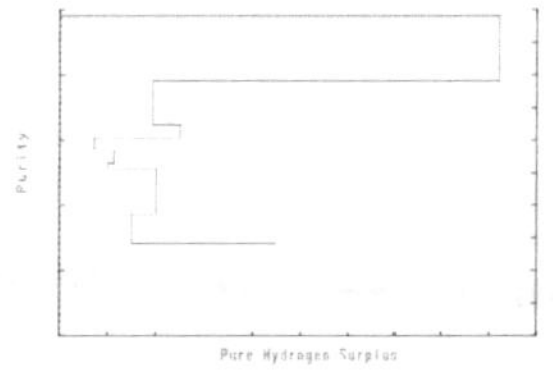

Hydrogen surplus cascade for optimised case considering realistic constraints (Realistic target)

Realistic target will be used as a bench mark

34

Simulation with Impurity Consideration

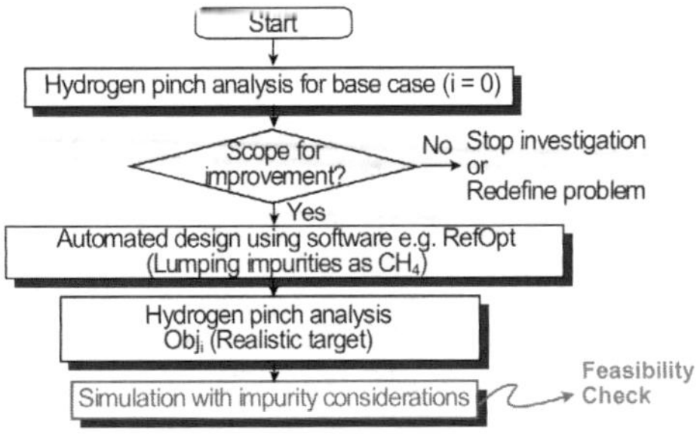

35

Feasibility Check for Optimised Network

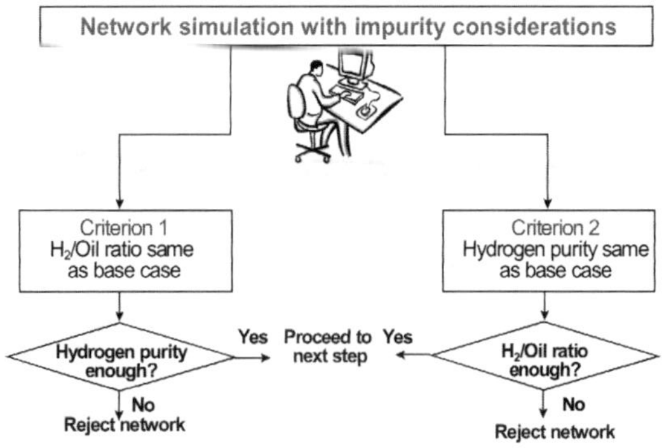

Remark: At least one criterion will give a feasible solution

36

Hydrogen Pinch Analysis for Simulated Network with Impurity Considerations

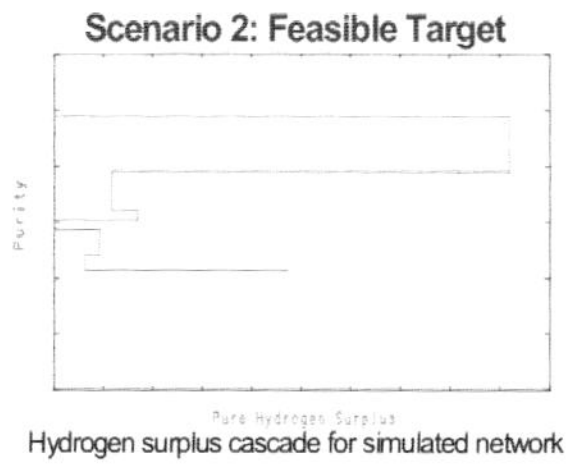

Hydrogen surplus cascade for simulated network

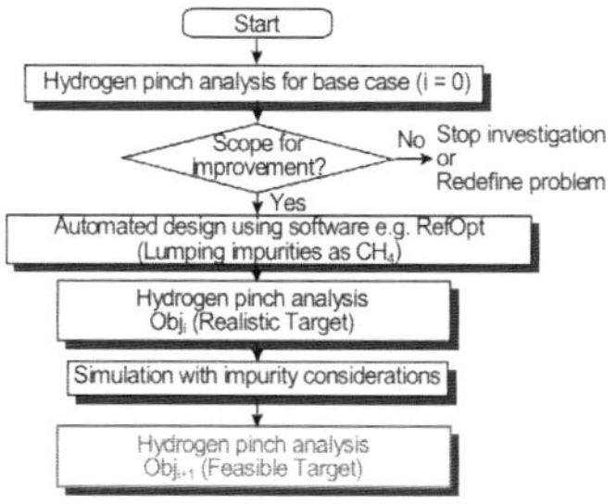

37

Comparison between Scenario 1 and Scenario 2

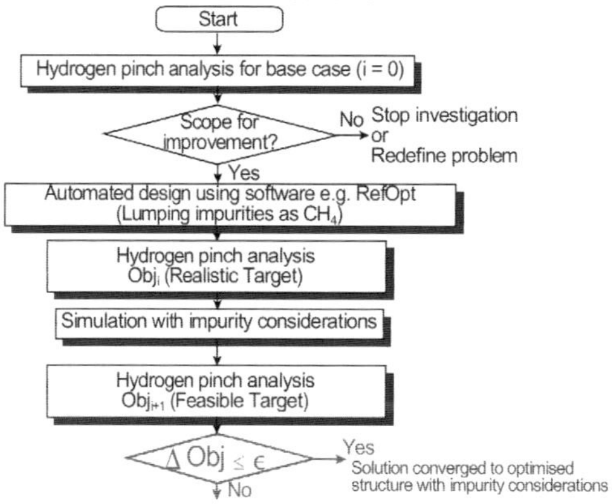

38

Update Data from Simulation

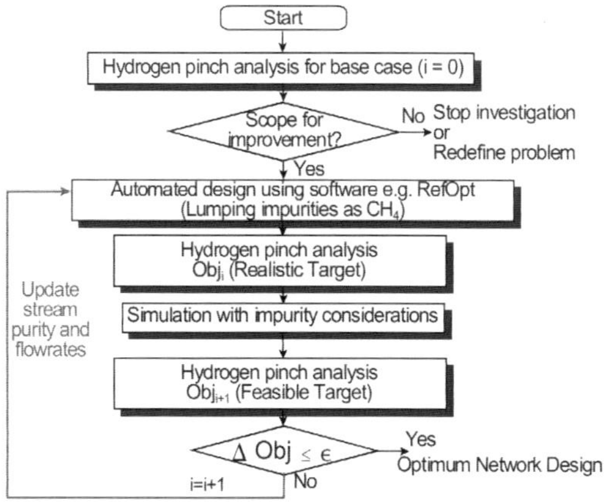

39

Overall Procedure

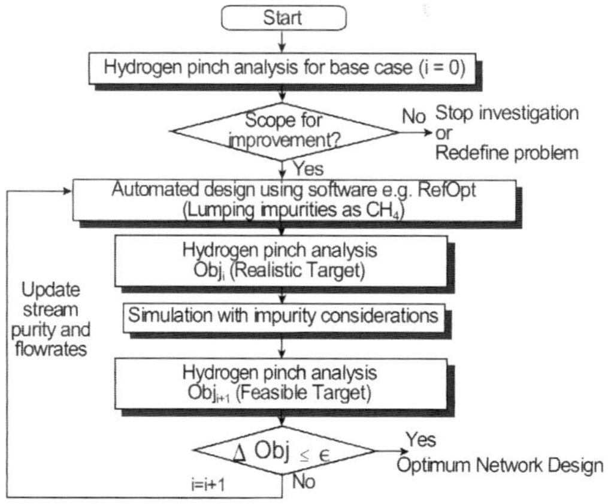

40

Summary

- A systematic procedure has been proposed to incorporate impurity considerations in hydrogen network management
  - The scope of optimisation in network is checked by hydrogen pinch analysis
  - Automated design is used to find optimised network with lumping impurities as CH_4
  - Simulation is used to incorporate the impact of impurities

41

5. Case Study

Objective: Optimum hydrogen network design with impurity considerations

42

Hydrogen Network for Base Case

Total hydrogen input = 125.969 MMscfd

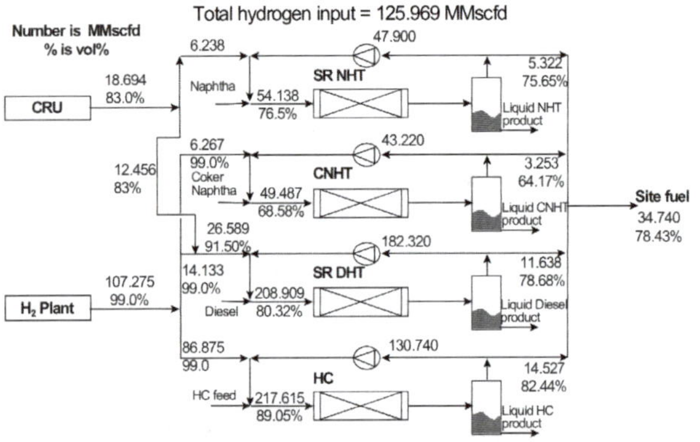

43

Hydrogen Surplus Cascade for Base Case

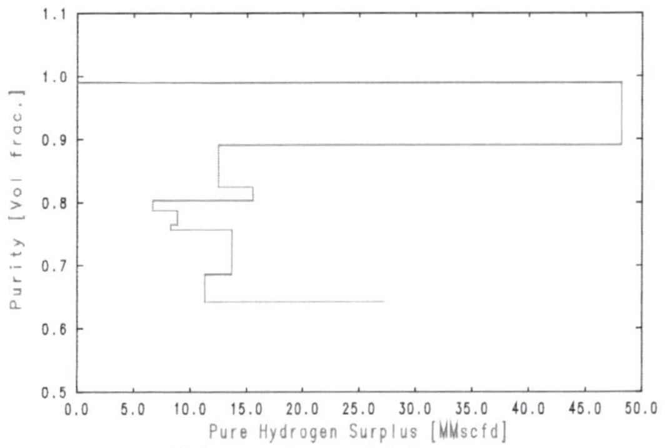

Hydrogen surplus cascade for base case

44

Check for Scope of Improvement in Base Case

Hydrogen pinch analysis shows scope for improvement in base case network

Proceed to the next step

45

Optimised Hydrogen Network (Lumping impurities as CH_4)

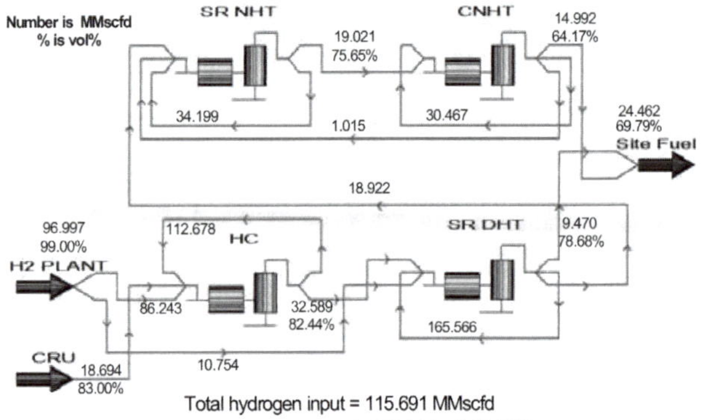

Total hydrogen input = 115.691 MMscfd

Saving in H_2 plant flow = 10.278 MMscfd (8.16%)

46

Hydrogen Pinch Analysis for Optimised Network with Lumping Impurities as CH_4

Scenario 1: Realistic Target

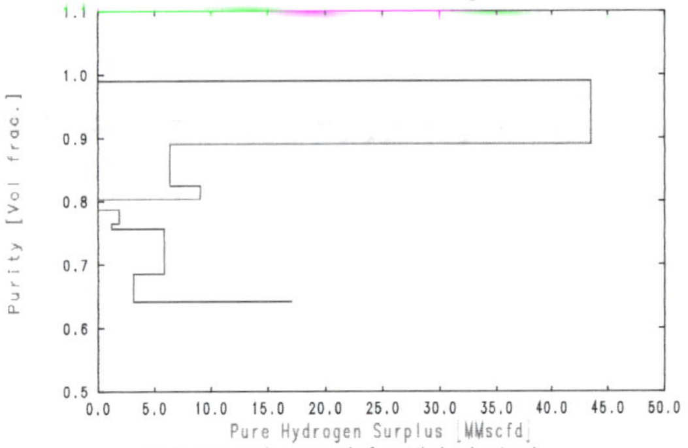

Hydrogen surplus cascade for optimised network

47

Is this optimised hydrogen network a feasible one, when considering impurities?

Let's check this network by simulation...

48

After Simulating Optimised Network

Total hydrogen input for Case 1 and Case 2 = 115.691 MMscfd

	Total inlet flowrate, MMscfd		Purity, vol%		H_2/Oil ratio, scfd/bpd		Hydrogen partial pressure, bar (a)	
Unit	Case 1	Case 2	Case 1	Case 2	Case 1	Case 2	Case 1	Case 2
SRNHT	54.136	55.026	76.49	75.61	1150	1156	15.30	15.12
CNHT	49.487	50.702	68.58	71.02	2263	2401	24.00	24.86
SR DHT	208.909	216.217	80.32	77.62	2796	2797	24.09	23.29
HC	217.615	221.577	89.05	87.46	5537	5537	83.71	82.22
SITE FUEL	24.462	24.462	69.79	69.79				

Case 1: Optimised network with impurities lumped as CH_4
Case 2: Simulated optimised network with impurity considerations

Remarks: H_2/Oil ratio change for SRNHT, CNHT, SRDHT
Hydrogen partial pressure change for all units

Remarks

- Previous approaches assume the same hydrogen partial pressures and H_2/Oil ratios in optimised network as in the base case
- Hydrogen partial pressures as well as H_2/Oil ratios change with different impurity composition in optimised hydrogen network

Deal with this issue systematically

Feasibility Check for Optimised Network
Criterion 1
(H_2/Oil ratio same as base case)

Hydrogen plant flow
Case 1 = 96.997 MMscfd
Case 3 = 96.885 MMscfd

	Total inlet flowrate, MMscfd		Purity, vol%		H_2/Oil ratio, scfd/bpd		Hydrogen partial pressure, bar (a)	
Unit	Case 1	Case 3	Case 1	Case 3	Case 1	Case 3	Case 1	Case 3
SRNHT	54.136	54.817	76.49	75.63	1150	1152	15.30	15.13
CNHT	49.487	47.685	68.58	71.19	2263	2263	24.00	24.92
SR DHT	208.909	216.126	80.32	77.61	2796	2796	24.09	23.28
HC	217.615	221.576	89.05	87.46	5537	5537	83.71	82.22

Case 1: Optimised network with impurities lumped as CH_4
Case 3: H_2/Oil ratio same as base case in simulated optimised network with impurity considerations

Remarks: CRU flowrate is same for Case 1 and Case 3
Hydrogen partial pressure decreases for SRNHT, SRDHT and HC
Hydrogen plant flow almost the same as Case 1

Hydrogen Pinch Analysis for Simulated Network with Impurity Considerations

(H_2/Oil ratio same as base case)

Scenario 2: Feasible Target

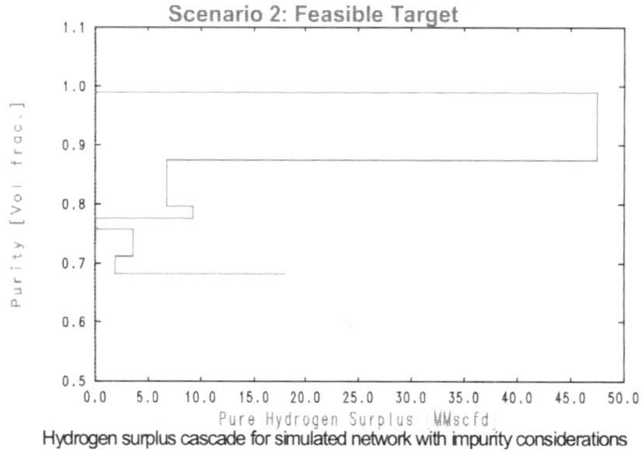

Hydrogen surplus cascade for simulated network with impurity considerations

Comparison between Scenario 1 and Scenario 2

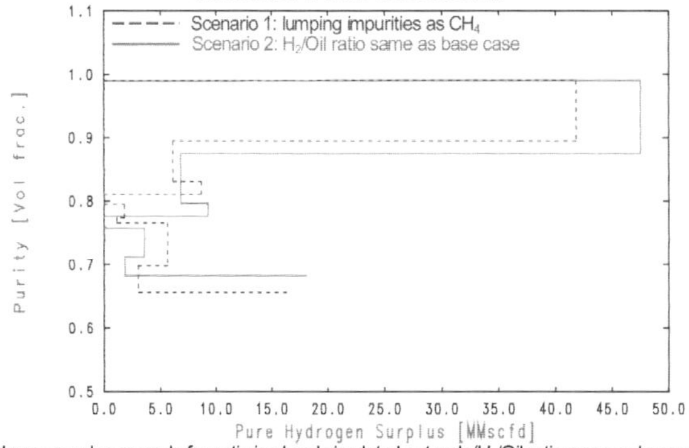

Hydrogen surplus cascade for optimised and simulated network (H_2/Oil ratio same as base case)

Remark: Comparison shows no scope for improvement in Scenario 2, therefore end of investigation

Feasibility Check for Optimised Network
Criterion 2
(Hydrogen partial pressure same as base case)

Hydrogen plant flow
Case 1 = 96.997 MMscfd
Case 4 = 102.640 MMscfd

	Total inlet flowrate, MMscfd		Purity, vol%		H_2/Oil ratio, scfd/bpd		Hydrogen partial pressure, bar (a)	
Unit	Case 1	Case 4	Case 1	Case 4	Case 1	Case 4	Case 1	Case 4
SRNHT	54.136	54.616	76.49	76.75	1150	1164	15.30	15.35
CNHT	49.487	52.413	68.58	68.58	2263	2396	24.00	24.00
SR DHT	208.909	215.751	80.32	80.38	2796	2890	24.09	24.12
HC	217.615	227.456	89.05	89.09	5537	5790	83.71	83.75

Case 1: Optimised network with impurities lumped as CH_4
Case 4: Hydrogen purity same as base case in simulated optimised network with impurity considerations

Remarks: CRU flowrate is same for Case 1 and Case 4
Hydrogen/Oil ratio increases for all units
Hydrogen plant flow increases
Previous approach over predicted savings in hydrogen plant flow

Hydrogen Pinch Analysis for Simulated Network with Impurity Considerations

(Hydrogen partial pressure same as base case)

Scenario 2: Feasible Target

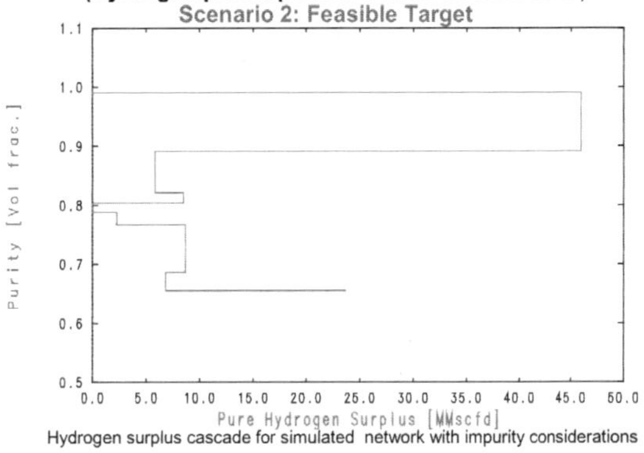

Hydrogen surplus cascade for simulated network with impurity considerations

55

Comparison between Scenario 1 and Scenario 2

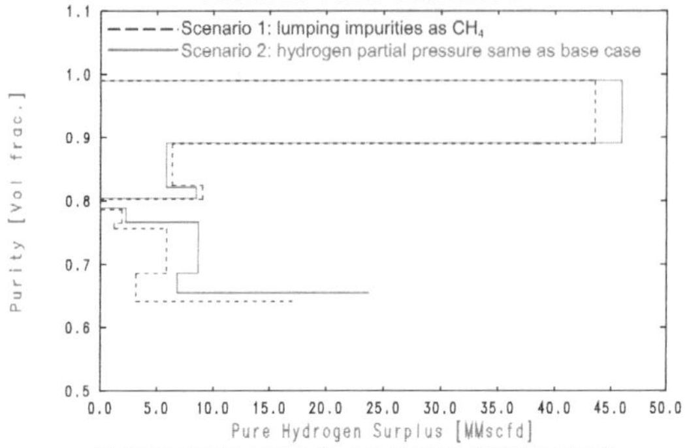

Hydrogen surplus cascade for optimised and simulated network (Hydrogen partial pressure same as base case)

Remark: Scenario 2 is same as Scenario 1, therefore problem converges

56

Results Comparison

Unit	H_2/Oil ratio, scfd/bpd				Hydrogen partial pressure, bar (a)			
	Case 1	Case 2	Case 3	Case 4	Case 1	Case 2	Case 3	Case 4
SRNHT	1150	**1156**	1152	1164	15.30	**15.12**	15.13	15.35
CNHT	2263	**2401**	2263	2396	24.00	**24.86**	24.92	24.00
SR DHT	2796	**2797**	2796	2890	24.09	**23.29**	23.28	24.12
HC	5537	**5537**	5537	5790	83.71	**82.22**	82.22	83.75

	Base Case	Case 1	Case 2	Case 3	Case 4
Total hydrogen input, MMscfd	**125.969**	115.691	**115.691**	115.679	121.33

Case 1: Optimised network with impurities lumped as CH_4
Case 2: Simulated optimised network with impurity considerations
Case 3: H_2/Oil ratio same as base case in simulated optimised network with impurity considerations
Case 4: Hydrogen purity same as base case in simulated optimised network with impurity considerations

Remarks: Case 3 is optimum network, if drop in hydrogen partial pressure is acceptable
Case 4 is optimum network, at the expense of more hydrogen input

57

Summary

- Hydrogen pinch analysis provides ultimate targets for minimum hydrogen consumption
- The network structure derived from hydrogen pinch analysis needs to be validated by taking into account of impurity distribution
- A systematic methodology has been shown to incorporate impurity considerations in hydrogen network management
  - Gives more realistic solution than previous hydrogen pinch approach
  - Hydrogen partial pressure as well as H_2/Oil ratio can be maintained up to acceptable levels

58

6. Conclusions

59

Conclusions

- Sulphur and Nitrogen impurities can be dealt by simulation
- Lumping impurities as CH_4 may generate unrealistic targets
  - Light hydrocarbon impurities strongly affect process purge purity as well as H_2/Oil ratio
- A new methodology has been proposed to incorporate impurities into hydrogen network management
  - Plant data as well as correlations (e.g. HPI correlations) can be used to model hydrogen consumer
  - Optimisation by hydrogen pinch analysis
  - Impact of impurities is considered by simulation

60

69a Future Directions in Membrane Technology

Richard D Noble

Membrane separations are based on the use of a selective barrier between phases. The membrane material, configuration and driving force across the membrane are not specified and can be chosen for a specific application. Future directions in membrane technology will include new materials, configurations and driving forces.

Recent advances in zeolite membrane research show high potential to meet and exceed the performance of other membrane types and overcome the problems associated with the different technologies. Zeolites are crystalline materials with pores of molecular dimensions and extremely narrow pore size distributions. The crystal structures are formed by tetrahedra with silicon or aluminum central atoms surrounded by four oxygen atoms. A fraction of the central atoms can also be substituted by elements like P, Ga, Ge, and B. Each tetrahedron is linked to four others through oxygen bridges. When all the cations in the framework of the crystal are Si^{4+} and the framework is uninterrupted, the structure is neutral. However, when some silicon atoms are replaced with Al^{3+} (for example), the charge deficit in the framework must be balanced by positive ions such as Na^{+} or H^{+}. These ions can be exchanged after synthesis to change adsorption, permeation, and selectivity behaviors. Zeolites can have one, two, and three-dimensional pore systems. Separation selectivity is based on either molecular sieving where only molecules can penetrate the membrane that are sufficiently small to enter the pores, differences in diffusion rates through the pores, or preferential adsorption.

Zeolite membranes are formed by synthesizing dense layers of intergrown zeolite crystals on a porous support surface (gamma-alumina, alpha-alumina, or stainless steel) or within the pores of a support. The porous support can be thick but with large pores (0.2 to 0.5 microns), so mechanical stability is excellent without introducing additional mass transfer resistance. Tubular supports can be used to provide a high surface to volume ratio, scale-up in modular form, and the ability to continuously separate.

Because the separation layer is based on inorganic oxides with an underlying support of porous ceramic or metal, these membranes are usable in high-pressure environments and are stable to temperatures of at least 750 K even in chemically corrosive conditions. Moreover, if they become fouled with organic decomposition products, heating them in air at 725 K can regenerate them. These features provide the ability to operate the separation at or near the processing conditions without any need for costly pressure changes or cooling and subsequent reheating.

Recent research has focused on room temperature Ionic Liquids (ILs) as novel, benign solvent replacements for volatile organic compounds traditionally used in organic synthesis and liquid/liquid separations, including liquid membranes. Large, asymmetric, organic cations and various anions make up the ILs, thereby creating non-volatile, non-flammable liquids capable of solubilizing a variety of solutes. These ionic solvents consist entirely of ions, strongly resembling ionic melts obtained by heating metallic salts such as NaCl to high temperatures, but are liquid at much lower temperatures. Many ILs are liquids over a wide temperature range with some known melting points as low as -96 oC and some liquid ranges in excess of 300 oC. High coulombic forces constrain the IL constituents and thus, the ILs exert practically no vapor pressure above the liquid surface. These features have led to current investigations of ILs as alternative media for a variety of applications that use organic solvents.

The cation of many ionic liquids contains a ring structure with one or more alkyl groups, which can be modified to adjust the solution properties. The anion structure is also available for modification. ILs combine a number of unique properties into one solvent system. ILs have high thermal stability, high ionic conductivity, negligible vapor pressure and are non-flammable. Unlike traditional organic media, IL properties may be adjusted via chemical alteration of the cation or anion to produce application specific compounds. The potential exists for complexing agents to be one of the ions giving a maximum loading of a complexing agent in the solvent, much greater than the loadings typical in standard solvent systems. Complexing agents can also be added or doped into the IL.

Composite structures can be formed using a polymer for mechanical strength and incorporating ionic liquids and/or zeolites into the matrix to tune the properties of the resulting membrane. These mixed matrix membranes can be used for a variety of gas separations.

An additional configuration is the refinery on a chip concept. This downsizing has the potential to produce chemical products on demand. An important element for this concept and similar microfluidic applications is a fluid pump that can also be scaled down to the size and flowrates needed. We have recently developed a membrane-based microfluidic pump that has no moving parts. The pump make use of redox reactions in compartments separated by an ion-selective membrane to provide the driving force for the fluid flow.

69b Commercialization of Membrane Technology for Refining Application

Xinjin Zhao

Davison Membranes pioneered many commercial membrane applications and has been a leading player in commercializing membrane-based solutions for refining applications ranging from natural gas separation to solvent recovery and gasoline sulfur reduction. Although the development of the membrane itself has attracted the most attention, there is generally a large gap between the identification of a working membrane to the development of a commercially viable process. In this paper, I will give an overview of the steps involved in the commercialization of membrane-based processes by using successful commercial examples. For example, MaxDeWax™ for lube solvent recovery, the world largest membrane refining applications jointly commercialized by Davison Membranes and ExxonMobil, has been in successful operation in ExxonMobil Beaumont refining since 1997. We have also developed processes for refining hydrogen recovery. More recently, we have commercialized S-Brane, an innovative separation process that selectively removes sulfur-containing molecules from FCC or other naphtha streams. S-Brane technology not only significantly reduces the overall capital and operating cost for clean fuel compliance, but also provides a mean for preserving octane that often occurs in hydrotreating-based technology.

69c Anti-fouling Membranes for Water Treatment

Jame C Li, Norman N Li, Richard Q Song

Fouling of reverse osmosis membranes has been a serious problem in water treatment. To overcome this problem, a variety of modifications of membrane surface and formulations may be used. The major methods of modification of membrane surfaces are increasing hydrophilicity, imparting surface charges, reducing roughness of membrane surface, and formulating biomimetic surfaces. In addition, the formulation of the thin film portion of the membranes can be modified to impart the anti-fouling characteristics to the membranes.

The effectiveness of these methods willl be described. Our own research results will be discussed.

69d Stable Multi-membrane Module Prepared by Counter Diffusion CVD for High Temperature Hydrogen Separation

Hitoshi Aida, Mikihiro Nomura, Shin-ichi Nakao, Suraj Gopalakrishnan, Takashi Sugawara

Porous inorganic membranes are widely considered over polymeric membranes for their resistance to physical and chemical degradation. Last decade showed much interest towards the development of hydrogen selective inorganic membranes with considerable attention focused towards silica membranes. Silica membranes are usually in the form of silica layers placed over rigid and porous ceramic supports such as Vycor glass, alumina, etc., using sol-gel and chemical vapor deposition (CVD). The hydrogen permeance values of these membranes are reported over a wide range (10^{-7} to 10^{-9} mol m^{-2} s^{-1} Pa^{-1}) and the selectivity for nitrogen ranges up to 3000. In the CVD process, suitable precursors were allowed to react near the supporting mesoporous/microporous substrate to produce a solid deposit inside/over these substrates resulting in complete or partial pore plugging. Two types of configurations are possible here; one is the one-sided geometry in which the reactants are made to react in the gas phase or near the surface of the substrate, producing an even deposit on the surface of the porous substrate. The second type is the opposing reactant geometry or counter diffusion CVD in which the reactants are fed through opposite sides of a porous support, which reacts inside the pores to form a deposit on the pore wall. This reduces the pore size and finally the reaction stops automatically when the reactants could no longer pass through these narrow pores. Thus a homogeneous micropore size distribution could be achieved using this method. Few authors have already reported the preparation and properties of singular membranes using the above method. In this presentation, we will explain the fabrication and permeation properties of a multi-membrane module prepared by counter diffusion CVD method using tetramethylorthosilicate (TMOS) and O_2 as precursors. In this process, the reaction takes place simultaneously inside the supports arranged parallel to each other.

Single component gas permeation studies were performed to evaluate this multi-membrane module. Permeation of larger molecules did not vary much within the temperature range of 373-873K. Activated transport displayed by the hydrogen molecules in the above range (Ea=22.9 kJ/mol) points to the narrow nature of the membrane pores. H2 permeance through this module was 5.02 x 10^{-8} mol m^{-2} s^{-1} Pa^{-1} yielding very high selectivity ratio with N_2 (>1000). This module also maintained a steady permeance ratio of 500 at 76 kPa of steam at 773K.

Novel Polymeric-Metallic Composite Membranes for CO_2 Separations at Elevated Temperatures

J. S. Young[1], K.A. Berchtold[1], E. S. Peterson[2], A. R. Greenberg[3], J. Acquaviva[4], F. Onorato[4], S. Hopkins[4]

[1]Los Alamos National Laboratory, Los Alamos, NM 87544
[2]Idaho National Engineering and Environmental Laboratory, Idaho Falls, Idaho, 83415
[3]Department of Mechanical Engineering, University of Colorado, Boulder, CO, 80309
[4]Pall Corporation, 2200 Northern Blvd,East Hills, NY 11548

Prepared for Presentation at AIChE 2005 Spring National Meeting, April 10-15, 2005, Advances in Membrane Separation Technologies, Unpublished

INTRODUCTION

Separation of CO_2 from process streams is readily accomplished via standard separation techniques such as amine scrubbing and pressure-swing adsorption. However, these techniques require low temperatures and produce a low-pressure CO_2 stream, resulting in a significant energy penalty for separating CO_2. Polymer-based membrane separations, on the other hand, are less energy intensive, requiring no phase change in the process and typically provide low-maintenance operations. Polymers exhibit a high selectivity due to the combination of size-based and solubility-based molecular separation, a dual-mode mechanism not shared by size-driven, inorganic membrane separations. Polymer membranes have been used successfully in a number of industrial applications including the production of high-purity nitrogen, gas dehydration, removal of acid gases, and recovery of hydrogen from process streams for recycle(Koros and Fleming 1993; Hagg 1998; Koros and Mahajan 2000). Use of polymer membranes in a synthesis gas or flue gas separation could become economically viable if a membrane that was thermally, chemically and mechanically stable at high temperature and high pressure were available. Unfortunately, the commercially available polymeric materials currently employed are not stable in demanding environments to the degree required. In general, as the glass transition temperature of the polymer is approached, membrane selectivity is significantly reduced and flux decline due to membrane compaction (creep) is increased. In addition, current membrane materials are often subject to chemical degradation by minor components in the process stream, a problem that is exacerbated by elevated temperature. Consequently, there is a compelling need for membrane materials that can operate under more extreme environmental conditions for extended periods of time while providing an acceptable level of performance.

Alternatives to polymer membranes are the inorganic membranes which include zeolites(Tuan, Falconer *et al.* 2000), carbon molecular sieves and selective surface-flow (SSF) membranes(Larbot, Fabre *et al.* 1988; Rao and Sircar 1993; Rao and Sircar 1993; Anand, Langsam *et al.* 1997; Armor 1998; Prabhu and Oyama 2000; Lin 2001; Lee 2002; Vu and Koros 2002). These membranes currently suffer from lack of reproducibility, densification due to humidity and elevated temperatures and high-cost fabrication. Given these limitations, mixed matrix membranes and crosslinked membranes have been suggested as viable development pathways of the future(Koros and Mahajan 2000). Mixed matrix membranes are composed of carbon molecular sieves or zeolites imbedded in a polymer matrix(Mahajan and Koros 2002). Although permeabilities and selectivities have been increased above the traditional upper bound for polymers, application temperatures are still limited by the polymer matrix. Alternatively, crosslinked membranes can reduce the plasticization effects at elevated pressures(Staudt-Bickel and Koros 1999) as well as increase membrane selectivity (Kita, Inada *et al.* 1994; Liu, Pan *et al.* 1999); however, permeability is adversely affected.

Hence, the development of high performance polymer membranes is still very much a viable approach to filling the need for high temperature membrane separations. The current study describes the development of membranes with a polybenzimidazole (PBI) selective layer. PBI possesses excellent chemical resistance, a very high glass transition temperature (450°C), good mechanical properties and excellent processability. Our approach also employs a unique porous metal substrate to provide mechanical support for the PBI selective layer in a manner similar to the ceramic supports for polyimide selective layers that have previously been developed (Rezac and Koros 1992). Given the promising results that we recently obtained for small disks of the PBI-metal composite, we have fabricated significantly larger membrane areas (≈67 vs 0.5 cm^2) in the form of tubular membranes in a stainless steel housing. This paper describes the initial results obtained for these PBI-metal composite tubular membranes.

MATERIALS AND METHODS

Composite Membrane Formation

The composite membranes utilized for this study are composed of a dense polybenzimidazole (PBI) layer supported by a tubular porous stainless steel substrate (Pall Corporation's AccuSep® media). The tubular supports were coated with PBI using methods proprietary to Pall Corporation. Figure 1 illustrates a coated tube along side the module housing.

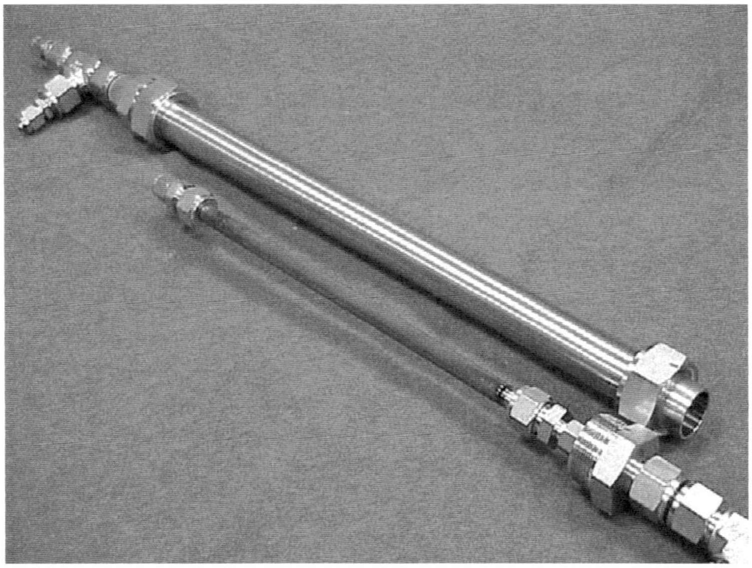

Figure 1: PBI coated AccuSep® media tube courtesy of Pall Corporation.

Permeation Experiments

Membrane permeance is a function of the diffusivity of the penetrating molecule and its solubility in the polymer. For a specific penetrant/membrane interaction, permeance (GPU, 10^{-6}scc/s/cm^2/cm Hg), can be calculated as follows:

$$GPU = \frac{\dot{v}*10^6}{\Delta P} \quad (1)$$

where $\dot{v}$ is the gas flux (cm^3/s/cm^2), and ΔP (cm Hg) is the pressure driving force across the membrane. The ideal selectivity $\alpha_{A/B}$ is the ratio of the permeances of the gases as determined from single-gas tests at a given temperature and equal pressure driving force.

Figure 2 illustrates the configuration of the test rig used for performing the single gas tests on a single tube module (membrane area of 66.84 cm^2, permeate volume of 28 ml, and feed volume of 138 ml.). Single-gas testing consisted of introducing either H_2, CO_2, CH_4, or N_2 with a minimum purity of 99.99% to a pressure of ~2600 torr with the permeate side under vacuum. Two types of tests were performed: pressure rise and pressure decay. The pressure-rise test was conducted by blocking off the permeate side from vacuum, and the pressure-decay test was performed by blocking off the feed side from the feed tank. The pressure data versus time were collected using LabView™. The test protocol consisted of performing a series of tests at a given temperature, increasing the temperature by 100 °C and performing another series of tests, then reducing the temperature to the original value and re-running the test series to verify that the membrane performance was not degraded as a result of the higher temperature testing. Tests were performed from ambient conditions, ~25 °C, to 400 °C.

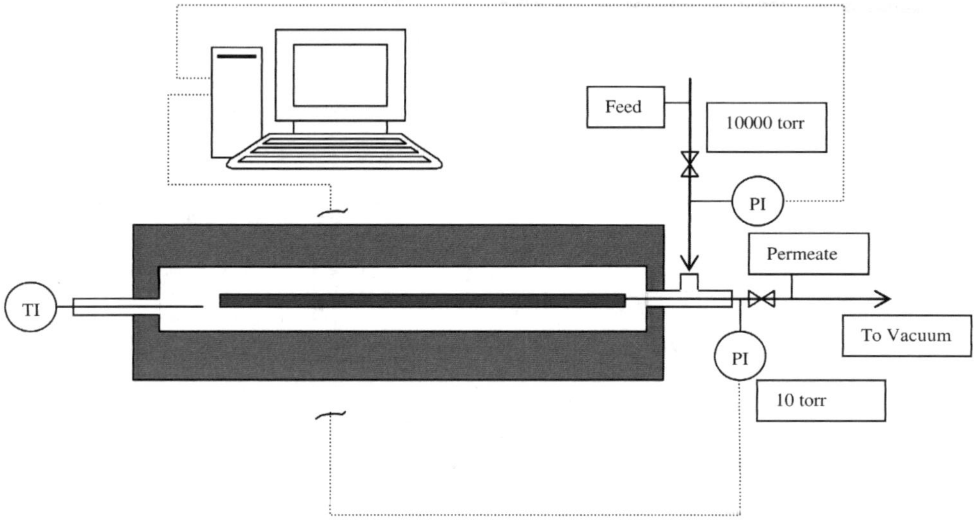

Figure 2: Illustration of test rig for the single tube module. The module was housed in an oven, and feed was introduced on the OD of the polymer coated stainless steel tube. Gas permeated across the membrane and could be vented to vacuum when necessary.

RESULTS AND DISCUSSION

Figure 3 summarizes the permeance data versus inverse temperature for all gases tested. For these initial studies, membrane thickness was not minimized, hence we expect in the future to significantly improve upon the permeances reported here. While not explicitly shown, the membrane maintained permeability and selectivity through repeated thermal cycles. Interestingly, hydrogen has the steepest slope versus temperature indicating that the selectivity between hydrogen and any other gas improves as the temperature increases! This phenomena was seen across multiple samples and formats of the composite membrane. Figure 4 illustrates selectivity as a function of inverse temperature. At 400 °C, the PBI/porous stainless steel support has almost an order of magnitude greater selectivity for H_2 over CO_2 than polymer membranes of comparable permeability(Robeson, Burgoyne *et al.* 1994). The selectivities of CO_2 and methane or CO_2 and N_2 decrease with increasing temperature due presumably to the larger negative impact of increased temperature on CO_2 solubility versus that of CH_4 or N_2.

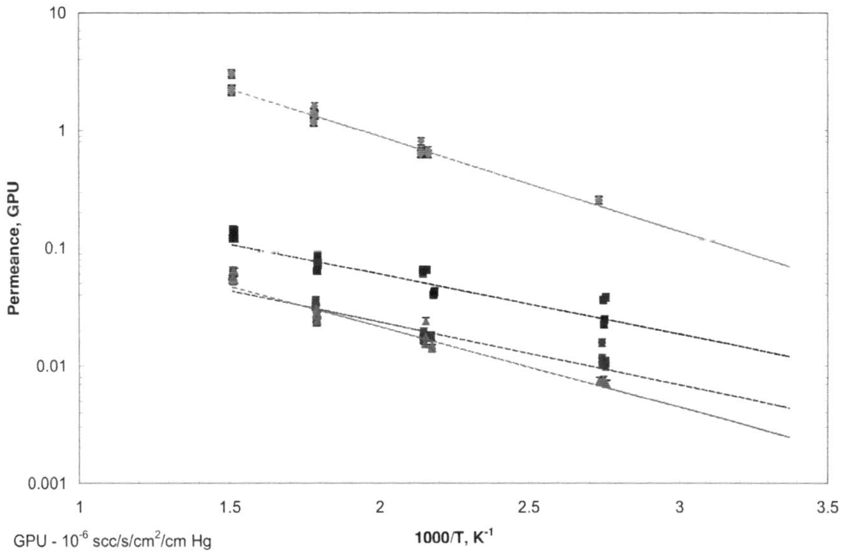

Figure 3: Summary of permeance versus inverse temperature for the gases hydrogen (♦), carbon dioxide (■), methane (●), and nitrogen (▲). The symbols are the experimental data points while the solid lines are fits of the data.

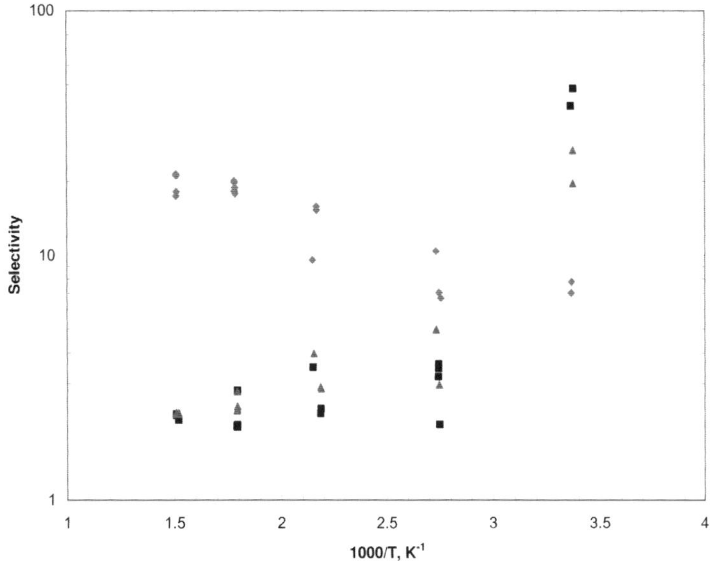

Figure 4. Selectivity for H_2/CO_2 (♦), CO_2/CH_4 (■), and CO_2/N_2 (▲) as a function of inverse temperature.

A comparison between the data obtained using small, flat, circular disks (0.48 cm^2) versus the single tube module (66.84 cm^2) is shown in Figure 5. Results indicate that the performance of the module in terms of both permeance and selectivity is not significantly different from that of the disks. The effect of temperature on the permeance is reduced for the module compared to that of the test disks. The reason for this behavior most likely lies in the uniformity with which the larger module area could be heated versus that of the disk. In addition, the potential for minute membrane defects increases as the membrane area is increased. This will affect the overall selectivity and reduce the effect of temperature on individual gas permeance. We are continuing efforts to improve the composite structure by reducing film thickness to significantly increase permeance while maintaining selectivity.

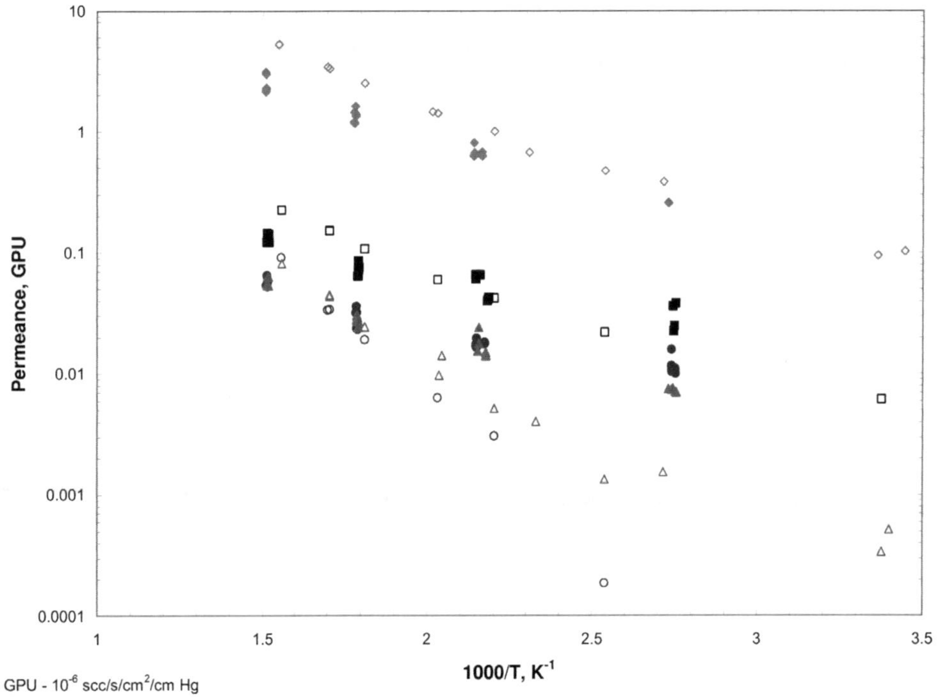

Figure 5: Comparison of H_2 (♦), CO_2 (■), CH_4 (●), and N_2 (▲) permeances obtained using test disks (0.48 cm^2) versus the single tube module (66.84 cm^2); solid symbols are module data while open symbols are disk data.

CONCLUSIONS

The feasibility of fabricating a tubular composite PBI/stainless steel support membrane has been demonstrated. This composite membrane has been tested using single gases (H_2, CO_2, CH_4, and N_2) from 25 to 400 °C using a protocol that involves repeated thermal cycling. The membrane maintained permeance and selectivity through multiple thermal cycles. The membrane area of this tubular geometry represents a scale-up of greater than two orders of magnitude over previous test specimens. The performance characteristics of this composite membrane are encouraging, particularly for H_2/CO_2 separations. In addition, the response of the module is very similar to that of the test disks, which provides strong evidence of the scalability of this composite membrane. This latter factor is critical for ultimate utilization of this membrane in commercial-scale gas separations.

ACKNOWLEDGEMENTS

The authors gratefully acknowledge funding for this project from the Department of Energy, Office of Fossil Energy, administered through the National Energy and Technology Laboratory, Morgantown, WV. (Project FE-01-0003). Los Alamos National Laboratory is operated by the University of California for the United States DOE under Contract W-7405-ENG-36.

REFERENCES

Anand, M., M. Langsam, et al. (1997). "Multicomponent gas separation by selective surface flow (SSF) and poly-trimethylsilylpropyne (PTMSP) membranes." Journal of Membrane Science **123**: 17-25.

Armor, J. N. (1998). "Applications of catalytic inorganic membrane reactors to refinery products." Journal of Membrane Science **147**: 217-233.

Hagg, M.-B. (1998). "Membranes in Chemical Processing: A Review of Applications and Novel Developments." Separation and Purification Methods **27**(1): 51-168.

Kita, H., T. Inada, et al. (1994). "Effect of Photo-Cross-Linking on Permeability and Permselectivity of Gases through Benzophenone-Containing Polyimide." Journal of Membrane Science **87**(1-2): 139-147.

Koros, W. J. and G. K. Fleming (1993). "Membrane-based gas separation." Journal of Membrane Science **83**: 1-80.

Koros, W. J. and R. Mahajan (2000). "Pushing the limits on possibilities for large scale gas separation: which strategies." Journal of Membrane Science **175**: 181-196.

Larbot, A., J. P. Fabre, et al. (1988). "Inorganic Membranes Obtained by Sol-Gel Techniques." Journal of Membrane Science **39**: 203-212.

Lee, D. (2002). "Gas permeation characteristics of a hydrogen selective supported silica membrane." Journal of Membrane Science **210**(2): 291.

Lin, Y. S. (2001). "Microporous and dense inorganic membranes: current status and prospective." Separation and Purification Technology **25**: 39-55.

Liu, Y., C. Pan, et al. (1999). "Effect of crosslinking distribution on gas permeability and permselectivity of crosslinked polyimides." European Polymer Journal **35**(9): 1739-1741.

Mahajan, R. and W. J. Koros (2002). "Mixed Matrix Membrane Materials With Glassy Polymers. Part 1." Polymer Engineering and Science **42**(7): 1420-1431.

Prabhu, A. and S. Oyama (2000). "Highly hydrogen selective ceramic membranes: application to the transformation of greenhouse gases." Journal of Membrane Science **176**(2): 233-248.

Rao, M. B. and S. Sircar (1993). "Nanoporous carbon membrane for gas separation." Gas Separation and Purification **7**(4).

Rao, M. B. and S. Sircar (1993). "Nanoporous carbon membranes for separation of gas mixtures by selective surface flow." Journal of Membrane Science **85**: 253-264.

Rezac, M. E. and W. J. Koros (1992). "Preparation of Polymer-Ceramic Composite Membranes with Thin Defect-Free Separating Layers." Journal of Applied Polymer Science **46**(11): 1927-1938.

Robeson, L. M., W. F. Burgoyne, et al. (1994). "High performance polymers for membrane separation." Polymer **35**(23): 4970-4978.

Staudt-Bickel, C. and W. J. Koros (1999). "Improvement of CO2/CH4 separation characteristics of polyimides by chemical crosslinking." Journal of Membrane Science **155**: 145-154.

Tuan, V. A., J. L. Falconer, et al. (2000). "Isomorphous substitution of Al, Fe, B, and Ge into MFI-zeolite membranes." Microporous and Mesoporous Materials **41**(1-3): 269-280.

Vu, D. Q. and W. J. Koros (2002). "High Pressure CO_2/CH_4 Separation Using Carbon Molecular Sieve Hollow Fiber Membranes." Industrial and Engineering Chemistry Research **41**(3): 367-380.

69f Carbon Molecular Sieve Membranes for Gas and Vapor Separations

Paul K T Liu, Richard J Ciora

Polymeric based gas/vapor separation membranes suffer from a number of drawbacks that have limited their penetration in the refining industry to a few applications. To overcome this problem, we have been developing carbon molecular sieve based membranes supported on tubular inorganic substrates. These membranes can be fabricated in a variety of pore sizes and display unique performance and excellent material stability. In this presentation, applications from hydrogen recovery and purification to solvent dehydration with these versatile membranes will be discussed. Results will be presented that demonstrate the high performance and stability of these membranes in a range of aggressive thermal, hydrothermal and chemical environments. Further, scale-up of these membranes to pre-commercial scale multi-tube elements has begun and results from preliminary pilot scale tests will be presented.

American Institute of Chemical Engineers
3 Park Avenue
New York, New York 10016-5991

ISBN: 978-1-60423-473-2